Lecture Notes in Control and Information Sciences 443

Editors

Series Advisory Board

For further volumes:
http://www.springer.com/series/642

Zheng-Guang Wu, Hongye Su, Peng Shi,
and Jian Chu

Analysis and Synthesis of Singular Systems with Time-Delays

Dr. Zheng-Guang Wu
Institute of Cyber-Systems
and Control
Zhejiang University
Hangzhou
China

Prof. Hongye Su
Institute of Cyber-Systems
and Control
Zhejiang University
Hangzhou
China

Prof. Peng Shi
School of Electrical and Electronic
Engineering
The University of Adelaide
Adelaide
Australia

and

College of Engineering and Science
Victoria University
Melbourne
Australia

Prof. Jian Chu
Institute of Cyber-Systems
and Control
Zhejiang University
Hangzhou
China

ISSN 0170-8643 ISSN 1610-7411 (electronic)
ISBN 978-3-642-37496-8 ISBN 978-3-642-37497-5 (eBook)
DOI 10.1007/978-3-642-37497-5
Springer Heidelberg New York Dordrecht London

Library of Congress Control Number: 2013934981

Printed on acid-free paper

Springer is part of Springer Science+Business Media (www.springer.com)

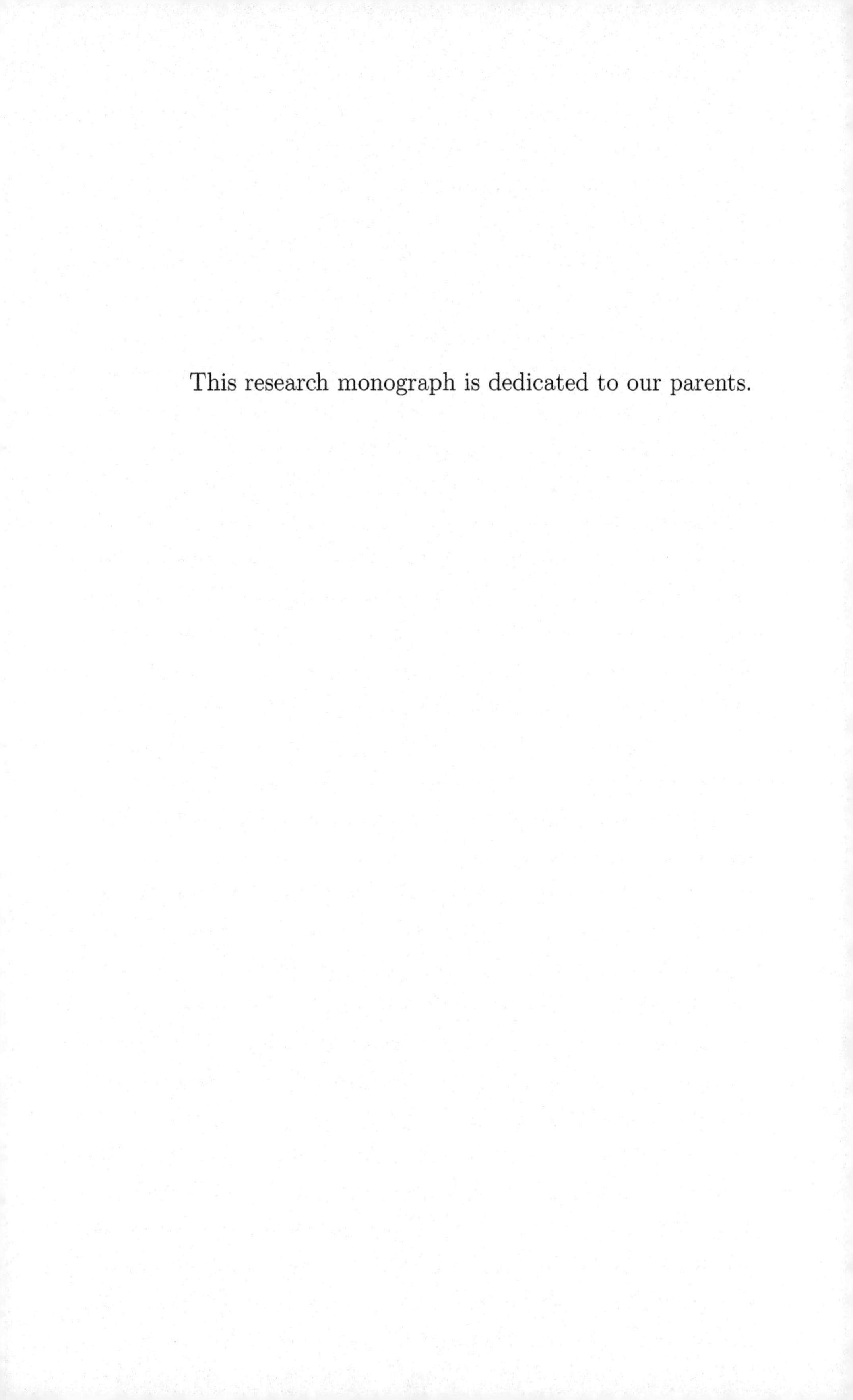

This research monograph is dedicated to our parents.

Preface

In the past decade, singular systems with time delays have attracted tremendous interest from the systems and control areas. The reason is mainly twofold: 1) singular systems have come to play an important role in modelling practical systems, e.g. power systems, biological systems, economic systems, and 2) time delay frequently occurs in a great number of dynamical systems including electrical networks, biological systems and nuclear reactor, and is a source of the generation of some complex and bad behaviours (oscillations, instability, poor performance). On the other hand, as an important class of random systems, Markov jump systems can better describe many practical systems, especially when random abrupt changes, which are induced by a variety of reasons such as unexpected events, changes in subsystem interconnections, sudden environmental disturbance and so on, occur in the systems. All kinds of techniques have been presented to study the analysis and synthesis problems for Markov jump systems, as well as the applications of such type of systems. In the past few years, the study on singular time-delay systems with Markov jumping parameters has also become the subject of extensive research that has received a lot of attention.

This book provides the recent advances in analysis and synthesis of singular time-delay systems with or without Markov jumping parameters. First, a review on the latest developments in the literature is presented in Chapter 1. Then this book will provide two parts:

Part I: Some developments on analysis and synthesis of singular systems with time delays. There are seven chapters in this part. Specifically, the robust stabilization problem is considered for discrete-time singular systems with time delay and norm bounded parametric uncertainties in Chapter 2. The problems of H_∞ control and reliable passive control are addressed for singular time-delay systems in Chapter 3 and Chapter 4, respectively. In Chapter 5, the problem of dissipativity analysis is investigated for continuous-time and discrete-time singular systems with time-varying delays. Chapter 6 and Chapter 7 study the problems of $\mathscr{L}_2 - \mathscr{L}_\infty$ and H_∞ filtering for singular systems with time delays, respectively. In Chapter 8, the problem of H_∞

filter design is discussed for discrete-time networked singular systems subject to multiple stochastic time-varying communication delays and probabilistic missing measurements.

Part II: Some developments on analysis and synthesis of singular Markov jump systems with time delays. There are six chapters in this part. Specifically, the problem of delay-dependent stability analysis is considered for discrete-time singular Markov jump systems with time-varying delay and partially unknown transition probabilities in Chapter 9. The H_∞ control for singular Markov jump systems with time delays is presented in Chapter 10. In Chapter 11, the problem of delay-dependent passivity analysis is studied for singular Markov jump systems with time delays. The mode-independent l_2-l_∞ filter design for discrete-time singular Markov jump systems with time-varying delays in given in Chapter 12. Both mode-independent and mode-dependent H_∞ filters are designed for singular Markov jump systems with time delays in Chapter 13 and Chapter 14, respectively.

Zhejiang University, China — Zheng-Guang Wu
Zhejiang University, China — Hongye Su
The University of Adelaide,
and Victoria University, Australia — Peng Shi
Zhejiang University, China — Jian Chu
February 2013

Symbols and Acronyms

$\mathbb{R}$	field of real numbers
$\mathbb{R}^n$	n-dimensional real Euclidean space
$\mathbb{R}^{m\times n}$	space of all $m \times n$ real matrices
I	identity matrix
0	zero matrix
$A > 0$	symmetric positive definite
$A \geqslant 0$	symmetric positive semi-definite
$A < 0$	symmetric negative definite
$A \leqslant 0$	symmetric negative semi-definite
$A^{1/2}$	symmetric square root of $A \geqslant 0$, i.e., $A^{1/2}A^{1/2} = A$
A^{T}	transpose of matrix A
A^{-1}	inverse of matrix A
A^{+}	Moore-Penrose inverse of matrix A
$\mathrm{rank}(A)$	rank of matrix A
$\rho(A)$	spectral radius of matrix A
$\sigma_{\max}(A)$	maximum singular value of matrix A
$\sigma_{\min}(A)$	minimum singular value of matrix A
$\lambda_{\max}(A)$	maximum eigenvalue of matrix A
$\lambda_{\min}(A)$	minimum eigenvalue of matrix A
$\det(A)$	determinant of matrix A
$\deg(\cdot)$	degree of a polynomial
$\mathrm{diag}\{\cdots\}$	block-diagonal matrix
$\lVert \cdot \rVert$	Euclidean norm of a vector and its induced norm of a matrix
sup	supremum
inf	infimum
$\lVert f(t) \rVert_d$	$\sup_{-d\leqslant s\leqslant 0} \lVert f(s) \rVert$
$\mathbb{N}[a,\, b]$	$\{a, a+1, \cdots, b-1, b\}$, where a and b are integers with $a < b$
$\mathscr{L}_2[0,\infty)$	space of square integrable functions on $[0,\infty)$
$l_2[0,\infty)$	space of square summable infinite sequence on $[0,\infty)$
Ω	sample space
$\mathcal{F}$	σ-algebra of subsets of Ω
$\mathcal{P}$	probability measure on $\mathcal{F}$
$\Pr\{x\}$	probability of x
$\Pr\{x\vert y\}$	probability of x given y
$\mathscr{E}\{x\}$	expectation of x
$\mathscr{E}\{x\vert y\}$	expectation of x conditional on y
LMI	linear matrix inequality
BRL	bounded real lemma
SSs	singular systems
MJSs	Markov jump systems
SMJSs	singular Markov jump systems

Contents

Part II: Singular Markov Jump Systems (SMJSs) with Time-Delays

1

Introduction

1.1 Singular Systems (SSs) with Time-Delays

During the past several decades, a great deal of research interest has been devoted to the study on analysis and synthesis of SSs. This stems from the fact that SS models are very suitable to describe a lot of practical systems including large-scale interconnected systems [96], mechanical engineering systems [60], power systems [121], chemical processes [72], circuit systems [132], and economic systems [97]. In many references, SSs are also named as differential-algebraic equation systems, implicit systems, descriptor systems, semistate systems, and generalized state-space systems (or generalized systems). Among these names, the name descriptor systems can better provide a natural description of the engineering economic systems [132] and thus is also very popular in the literature.

It has be shown that SSs are essentially different from state-space systems. One of the most important differences between the two types of systems is that singular systems usually contain three kinds of modes, that is, finite dynamic modes, infinite dynamic modes and non-dynamic modes, while state-space systems only have finite dynamic modes [21]. It should be pointed out that for a singular system, owing to infinite dynamic modes and non-dynamic modes, the existence and uniqueness of its solution cannot be always ensured and it may also have undesired impulsive behavior. Thus, when dealing with the analysis and synthesis of SSs, we not only need to consider stability, which is the most fundamental problem of systems, but also have to investigate regularity and non-impulsiveness (for continuous-time case) or causality (for discrete-time case) simultaneously, whereas for state-space systems the latter two issues do not arise due to the fact that the regularity and absence of impulses or causality are guaranteed definitely in state-space systems [178]. It is therefore the study of SSs is much more complicated and interesting than that of state-space systems.

Considering the importance of SSs both in practice and in theory, a lot of researchers from a variety of engineering areas have paid considerable attention to the study of SSs. In particular, in the last two decades there

Z.-G. Wu et al.: *Anal. & Synth. of Singular Syst. with Time-Delays*, LNCIS 443, pp. 1–9.
DOI: 10.1007/978-3-642-37497-5_1

has been a surge of research activities and thus extensive results have been presented by all kinds of methods in the literature, and a lot of fundamental and important notions and results in systems and control theory based on state-space systems have been successfully extended to SSs including controllability, observability, stability and stabilization, guaranteed cost control, H_∞ control and filtering, model reduction, sliding mode control and so on. For more details on the related results, we refer the readers to [18, 80, 107, 119, 130, 141, 142, 145, 146, 177, 181, 185, 189, 191, 198, 208] and the references therein.

It is well known that time delay, which is frequently encountered in various engineering systems such as nuclear reactors, rolling mills, hydraulic systems, manufacturing processes, and so on [49, 52], is often a source of the generation of oscillation and a source of instability of control systems [71]. Therefore, it is naturally important and necessary to study time-delay systems, which have been a fascinating subject of research attracting constant attention from the mathematical and control communities during the past several decades. A great number of approaches have been proposed to study the analysis, control and filtering problems for time-delay systems in order to meet the needs of practical engineering. Among these approaches, the linear matrix inequality (LMI) method appears to be very popular, the advantage of which lies in the fact that LMIs can be converted to a convex optimization problem which can be handled efficiently via standard numerical software [10]. The existing results on time-delay systems can be classified into two types, that is, delay-independent results and delay-dependent results; the former does not include any information on the size of delay while the latter employs such information. Generally speaking, the delay-independent results are regarded as more conservative than the delay-dependent results, especially when the time delay is comparatively small. Therefore, delay-dependent problems on time-delay systems have received visible research attention and a wealth of approaches have been proposed such as model transformation approach [77], bounding techniques (Park inequality [115], Moon inequality [113], Jensen inequality [49, 54]), descriptor system approach [36, 37], discretized Lyapunov functional method [48, 49], free-weighting matrix approach [57–59, 150, 151], delay partitioning technique [73, 117], etc. A survey of the recent results on time-delay systems can be found in [180].

It is not surprising that, in the past ten years, singular time-delay systems have been explored extensively by lots of researchers. Up to now, various methods and techniques have been developed for the study of SSs with time-delays in a wide range of research topics. It is worth pointing out that the development of singular time-delay systems is also mainly based on the well known Lyapunov theory, but unfortunately the traditional Lyapunov theorem for state-space systems with time-delays cannot be directly applied to singular time-delay systems [52, 81]. Based on the Barbalat lemma [47], a singular version of Lyapunov Theorem has been proposed in [35], based on which sufficient conditions for delay-dependent/delay-independent stability and for robustness of stability with respect to small delays have been given by LMIs. In [38], the singular version of Lyapunov Theorem has been further adopted to SSs with time-delays, and the H_∞ filtering, the state-feedback and the output-feedback H_∞ control problems have been solved. The singular version of Lyapunov Theorem has also been used to study fuzzy singular time-delay systems in the extended Takagi-Sugeno (T-S) fuzzy model in [81], where sufficient conditions have been derived for the stability and stabilization. However, in [35, 38], the decomposition and transformation of original system coefficient matrices are required and the derived results are not formulated in terms of the coefficient matrices of the original system, which make the analysis and design procedures complex and unreliable. In [175], the problems of robust stability and stabilization for uncertain SSs with state delay have been investigated and the following notable stable condition has been proposed.

Lemma 1.1. *System*

$$\begin{cases} E\dot{x}(t) = Ax(t) + A_d x(t-d), \\ x(t) = \phi(t), \;\; t \in [-d, 0] \end{cases} \tag{1.1}$$

is regular, impulse free and stable if there exist matrices $Q > 0$ and P such that

$$EP^{\mathrm{T}} = PE^{\mathrm{T}} \geqslant 0 \tag{1.2}$$

$$\begin{bmatrix} AP^{\mathrm{T}} + PA^{\mathrm{T}} + Q & A_d P^{\mathrm{T}} \\ * & -Q \end{bmatrix} < 0 \tag{1.3}$$

Based on the above lemma, necessary and sufficient conditions for generalized quadratic stability and generalized quadratic stabilization have been presented in terms of LMIs, respectively. An obvious and important merit of the results given in [175] is that all the given results are expressed by using the original system matrices and thus no system decomposition is required, which can avoid certain numerical problems arising from decomposition of matrices and thus make the analysis and design procedures relatively simple and reliable. It is worth pointing out that Lemma 1.1 and the techniques adopted in [175] play important roles in the study of SSs with time delays, and have been routinely used in the literature. Up to now, various approaches have been developed to the analysis and synthesis of SSs with time delays in a wide range of related research topics, including stability and stabilization [30, 51, 81, 158, 218], guaranteed cost control [70, 140, 194], sliding mode control [23, 53, 149], H_∞ control [29, 166, 173, 186, 187], dissipativity analysis [34, 154], H_∞ filtering [86, 93, 94, 201, 219], dissipative filtering [33], and fault detection [13].

The past few decades have witnessed a significant progress on the theory of Markov jump systems (MJSs), which is arguably one of the most active research areas in systems and control. The reason is mainly that as a special class of hybrid systems, MJSs have great ability to model the dynamic systems whose structure is subject to random abrupt variation mainly due to, for example, changing in subsystem interconnections, random component failures or repairs, sudden environmental changes, uncontrolled configuration changes, and sudden variations of the operating points of a nonlinear system [20]. As a result, many essential results have been reported on the theory of analysis, control and filtering of MJSs [2, 17, 19, 22, 27, 66, 68, 88, 95, 98, 110, 114, 126, 136, 147, 152, 165, 174, 203, 206], and the important applications of such type of systems can be found in diverse fields, e.g. robot manipulations, economic systems, networked control systems, multi-agent control systems, aircraft control systems, fault tolerant systems, power systems [1, 62, 127].

In recent years, some attention has also been paid to SMJSs and a few results have been reported by various techniques. For example, in [162, 164, 169, 171, 178], necessary and sufficient conditions for stochastic admissibility have been obtained in terms of LMIs for continuous-time and discrete-time SMJSs. It should be mentioned that the results reported in [169, 171, 178] require the complete knowledge of transition probabilities (rates) of the Markov chain (process) that describes the switching of the system modes. However, in fact, it is very hard and expensive to obtain precisely all the transition probabilities (rates) even for a simple system [8]. To overcome the deadlock, [8] and [164] have discussed the case of unknown transition probabilities, and thus the proposed results are more general and powerful than these of [169, 171, 178]. The problems of guaranteed cost control and H_∞ control for SMJSs have been addressed in [4, 74, 178]. In [211], the observer design problem has been investigated for a class of SMJSs with nonlinear perturbations, and the full-order and reduced-order observers have been designed such that the dynamics of each estimation error is guaranteed to be globally exponentially stable in the mean square. The problem of fault detection filter design for SMJSs with intermittent measurements has been addressed in [197], where a fault detection filter has been designed such that the residual system is stochastically admissible and satisfies some expected performances. Nevertheless, compared to the rich results for MJSs and SMJSs, the study of SMJSs with time delay has received much less attention. When time delays appear in SMJSs, the problems of stability and stabilization have been discussed for continuous-time cases in [9, 160] and discrete-time cases in [104, 216], respectively. In [133], the guaranteed cost control has been considered for SMJSs with time-varying delay, and an existence condition of the guaranteed cost state feedback controller has been derived. The H_∞ control problem has been investigated for SMJSs with time delay in [85, 105, 159, 215], where several BRLs have been given based on LMI approach and the design methods of desired controller have also been propsoed. The results on filtering for SMJSs with time delays can be found in [84, 156, 170]. In [148], the problem of sliding mode control has been discussed for SMJSs with time delay and an SMC law has been given to drive the system trajectories onto the predefined switching surface.

1.2 Book Organization

So far a large number of important and interesting results have been proposed for singular time-delay systems with or without Markov jumping parameters. However, there lacks of a monograph to provide the up-to-date advances in the literature. Thus, the main purpose of this book is to fill such gap by providing some recent developments in the analysis and synthesis issues for SSs with time delay and SMJSs with time delay. The materials adopted in the book are mainly based on research results of the authors.

Besides this short Introduction, this book is organized into two main parts; i.e., Part I: SSs with time delays and Part II: SMJSs with time delays.

Some developments on analysis and synthesis of SSs with time delays is presented in Part I, which starts with Chapter 2 and consists of seven relevant but independent chapters as follows.

Chapter 2 studies the robust stabilization problem for uncertain discrete-time SSs with time delays. In terms of strict LMI and a finite sum inequality, a delay-dependent criterion is obtained for the nominal systems to be admissible. Based on the criterion, a state feedback controller, which guarantees that, for all uncertainties, the resulting closed-loop system is regular, causal and stable, is constructed. An explicit expression for the desired controller is also given. The obtained results include both delay-independent and delay-dependent cases.

Chapter 3 investigates the problem of delay-dependent H_∞ control for SSs with state delay. In terms of LMI approach, a delay-dependent BRL is presented to ensure the system to be regular, impulse free and stable with H_∞ performance condition via an augmented Lyapunov functional. Based on the BRL obtained, the delay-dependent condition for the existence of H_∞ state feedback controller is presented via strict LMI. An explicit expression for the desired state feedback controller is also given.

Chapter 4 is devoted to the problem of reliable passive control for SSs with time-varying delays. The aim of the addressed reliable passive control problem is to design a state feedback controller such that, for all possible actuator failures, the resultant closed-loop system is regular, impulse-free, exponentially stable, and passive. A delay-dependent condition is established to guarantee the considered system to be regular, impulse-free, exponentially stable, and passive. Based on the derived condition, the reliable passive control problem is solved, and an explicit expression for the desired controller is given.

Chapter 5 addresses the dissipativity analysis problem for continuous-time/discrete-time SSs with time-varying delays. Some delay-dependent criterion are established to guarantee the dissipativity of the underlying systems using the delay partitioning technique. All the results given are not only dependent upon the time delay, but also dependent upon the number of delay partitions.

Chapter 6 focuses on the problem of delay-dependent $\mathscr{L}_2 - \mathscr{L}_\infty$ filtering for SSs with time delays. Attention is focused on the design of full-order filter that guarantees the delay-dependent exponentially admissible and a prescribed noise attenuation level in $\mathscr{L}_2 - \mathscr{L}_\infty$ sense for the filtering error dynamics. The desired filter can be constructed by solving certain LMIs.

Chapter 7 is concerned with the delay-dependent H_∞ filtering problem for SSs with time-varying delays in a range. In terms of LMI approach, the delay-range-dependent BRLs are proposed, which guarantee the considered system to be regular, impulse free and exponentially stable while satisfying a prescribed H_∞ performance level. The sufficient conditions are proposed for the existence of linear H_∞ filter.

Chapter 8 deals with the problem of H_∞ filter design for discrete-time networked SSs with both multiple stochastic time-varying communication delays and probabilistic missing measurements. Two kinds of stochastic time-varying communication delays, namely stochastic discrete delays and stochastic distributed delays, are simultaneously considered. The purpose of the addressed filtering problem is to design a filter such that, for the admissible random measurement missing and communication delays, the filtering error dynamics is asymptotically stable in the mean square with a prescribed H_∞ performance index. In terms of LMI method, a sufficient condition is established that ensures the asymptotical stability in the mean square with a prescribed H_∞ performance index of the filtering error dynamics and then the filter parameters are characterized by the solution to a LMI.

Some developments on analysis and synthesis of SMJSs with time delays presented in Part II, which starts with Chapter 9 and consists of six relevant but independent chapters as follows.

Chapter 9 considers the problem of delay-dependent stability analysis for a class of discrete-time SMJSs with time-varying delay. The transition probabilities in Markov chain are assumed to be partially unknown. In terms of LMI approach, the delay-dependent criteria are proposed to ensure the underlying system to be regular, causal and stochastically stable.

Chapter 10 studies the problem of delay-dependent H_∞ control for SMJSs with time delays. The aim of the problem is to design a state feedback controller, which guarantees that the resultant closed-loop system is not only regular, impulse free and stochastically stable, but also satisfies a prescribed H_∞ performance level for all delays no larger than a given upper bound in terms of LMI approach. A strict LMI condition is developed to guarantee the existence of the desired state feedback controller. An explicit expression for the desired controller is also given.

In Chapter 11, the problem of delay-dependent passivity analysis is studied for SMJSs with time-varying/time-invariant delays. By use of the delay partitioning method, some delay-dependent passivity conditions are derived for the considered systems via novel Lyapunov functionals including mode-dependent double integral terms. Some stochastic stability criteria are also given. All the results reported in this paper not only depend upon the time delays, but also depend upon their partitioning, which aims at reducing the conservatism.

Chapter 12 is devoted to the problem of mode-independent $l_2 - l_\infty$ filter design for discrete-time SMJSs with time-varying delays. By using the delay partitioning technique, a delay-dependent condition is established to guarantee the filtering error systems to be stochastically admissible and achieve a prescribed $l_2 - l_\infty$ performance index. Based on the derived condition, the full-order and reduced-order filters with mode-independent characterization are designed in a unified framework. The corresponding filter parameters can be obtained by solving a set of LMIs.

Chapter 13 studies the problem of mode-independent H_∞ filtering for SMJSs with time delay. In terms of LMI approach, a delay-dependent BRL is proposed for the considered system to be stochastically admissible while achieving the prescribed H_∞ performance condition. Based on the BRL and under partial knowledge of the jump rates of the Markov process, both delay-dependent and delay-independent sufficient conditions that guarantee the existence of the desired filter are presented. The explicit expression of the desired filter gains is also characterized by solving a set of strict LMIs.

Chapter 14 presents the synthesis design procedure of a mode-dependent H_∞ filter for SMJSs with time delays. A delay-dependent BRL is given to ensure that the system achieves mean-square exponentially admissible and guarantees a prescribed H_∞ performance index in terms of LMI approach. Based on the BRL, the H_∞ filtering problem is solved and the desired filter can be constructed by solving the corresponding LMIs.

1.3 Some Lemmas

Before ending this chapter, we give the following lemmas, which will be used throughout this book.

Lemma 1.2. *[118] Given matrices Ω, Γ and Ξ with appropriate dimensions and with Ω symmetrical, then*

$$\Omega + \Gamma F \Xi + \Xi^{\mathrm{T}} F^{\mathrm{T}} \Gamma^{\mathrm{T}} < 0 \tag{1.4}$$

holds for any F satisfying $F^{\mathrm{T}} F \leqslant I$, if and only if there exists a scalar $\varepsilon > 0$ such that

$$\Omega + \varepsilon \Gamma \Gamma^{\mathrm{T}} + \varepsilon^{-1} \Xi^{\mathrm{T}} \Xi < 0. \tag{1.5}$$

Lemma 1.3. *[10, 178] The matrix inequality*

$$\begin{bmatrix} Z_1 & Z_2 \\ * & Z_3 \end{bmatrix} \geqslant 0 \tag{1.6}$$

holds if and only if

$$Z_1 \geqslant 0,\ Z_1 - Z_2 Z_3^{+} Z_2^{\mathrm{T}} \geqslant 0,\ Z_2(I - Z_3 Z_3^{+}) = 0. \tag{1.7}$$

Lemma 1.4. *(Jensen Inequality)[49, 125]: For any matrix $W > 0$ with appropriate dimension, scalars γ_1 and γ_2 satisfying $\gamma_2 > \gamma_1$, and a vector function $\omega : [\gamma_1, \gamma_2] \to \mathbb{R}^n$ such that the integrations concerned are well defined, then*

$$(\gamma_2 - \gamma_1) \int_{\gamma_1}^{\gamma_2} \omega(\alpha)^{\mathrm{T}} W \omega(\alpha)\, \mathrm{d}\alpha \geqslant \left[\int_{\gamma_1}^{\gamma_2} \omega(\alpha)\, \mathrm{d}\alpha \right]^{\mathrm{T}} W \left[\int_{\gamma_1}^{\gamma_2} \omega(\alpha)\, \mathrm{d}\alpha \right]. \quad (1.8)$$

Lemma 1.5. *(Discretized Jensen Inequality)[67]: For any matrix $W > 0$ with appropriate dimension, integers γ_1 and γ_2 satisfying $\gamma_2 > \gamma_1$, and vector function $\omega : \mathbb{N}[\gamma_1, \gamma_2] \to \mathbb{R}^n$ such that the sums concerned are well defined, then*

$$(\gamma_2 - \gamma_1 + 1) \sum_{\alpha=\gamma_1}^{\gamma_2} \omega(\alpha)^{\mathrm{T}} W \omega(\alpha) \geqslant \sum_{\alpha=\gamma_1}^{\gamma_2} \omega(\alpha)^{\mathrm{T}} W \sum_{\alpha=\gamma_1}^{\gamma_2} \omega(\alpha). \quad (1.9)$$

Lemma 1.6. *For any matrices T_1, Y_1, Y_2, E and $Z > 0$ with appropriate dimensions, an integer $d > 0$, and a vector function $x : [k-d, k] \to \mathbb{R}^n$ such that the sums concerned are well defined, then*

$$\begin{aligned} -\sum_{\alpha=k-d}^{k-1} \eta(\alpha)^{\mathrm{T}} Z \eta(\alpha) \leqslant{}& 2\xi(k)^{\mathrm{T}} \begin{bmatrix} Y_1 E \\ Y_2 E \\ T_1 E \end{bmatrix} [x(k) - x(k-d)] \\ &+ d\xi(k)^{\mathrm{T}} \begin{bmatrix} Y_1 \\ Y_2 \\ T_1 \end{bmatrix} Z^{-1} \begin{bmatrix} Y_1 \\ Y_2 \\ T_1 \end{bmatrix}^{\mathrm{T}} \xi(k). \end{aligned} \quad (1.10)$$

where

$$\begin{aligned} \eta(k) &= Ex(k+1) - Ex(k), \\ \xi(k) &= \left[x(k)^{\mathrm{T}}\ \eta(k)^{\mathrm{T}}\ x(k-d)^{\mathrm{T}} \right]^{\mathrm{T}}. \end{aligned}$$

Proof. See Appendix.

Lemma 1.7. *For any given matrix $\begin{bmatrix} Z_1 & Z_2 \\ * & Z_3 \end{bmatrix} > 0$ with appropriate dimension, a scalar $d > 0$, and a vector function $x : [t-d, t] \to \mathbb{R}^n$ such that the integrations concerned are well defined, then*

$$\begin{aligned} -d \int_{t-d}^{t} & \begin{bmatrix} x(\alpha) \\ E\dot{x}(\alpha) \end{bmatrix}^{\mathrm{T}} \begin{bmatrix} Z_1 & Z_2 \\ * & Z_3 \end{bmatrix} \begin{bmatrix} x(\alpha) \\ E\dot{x}(\alpha) \end{bmatrix} \mathrm{d}\alpha \\ & \leqslant \begin{bmatrix} x(t) \\ x(t-d) \\ \int_{t-d}^{t} x(\alpha)\mathrm{d}\alpha \end{bmatrix}^{\mathrm{T}} \begin{bmatrix} -E^{\mathrm{T}} Z_3 E & E^{\mathrm{T}} Z_3 E & -E^{\mathrm{T}} Z_2^{\mathrm{T}} \\ E^{\mathrm{T}} Z_3 E & -E^{\mathrm{T}} Z_3 E & E^{\mathrm{T}} Z_2^{\mathrm{T}} \\ -Z_2 E & Z_2 E & -Z_1 \end{bmatrix} \begin{bmatrix} x(t) \\ x(t-d) \\ \int_{t-d}^{t} x(\alpha)\mathrm{d}\alpha \end{bmatrix}. \end{aligned} \quad (1.11)$$

Proof. See Appendix.

Lemma 1.8. *For any matrix* $\begin{bmatrix} M & S \\ * & M \end{bmatrix} \geqslant 0$ *with appropriate dimension, integers* d_1, d_2 *and* $d(k)$ *satisfying* $d_1 \leqslant d(k) \leqslant d_2$, *and a vector function* $x : \mathbb{N}[k-d_2, k-d_1] \to \mathbb{R}^n$ *such that the sums concerned are well defined, then*

$$-d_{12} \sum_{\alpha=k-d_2}^{k-d_1-1} \zeta(\alpha)^{\mathrm{T}} M \zeta(\alpha) \leqslant \varpi(k)^{\mathrm{T}} \Omega \varpi(k), \tag{1.12}$$

where $d_{12} = d_2 - d_1$, $\zeta(\alpha) = x(\alpha+1) - x(\alpha)$ *and*

$$\varpi(k) = \left[x(k-d_1)^{\mathrm{T}} \; x(k-d(k))^{\mathrm{T}} \; x(k-d_2)^{\mathrm{T}} \right]^{\mathrm{T}},$$

$$\Omega = \begin{bmatrix} -M & M-S & S \\ * & -2M+S+S^{\mathrm{T}} & -S+M \\ * & * & -M \end{bmatrix}.$$

Remark 1.9. Lemma 1.8 is a special case of the lower bounds lemma of [116], which is presented in a form more convenient for the application in this book. It is noted that if $S = 0$, the finite sum inequality (1.12) reduces to [129, inequality (11)]. Thus, inequality (1.12) is tighter than the existing one.

Lemma 1.10. *[125, 199] Suppose that a positive continuous function* $f(t)$ *satisfies*

$$f(t) \leqslant \zeta_1 \sup_{t-d \leqslant s \leqslant t} f(s) + \zeta_2 e^{-\varepsilon t}, \tag{1.13}$$

where $\varepsilon > 0$, $0 < \zeta_1 < 1$, $0 < \zeta_1 e^{\varepsilon d} < 1$ $\zeta_2 > 0$ *and* $d > 0$, *then*

$$f(t) \leqslant e^{-\varepsilon t} \sup_{-d \leqslant s \leqslant 0} f(s) + \frac{\zeta_2 e^{-\varepsilon t}}{1 - \zeta_1 e^{\varepsilon d}}. \tag{1.14}$$

Lemma 1.11. *Suppose that a positive continuous stochastic function* $f(t)$ *satisfies*

$$\mathscr{E}\{f(t)\} \leqslant \zeta_1 \mathscr{E}\{ \sup_{t-d \leqslant s \leqslant t} f(s)\} + \zeta_2 e^{-\varepsilon t}, \tag{1.15}$$

where $\varepsilon > 0$, $0 < \zeta_1 < 1$, $0 < \zeta_1 e^{\varepsilon d} < 1$ $\zeta_2 > 0$ *and* $d > 0$, *then*

$$\mathscr{E}\{f(t)\} \leqslant e^{-\varepsilon t} \mathscr{E}\{ \sup_{-d \leqslant s \leqslant 0} f(s)\} + \frac{\zeta_2 e^{-\varepsilon t}}{1 - \zeta_1 e^{\varepsilon d}}. \tag{1.16}$$

Remark 1.12. The proof of Lemma 1.11 can be carried out similarly to that of [199, Lemma 2]. It is noted that Lemma 1.11 is the stochastic case of Lemma 1.10.

Part I

Singular Systems (SSs) with Time-Delays

2

Robust Stabilization for Uncertain Discrete-Time SSs with State Delay

2.1 Introduction

The problems of delay-dependent stability analysis and stabilization for singular time-delay systems have received increasing attention. Based on LMI approach, several delay-dependent stability criteria have been given in [7, 38, 187]. In [128], the delay-dependent robust stabilization problem for uncertain singular time-delay systems has been investigated by using model transformation and bounding technique for cross terms, and several LMI conditions for the solvability of this problem have been obtained. The problem has also been discussed in [120], where neither model transformation nor bounding technique for cross terms is needed in the development of the results.

It should be pointed out that all of the above mentioned works have been developed in the context of continuous-time singular time-delay systems. It seems that the corresponding results for discrete-time case are relatively few. In [14, 15, 176, 188], the robust stability problem has been discussed for discrete-time SSs with state delay and uncertainty, and several robust stability and D-stability conditions have been established. But it should be pointed out that in the above mentioned results, some additional assumptions on the system are needed in the development of the results. Based on LMI method, the delay-dependent robust stability problem has been discussed for discrete-time SSs with state delay and norm bounded parametric uncertainties in [63, 64], and several delay-dependent and delay-independent stability criteria have been derived. Because the results of [63, 64] are formulated by LMI and no additional assumption on the system is necessary, the results proposed in [63, 64] are more desired and have computational advantages over [14, 15, 176, 188]. However, the delay-dependent robust stabilization problem has not been solved in the above mentioned papers.

Based on LMI approach and a finite sum inequality, the problem of robust stabilization for discrete-time SSs with time-delay and norm bounded parametric uncertainties is addressed in this chapter. The aim of the problem

Z.-G. Wu et al.: *Anal. & Synth. of Singular Syst. with Time-Delays*, LNCIS 443, pp. 13–25.
DOI: 10.1007/978-3-642-37497-5_2

is to design a state feedback controller which renders, for all uncertainties, the resulting closed-loop system is regular, causal and stable. The robust stability condition and robust stabilizability condition of the considered system are proposed by strict LMIs. These strict LMI-based conditions include delay-dependent and delay-independent results and all obtained results are formulated by the coefficient matrices of the original systems, which avoids the decomposition of the given system matrices. Two examples are provided to demonstrate that the results given in this chapter improve the existing ones.

2.2 Preliminaries

Consider the following uncertain discrete-time SS with state delay:

$$\begin{cases} Ex(k+1) = (A+\Delta A)x(k) + (A_d + \Delta A_d)x(k-d) \\ \qquad\qquad + (B+\Delta B)u(k), \\ x(k) = \phi(k),\ k \in \mathbb{N}[-d,\ 0], \end{cases} \tag{2.1}$$

where $x(k) \in \mathbb{R}^n$ is the state, $u(k) \in \mathbb{R}^m$ is the control input, $d > 0$ is an integer representing the time delay, and $\phi(k)$ is the compatible initial condition. The matrix $E \in \mathbb{R}^{n\times n}$ may be singular and it is assumed that $\operatorname{rank} E = r \leqslant n$. A, A_d and B are known real constant matrices with appropriate dimensions. ΔA, ΔA_d and ΔB are unknown matrices representing norm-bounded parametric uncertainties and are assumed to be of the following form,

$$\begin{bmatrix} \Delta A\ \Delta A_d\ \Delta B \end{bmatrix} = MF(k) \begin{bmatrix} N_1\ N_2\ N_3 \end{bmatrix}, \tag{2.2}$$

where M, N_1, N_2 and N_3 are known real constant matrices with appropriate dimensions, and $F(k) \in \mathbb{R}^{q\times k}$ is an unknown real and possibly time-varying matrix satisfying

$$F(k)^{\mathrm{T}}F(k) \leqslant I. \tag{2.3}$$

The nominal unforced system of (2.1) can be written as

$$\begin{cases} Ex(k+1) = Ax(k) + A_d x(k-d), \\ x(k) = \phi(k),\ k \in \mathbb{N}[-d, 0]. \end{cases} \tag{2.4}$$

Definition 2.1. *[21, 178]*

1. *The pair (E, A) is said to be regular if* $\det(zE - A) \not\equiv 0$,
2. *The pair (E, A) is said to be causal if* $\deg(\det(zE - A)) = \operatorname{rank} E$.

Definition 2.2. *[178]*

1. *System (2.4) is said to be regular and causal, if the pair (E, A) is regular and causal,*

2. System (2.4) is said to be stable if for any scalar $\varepsilon > 0$, there exits a scalar $\delta(\varepsilon) > 0$ such that, for any compatible initial conditions $\phi(k)$ satisfying $\sup_{-d\leqslant k\leqslant 0} \|\phi(k)\| \leqslant \delta(\varepsilon)$, *the solution $x(k)$ of system (2.4) satisfies $\|x(k)\| \leqslant \varepsilon$ for any $k \geqslant 0$, moreover $\lim_{k\to\infty} x(k) = 0$,*
3. System (2.4) is said to be admissible if it is regular, causal and stable.

Before closing this section, we present the following result which will be used in the proof of the further results.

Lemma 2.3. *System (2.4) is stable, if there exit scalars $\alpha > 0$ and $\beta > 0$, and a discrete-time functional $V(k) = V(x(k), x(k-1), \cdots, x(k-d))$ such that for any $x(k)$ satisfying system (2.4),*

$$0 \leqslant V(k) \leqslant \alpha \sup_{k-d\leqslant i\leqslant k} \|x(i)\|^2,$$
$$\Delta V(k) = V(k+1) - V(k) \leqslant -\beta\|x(k)\|^2. \tag{2.5}$$

Proof. From (2.5), we obtain that for any $k \geqslant 0$,

$$-V(0) \leqslant V(k+1) - V(0) = \sum_{i=0}^{k} \Delta V(i) \leqslant -\beta \sum_{i=0}^{k} ||x(i)||^2. \tag{2.6}$$

Hence, for any $x(k)$ satisfying system (2.4), we have

$$\sum_{i=0}^{k} ||x(i)||^2 \leqslant \frac{1}{\beta} V(0) \leqslant \frac{\alpha}{\beta} \sup_{-d\leqslant i\leqslant 0} \|\phi(i)\|^2. \tag{2.7}$$

Therefore, $\|x(k)\|$ is small enough for small enough $\sup_{-d\leqslant i\leqslant 0} \|\phi(i)\|$. Moreover, Series $\sum_{i=0}^{\infty} ||x(i)||^2$ converge, which implies that $\lim_{k\to\infty} x(k) = 0$. According to Definition 2.2, system (2.4) is stable. This completes the proof.

Remark 2.4. It is noted that Lemma 2.3 proposes a simple and effective method to determine the stability of system (2.4). It can be viewed as a discrete-time counterpart of [35, Lemma 1], which is used to demonstrate the stability for continuous-time SSs with time delay.

Remark 2.5. In [14, 178, 184], to establish the stability conditions for the same system, the considered system is converted into delay-free systems by augmentation approach. However, for large known delay, this method will lead to large-dimensional systems, which makes the proof procedure complicated and unreliable. Furthermore, for unknown delay or time-varying delay the augmentation method become difficult to apply. For Lemma 2.3, however, not only it is not necessary to convert the considered systems, but also the stability analysis problem for systems with unknown delay or time-varying delay can be solved. Hence, Lemma 2.3 is rather powerful and general to establish stability conditions for discrete-time SSs with time delay.

2.3 Main Results

2.3.1 Stability Analysis

The delay-dependent and delay-independent conditions will be established based on the LMI approach for system (2.1) with $u(k) = 0$ to be regular, causal and stable for all uncertainties. Initially, the following theorem is presented for system (2.4) to be regular, causal and delay-dependent stable.

Theorem 2.6. *System (2.4) is admissible, if there exist matrices* $P_1 > 0$, $Q > 0$, $Z > 0$, S, S_e S_d, P_2, P_3, P_4, Y_1, Y_2 *and* T_1 *such that*

$$\Lambda = \begin{bmatrix} \Xi_{11} & \Xi_{12} & \Xi_{13} & dY_1 \\ * & \Xi_{22} & \Xi_{23} & dY_2 \\ * & * & \Xi_{33} & dT_1 \\ * & * & * & -dZ \end{bmatrix} < 0, \tag{2.8}$$

where $R \in \mathbb{R}^{n\times(n-r)}$ *is any full column rank matrix satisfying* $E^{\mathrm{T}}R = 0$, *and*

$$\begin{aligned} \Xi_{11} &= P_2^{\mathrm{T}}(A-E) + (A-E)^{\mathrm{T}}P_2 + Y_1E + E^{\mathrm{T}}Y_1^{\mathrm{T}} + Q, \\ \Xi_{12} &= E^{\mathrm{T}}P_1 + SR^{\mathrm{T}} - P_2^{\mathrm{T}} + (A-E)^{\mathrm{T}}P_3 + E^{\mathrm{T}}Y_2^{\mathrm{T}}, \\ \Xi_{13} &= -Y_1E + P_2^{\mathrm{T}}A_d + E^{\mathrm{T}}T_1^{\mathrm{T}} + (A-E)^{\mathrm{T}}P_4, \\ \Xi_{22} &= S_eR^{\mathrm{T}} + RS_e^{\mathrm{T}} - P_3 - P_3^{\mathrm{T}} + P_1 + dZ, \\ \Xi_{23} &= -Y_2E + P_3^{\mathrm{T}}A_d - P_4 + RS_d^{\mathrm{T}}, \\ \Xi_{33} &= -Q - T_1E - E^{\mathrm{T}}T_1^{\mathrm{T}} + P_4^{\mathrm{T}}A_d + A_d^{\mathrm{T}}P_4. \end{aligned}$$

Proof. From (2.8), it is easy to show that

$$\begin{bmatrix} \Omega_1 & \tilde{S}_1\tilde{R}^{\mathrm{T}}\tilde{A}_d + \tilde{A}^{\mathrm{T}}\tilde{R}\tilde{S}_2^{\mathrm{T}} - \tilde{Y}\tilde{E} + \tilde{E}^{\mathrm{T}}\tilde{T}^{\mathrm{T}} \\ * & -\tilde{Q} + \tilde{S}_2\tilde{R}^{\mathrm{T}}\tilde{A}_d + \tilde{A}_d^{\mathrm{T}}\tilde{R}\tilde{S}_2^{\mathrm{T}} - \tilde{T}\tilde{E} - \tilde{E}^{\mathrm{T}}\tilde{T}^{\mathrm{T}} \end{bmatrix} < 0, \tag{2.9}$$

where

$$\begin{aligned} \Omega_1 &= \tilde{A}^{\mathrm{T}}\tilde{P}\tilde{A} - \tilde{E}^{\mathrm{T}}\tilde{P}\tilde{E} + \tilde{S}_1\tilde{R}^{\mathrm{T}}\tilde{A} + \tilde{A}^{\mathrm{T}}\tilde{R}\tilde{S}_1^{\mathrm{T}} + \tilde{Y}\tilde{E} + \tilde{E}^{\mathrm{T}}\tilde{Y}^{\mathrm{T}} + \tilde{Q}, \\ \tilde{E} &= \begin{bmatrix} E & 0 \\ 0 & 0 \end{bmatrix}, \ \tilde{A} = \begin{bmatrix} E & I \\ A-E & -I \end{bmatrix}, \ \tilde{A}_d = \begin{bmatrix} 0 & 0 \\ A_d & 0 \end{bmatrix}, \\ \tilde{P} &= \begin{bmatrix} P_1 & 0 \\ 0 & 0 \end{bmatrix}, \ \tilde{Y} = \begin{bmatrix} Y_1 & 0 \\ Y_2 & 0 \end{bmatrix}, \ \tilde{T} = \begin{bmatrix} T_1 & 0 \\ 0 & 0 \end{bmatrix}, \ \tilde{Q} = \begin{bmatrix} Q & 0 \\ 0 & dZ \end{bmatrix}, \\ \tilde{S}_1 &= \begin{bmatrix} S & P_2^{\mathrm{T}} \\ S_e & P_3^{\mathrm{T}} \end{bmatrix}, \ \tilde{S}_2 = \begin{bmatrix} S_d & P_4^{\mathrm{T}} \\ 0 & 0 \end{bmatrix}, \ \tilde{R} = \begin{bmatrix} R & 0 \\ 0 & I \end{bmatrix}. \end{aligned}$$

Since $\operatorname{rank}\tilde{E} = \operatorname{rank}E = r \leqslant n$, there exist nonsingular matrices M and N such that

$$\hat{E} = M\tilde{E}N = \begin{bmatrix} I_r & 0 \\ 0 & 0 \end{bmatrix}. \tag{2.10}$$

Denote

$$\hat{A} = M\tilde{A}N = \begin{bmatrix} A_{11} & A_{12} \\ A_{21} & A_{22} \end{bmatrix}, \tag{2.11a}$$

$$\hat{P} = M^{-T}\tilde{P}M^{-1} = \begin{bmatrix} P_{11} & P_{12} \\ P_{21} & P_{22} \end{bmatrix}, \tag{2.11b}$$

$$\hat{Y} = N^{\mathrm{T}}\tilde{Y}M^{-1} = \begin{bmatrix} Y_{11} & Y_{12} \\ Y_{21} & Y_{22} \end{bmatrix}, \tag{2.11c}$$

$$\hat{S} = N^{T}\tilde{S}_1 = \begin{bmatrix} S_1 \\ S_2 \end{bmatrix}, \tag{2.11d}$$

$$\hat{R} = M^{-T}\tilde{R} = \begin{bmatrix} 0 \\ I \end{bmatrix} H, \tag{2.11e}$$

where $H \in \mathbb{R}^{(2n-r)\times(2n-r)}$ is any nonsingular matrix. From (2.9), it is easy to see that

$$-\tilde{E}^{\mathrm{T}}\tilde{P}\tilde{E} + \tilde{S}_1\tilde{R}^{\mathrm{T}}\tilde{A} + \tilde{A}^{\mathrm{T}}\tilde{R}\tilde{S}_1^{\mathrm{T}} + \tilde{Y}\tilde{E} + \tilde{E}^{\mathrm{T}}\tilde{Y}^{\mathrm{T}} < 0. \tag{2.12}$$

Pre-multiplying and post-multiplying (2.12) by N^{T} and N, respectively, we have

$$\begin{bmatrix} \star & \star \\ \star & S_2H^{\mathrm{T}}A_{22} + A_{22}^{\mathrm{T}}HS_2^{\mathrm{T}} \end{bmatrix} < 0, \tag{2.13}$$

where "$\star$" represents matrices that are not relevant in the following discussion. From (2.13), we can obtain that A_{22} is nonsingular, which implies the pair $(\tilde{E}, \tilde{A})$ is regular and causal. Noting the fact that $\det(zE - A) = \det(z\tilde{E} - \tilde{A})$, we can easily see that the pair (E, A) are regular and causal. Hence, according to Definition 2.2, system (2.4) is regular and causal.

Next we will show that system (2.4) is stable. To the end, we propose the following Lyapunov functional:

$$V(k) = V_1(k) + V_2(k) + V_3(k), \tag{2.14}$$

where

$$\begin{aligned} V_1(k) =& x(k)^{\mathrm{T}}E^{\mathrm{T}}P_1Ex(k), \\ V_2(k) =& \sum_{\beta=-d}^{-1}\sum_{\alpha=k+\beta}^{k-1} [Ex(\alpha+1) - Ex(\alpha)]^{\mathrm{T}}Z[Ex(\alpha+1) - Ex(\alpha)], \\ V_3(k) =& \sum_{\alpha=k-d}^{k-1} x(\alpha)^{\mathrm{T}}Qx(\alpha). \end{aligned}$$

It is clear that $0 \leqslant V(k) \leqslant \alpha \sup_{k-d\leqslant i\leqslant k} \|x(i)\|^2$, where $\alpha = \lambda_{\max}(E^{\mathrm{T}}P_1E) + d\lambda_{\max}(Q) + 4d^2\lambda_{\max}(E^{\mathrm{T}}ZE)$. Since $\Delta V_1(k)$, $\Delta V_2(k)$ and $\Delta V_3(k)$ yield the relation

$$\begin{aligned}\Delta V_1(k) =& x(k+1)^{\mathrm{T}} E^{\mathrm{T}} P_1 E x(k+1) - x(k)^{\mathrm{T}} E^{\mathrm{T}} P_1 E x(k),\\ =& [Ex(k+1) - Ex(k)]^{\mathrm{T}} P_1 [Ex(k+1) - Ex(k)]\\ &+ 2x(k)^{\mathrm{T}} E^{\mathrm{T}} P_1 [Ex(k+1) - Ex(k)],\end{aligned} \tag{2.15}$$

$$\begin{aligned}\Delta V_2(k) \leqslant& d[Ex(k+1) - Ex(k)]^{\mathrm{T}} Z [Ex(k+1) - Ex(k)]\\ &- \sum_{\alpha=k-d}^{k-1} [Ex(\alpha+1) - Ex(\alpha)]^{\mathrm{T}} Z [Ex(\alpha+1) - Ex(\alpha)],\end{aligned} \tag{2.16}$$

$$\Delta V_3(k) = x(k)^{\mathrm{T}} Q x(k) - x(k-d)^{\mathrm{T}} Q x(k-d). \tag{2.17}$$

By Lemma 1.6, we have

$$\begin{aligned}\Delta V_2(k) \leqslant& d[Ex(k+1) - Ex(k)]^{\mathrm{T}} Z [Ex(k+1) - Ex(k)]\\ &+ 2\xi(k)^{\mathrm{T}} \begin{bmatrix} Y_1 E \\ Y_2 E \\ T_1 E \end{bmatrix} [x(k) - x(k-d)]\\ &+ d\xi(k)^{\mathrm{T}} \begin{bmatrix} Y_1 E \\ Y_2 E \\ T_1 E \end{bmatrix} Z^{-1} \begin{bmatrix} Y_1 E \\ Y_2 E \\ T_1 E \end{bmatrix}^{\mathrm{T}} \xi(k),\end{aligned} \tag{2.18}$$

where $\xi(k) = \begin{bmatrix} x(k)^{\mathrm{T}} & (Ex(k+1) - Ex(k))^{\mathrm{T}} & x(k-d)^{\mathrm{T}} \end{bmatrix}^{\mathrm{T}}$. On the other hand, we can find that, for any appropriately dimensioned matrices S, S_e S_d, P_2, P_3 and P_4, the following equalities hold:

$$\begin{aligned}\alpha(k) =& 2[x(k)^{\mathrm{T}} P_2^{\mathrm{T}} + [Ex(k+1) - Ex(k)]^{\mathrm{T}} P_3^{\mathrm{T}} + x(k-d)^{\mathrm{T}} P_4^{\mathrm{T}}]\\ &\times [(A-E)x(k) - [Ex(k+1) - Ex(k)] + A_d x(k-d)]\\ =& 0,\end{aligned} \tag{2.19}$$

$$\begin{aligned}\beta(k) =& 2[x(k)^{\mathrm{T}} S R^{\mathrm{T}} + [Ex(k+1) - Ex(k)]^{\mathrm{T}} S_e R^{\mathrm{T}} + x(k-d)^{\mathrm{T}} S_d R^{\mathrm{T}}]\\ &\times [Ex(k+1) - Ex(k)]\\ =& 0.\end{aligned} \tag{2.20}$$

Hence,

$$\begin{aligned}\Delta V(k) =& \Delta V_1(k) + \Delta V_2(k) + \Delta V_3(k) + \alpha(k) + \beta(k)\\ \leqslant& \xi(k)^{\mathrm{T}} \Pi \xi(k),\end{aligned} \tag{2.21}$$

where

$$\Pi = \begin{bmatrix} \Xi_{11} & \Xi_{12} & \Xi_{13} \\ * & \Xi_{22} & \Xi_{23} \\ * & * & \Xi_{33} \end{bmatrix} + d \begin{bmatrix} Y_1 E \\ Y_2 E \\ T_1 E \end{bmatrix} Z^{-1} \begin{bmatrix} Y_1 E \\ Y_2 E \\ T_1 E \end{bmatrix}^{\mathrm{T}}.$$

By Schur complement, $\Pi < 0$ can be obtained from (2.8). Therefore, $\Delta V(k) \leqslant -\beta \|x(k)\|^2$, where $\beta = -\lambda_{\max}(\Pi) > 0$. Thus, according to Lemma 2.3, system (2.4) is stable. This completes the proof.

Remark 2.7. It is noted that Theorem 2.6 proposes a strict LMI-based delay-dependent criterion guaranteeing the considered system to be admissible. In the case when $E = I$, that is, system (2.4) reduces to a state-space time-delay system, it is easy to show that Theorem 2.6 coincides with [41, Proposition 2]. Therefore, Theorem 2.6 can be viewed as an extension of existing results on state-space systems with time-delay to SSs with time-delay.

Remark 2.8. From the proof of Theorem 2.6, it is clear to see that the finite sum inequality (1.10) plays an important role in the derivation of the delay-dependent condition. Neither model transformation nor bounding technique for cross-terms is involved. Hence, the conservatism inherited from these ideas will no longer exit in Theorem 2.6.

Remark 2.9. In the computation of $\Delta V(k)$, two null terms $\alpha(k)$ and $\beta(k)$ are introduced to get less restriction solution. Hence, the free-weighting matrix approach, which has been widely used to deal with the delay-dependent problem for state-space time-delay systems, is extended to discrete-time SSs with time delay in Theorem 2.6. It is worth pointing out that $\beta(k)$ for state-space time-delay systems will vanish because of the nonsingularity of E.

The delay-dependent result for uncertain system (2.1) with $u(k) = 0$ can be described as the following theorem.

Theorem 2.10. *System* (2.1) *with* $u(k) = 0$ *is admissible, if there exist matrices* $P_1 > 0$, $Q > 0$, $Z > 0$, S, S_e, S_d, P_2, P_3, P_4, Y_1, Y_2 *and* T_1 *such that*

$$\begin{bmatrix} \Omega_{11} & \Xi_{12} & \Omega_{13} & dY_1 & P_2^{\mathrm{T}}M \\ * & \Xi_{22} & \Xi_{23} & dY_2 & P_3^{\mathrm{T}}M \\ * & * & \Omega_{33} & dT_1 & P_4^{\mathrm{T}}M \\ * & * & * & -dZ & 0 \\ * & * & * & * & -I \end{bmatrix} < 0, \tag{2.22}$$

where R, Ξ_{12}, Ξ_{22} *and* Ξ_{23} *are given in Theorem 2.6, and*

$$\begin{aligned} \Omega_{11} &= P_2^{\mathrm{T}}(A - E) + (A - E)^{\mathrm{T}}P_2 + Y_1E + E^{\mathrm{T}}Y_1^{\mathrm{T}} + Q + N_1^{\mathrm{T}}N_1, \\ \Omega_{13} &= -Y_1E + P_2^{\mathrm{T}}A_d + E^{\mathrm{T}}T_1^{\mathrm{T}} + (A - E)^{\mathrm{T}}P_4 + N_1^{\mathrm{T}}N_2, \\ \Omega_{33} &= -Q - T_1E - E^{\mathrm{T}}T_1^{\mathrm{T}} + P_4^{\mathrm{T}}A_d + A_d^{\mathrm{T}}P_4 + N_2^{\mathrm{T}}N_2. \end{aligned}$$

Proof. Using Schur complement, based on (2.22), we have

$$\Lambda + DD^{\mathrm{T}} + N^{\mathrm{T}}N < 0, \tag{2.23}$$

where Λ is the same as the left-hand side of (2.8), and

$$\begin{aligned} D &= \begin{bmatrix} M^{\mathrm{T}}P_2 & M^{\mathrm{T}}P_3 & M^{\mathrm{T}}P_4 & 0 \end{bmatrix}^{\mathrm{T}}, \\ N &= \begin{bmatrix} N_1 & 0 & N_2 & 0 \end{bmatrix}. \end{aligned}$$

By Lemma 1.2, it follows from (2.23) that

$$\begin{bmatrix} \Psi_{11} & \Psi_{12} & \Psi_{13} & dY_1 \\ * & \Xi_{22} & \Psi_{23} & dY_2 \\ * & * & \Psi_{33} & dT_1 \\ * & * & * & -dZ \end{bmatrix} < 0, \tag{2.24}$$

where

$$\begin{aligned}
\Psi_{11} &= P_2^{\mathrm{T}}(A+\Delta A-E)+(A+\Delta A-E)^{\mathrm{T}}P_2+Y_1E+E^{\mathrm{T}}Y_1^{\mathrm{T}}+Q,\\
\Psi_{12} &= E^{\mathrm{T}}P_1+SR^{\mathrm{T}}-P_2^{\mathrm{T}}+(A+\Delta A-E)^{\mathrm{T}}P_3+E^{\mathrm{T}}Y_2^{\mathrm{T}},\\
\Psi_{13} &= -Y_1E+P_2^{\mathrm{T}}(A_d+\Delta A_d)+E^{\mathrm{T}}T_1^{\mathrm{T}}+(A+\Delta A-E)^{\mathrm{T}}P_4,\\
\Psi_{23} &= -Y_2E+P_3^{\mathrm{T}}(A_d+\Delta A_d)-P_4+RS_d^{\mathrm{T}},\\
\Psi_{33} &= -Q-T_1E-E^{\mathrm{T}}T_1^{\mathrm{T}}+P_4^{\mathrm{T}}(A_d+\Delta A_d)+(A_d+\Delta A_d)^{\mathrm{T}}P_4.
\end{aligned}$$

According to Theorem 2.6, we have the desired result immediately. This completes the proof.

If the matrices, in (2.22), $Y_1=Y_2=T_1=0$ and $Z=\varepsilon I/d(\varepsilon\to 0)$, then Theorem 2.10 provides the delay-independent result, which is stated as follows.

Corollary 2.11. *System (2.1) with $u(k)=0$ is admissible, if there exist $P_1>0$, $Q>0$, S, S_e, S_d, P_2, P_3 and P_4 such that*

$$\begin{bmatrix} \Omega_{11} & \Omega_{12} & \Omega_{13} & P_2^{\mathrm{T}}M \\ * & \Omega_{22} & \Omega_{23} & P_3^{\mathrm{T}}M \\ * & * & \Omega_{33} & P_4^{\mathrm{T}}M \\ * & * & * & -I \end{bmatrix} < 0, \tag{2.25}$$

where $R\in\mathbb{R}^{n\times(n-r)}$ is any full column rank matrix satisfying $E^{\mathrm{T}}R=0$, and

$$\begin{aligned}
\Omega_{11} &= P_2^{\mathrm{T}}(A-E)+(A-E)^{\mathrm{T}}P_2+Q+N_1^{\mathrm{T}}N_1,\\
\Omega_{12} &= E^{\mathrm{T}}P_1+SR^{\mathrm{T}}-P_2^{\mathrm{T}}+(A-E)^{\mathrm{T}}P_3,\\
\Omega_{13} &= P_2^{\mathrm{T}}A_d+(A-E)^{\mathrm{T}}P_4+N_1^{\mathrm{T}}N_2,\\
\Omega_{22} &= S_eR^{\mathrm{T}}+RS_e^{\mathrm{T}}-P_3-P_3^{\mathrm{T}}+P_1,\\
\Omega_{23} &= P_3^{\mathrm{T}}A_d-P_4+RS_d^{\mathrm{T}},\\
\Omega_{33} &= -Q+P_4^{\mathrm{T}}A_d+A_d^{\mathrm{T}}P_4+N_2^{\mathrm{T}}N_2.
\end{aligned}$$

Remark 2.12. It is noted that a delay-independent condition is established in Corollary 2.11 for system (2.1) with $u(k)=0$ to be admissible for all uncertainties by strict LMI, which is contrast to that in [184], where a nonstrict LMI condition has been reported. Compared with a nonstrict LMI condition, testing a strict LMI condition can avoid some numerical problems arising from equality constraints. Thus, the result in this chapter is elegant than that of [184] from the mathematical point of view.

2.3.2 Stabilization

A state feedback controller $u(k) = Kx(k)$, $K \in \mathbb{R}^{m\times n}$ will be designed for system (2.1) such that the resultant closed-loop system is delay-dependent/delay-independent admissible for all uncertainties.

We first consider the nominal system of system (2.1) described by

$$\begin{cases} Ex(k+1) = Ax(k) + A_d x(k-d) + Bu(k), \\ \quad x(k) = \phi(k),\ k \in \mathbb{N}[-d, 0]. \end{cases} \tag{2.26}$$

In the following theorem, we make use of Theorem 2.6 to design the controller for the above system such that the resultant closed-loop system is delay-dependent admissible.

Theorem 2.13. *For given scalars ϵ_1, ϵ_2 and ϵ_3, system (2.26) is admissible, if there exist matrices $P_1 > 0$, $Q > 0$, $Z > 0$, S, S_e, S_d, P, X, Y_1, Y_2 and T_1 such that*

$$\Omega = \begin{bmatrix} \Xi_{11} & \Xi_{12} & \Xi_{13} & dY_1 \\ * & \Xi_{22} & \Xi_{23} & dY_2 \\ * & * & \Xi_{33} & dT_1 \\ * & * & * & -dZ \end{bmatrix} < 0, \tag{2.27}$$

where $R \in \mathbb{R}^{n\times(n-r)}$ is any full column rank matrix satisfying $ER = 0$, and

$$\begin{aligned} \Xi_{11} &= \epsilon_1 P^{\mathrm{T}}(A-E)^{\mathrm{T}} + \epsilon_1(A-E)P + \epsilon_1(BX + X^{\mathrm{T}}B^{\mathrm{T}}) + Y_1E^{\mathrm{T}} + EY_1^{\mathrm{T}} + Q, \\ \Xi_{12} &= EP_1 + SR^{\mathrm{T}} - \epsilon_1 P^{\mathrm{T}} + \epsilon_2(A-E)P + \epsilon_2 BX + EY_2^{\mathrm{T}}, \\ \Xi_{13} &= -Y_1E + \epsilon_1 P^{\mathrm{T}}A_d^{\mathrm{T}} + ET_1^{\mathrm{T}} + \epsilon_3(A-E)P + \epsilon_3 BX, \\ \Xi_{22} &= S_eR^{\mathrm{T}} + RS_e^{\mathrm{T}} - \epsilon_2 P^{\mathrm{T}} - \epsilon_2 P + P_1 + dZ, \\ \Xi_{23} &= -Y_2E^{\mathrm{T}} + \epsilon_2 P^{\mathrm{T}}A_d^{\mathrm{T}} - \epsilon_3 P + RS_d^{\mathrm{T}}, \\ \Xi_{33} &= -Q - T_1E^{\mathrm{T}} - ET_1^{\mathrm{T}} + \epsilon_3 P^{\mathrm{T}}A_d^{\mathrm{T}} + \epsilon_3 A_d P. \end{aligned}$$

Furthermore, if (2.27) is solvable, the desired controller gain is given as

$$K = XP^{-1}. \tag{2.28}$$

Proof. Applying the controller $u(k) = Kx(k)$ to system (2.26) yields

$$Ex(k+1) = (A + BK)x(k) + A_d x(k-d). \tag{2.29}$$

Noting the fact that $\det(zE - (A+BK)) = \det(zE^{\mathrm{T}} - (A+BK)^{\mathrm{T}})$, and the solutions to $\det(zE - (A+BK) - z^{-d}A_d) = 0$ are the same as those of $\det(zE^{\mathrm{T}} - (A+BK)^{\mathrm{T}} - z^{-d}A_d^{\mathrm{T}}) = 0$, as long as the regularity, causality and stability are concerned, we can see that the above system is equivalent to the following system,

$$E^{\mathrm{T}}\zeta(k+1) = (A+BK)^{\mathrm{T}}\zeta(k) + A_d^{\mathrm{T}}\zeta(k-d). \tag{2.30}$$

Replacing E by E^{T}, A by $(A+BK)^{\mathrm{T}}$, A_d by A_d^{T} in (2.8), and setting $P_2 = \epsilon_1 P$, $P_3 = \epsilon_2 P$, $P_4 = \epsilon_3 P$ and $X = KP$, we can obtain (2.27) immediately. This completes the proof.

The result of Theorem 2.13 can be extended to the uncertain case and the following result can be obtained.

Theorem 2.14. *For given scalars ϵ_1, ϵ_2 and ϵ_3, system (2.1) is admissible, if there exist a scalar $\varepsilon > 0$, and matrices $P_1 > 0$, $Q > 0$, $Z > 0$, S, S_e, S_d, P, X, Y_1, Y_2 and T_1 such that*

$$\begin{bmatrix} \Xi_{11}+\varepsilon MM^{\mathrm{T}} & \Xi_{12} & \Xi_{13} & dY_1 & \epsilon_1(N_1P+N_3X)^{\mathrm{T}} & \epsilon_1 N_2 P^{\mathrm{T}} \\ * & \Xi_{22} & \Xi_{23} & dY_2 & \epsilon_2(N_1P+N_3X)^{\mathrm{T}} & \epsilon_2 N_2 P^{\mathrm{T}} \\ * & * & \Xi_{33}+\varepsilon MM^{\mathrm{T}} & dT_1 & \epsilon_3(N_1P+N_3X)^{\mathrm{T}} & \epsilon_3 N_2 P^{\mathrm{T}} \\ * & * & * & -dZ & 0 & 0 \\ * & * & * & * & -\varepsilon I & 0 \\ * & * & * & * & * & -\varepsilon I \end{bmatrix} < 0, \tag{2.31}$$

where R, Ξ_{11}, Ξ_{12}, Ξ_{13}, Ξ_{22}, Ξ_{23} and Ξ_{33} are given in (2.27). Furthermore, if (2.31) is solvable, the desired controller gain is given in (2.28).

Proof. Replacing A, A_d and B with $A+\Delta A$, $A_d+\Delta A_d$ and $B+\Delta B$, respectively, (2.27) can be written as

$$\Omega + \Xi \begin{bmatrix} F(k) & 0 \\ 0 & F(k) \end{bmatrix} \Phi + \Phi^{\mathrm{T}} \begin{bmatrix} F(k) & 0 \\ 0 & F(k) \end{bmatrix}^{\mathrm{T}} \Xi^{\mathrm{T}} < 0, \tag{2.32}$$

where Ω is given in (2.27),

$$\Xi = \begin{bmatrix} M^{\mathrm{T}} & 0 & 0 & 0 \\ 0 & 0 & M^{\mathrm{T}} & 0 \end{bmatrix}^{\mathrm{T}},$$

$$\Phi = \begin{bmatrix} \epsilon_1(N_1P+N_3X) & \epsilon_2(N_1P+N_3X) & \epsilon_3(N_1P+N_3X) & 0 \\ \epsilon_1 N_2 P & \epsilon_2 N_2 P & \epsilon_3 N_2 P & 0 \end{bmatrix}.$$

By Lemma 1.2, it is easy to see that (2.32) holds for any $F(k)$ satisfying (2.3) if and only if there exist a scalar $\varepsilon > 0$ such that

$$\Omega + \varepsilon \Xi \Xi^{\mathrm{T}} + \varepsilon^{-1} \Phi^{\mathrm{T}} \Phi < 0. \tag{2.33}$$

Applying Schur complement, (2.33) is equivalent to (2.31). This completes the proof.

Remark 2.15. It can be seen that without any additional assumption on system matrices, the condition for solvability of the delay-dependent robust control problem for system (2.1) is formulated by strict LMI. Moreover, the design procedures of the desired controller involve no decomposition of system, which can get around certain numerical problems arising from decomposition of matrices and thus make the design procedures relatively simple and reliable.

If the matrices, in (2.31), $Y_1 = Y_2 = T_1 = 0$ and $Z = \rho I/d(\rho \to 0)$, Theorem 2.14 provides the following delay-independent result.

Corollary 2.16. *For given scalars ϵ_1, ϵ_2 and ϵ_3, system (2.1) is admissible, if there exist a scalar $\varepsilon > 0$, and matrices $P_1 > 0$, $Q > 0$, $Z > 0$, S, S_e, S_d, P and X such that*

$$\begin{bmatrix} \Xi_{11} & \Xi_{12} & \Xi_{13} & \epsilon_1(N_1P + N_3X)^{\mathrm{T}} & \epsilon_1 N_2 P^{\mathrm{T}} \\ * & \Xi_{22} & \Xi_{23} & \epsilon_2(N_1P + N_3X)^{\mathrm{T}} & \epsilon_2 N_2 P^{\mathrm{T}} \\ * & * & \Xi_{33} & \epsilon_3(N_1P + N_3X)^{\mathrm{T}} & \epsilon_3 N_2 P^{\mathrm{T}} \\ * & * & * & -\varepsilon I & 0 \\ * & * & * & * & -\varepsilon I \end{bmatrix} < 0, \tag{2.34}$$

where $R \in \mathbb{R}^{n\times(n-r)}$ is any full column rank matrix satisfying $ER = 0$, and

$$\begin{aligned} \Xi_{11} &= \epsilon_1 P^{\mathrm{T}}(A-E)^{\mathrm{T}} + \epsilon_1(A-E)P + \epsilon_1(BX + X^{\mathrm{T}}B^{\mathrm{T}}) + Q + \varepsilon MM^{\mathrm{T}}, \\ \Xi_{12} &= EP_1 + SR^{\mathrm{T}} - \epsilon_1 P^{\mathrm{T}} + \epsilon_2(A-E)P + \epsilon_2 BX, \\ \Xi_{13} &= \epsilon_1 P^{\mathrm{T}} A_d^{\mathrm{T}} + \epsilon_3(A-E)P + \epsilon_3 BX, \\ \Xi_{22} &= S_e R^{\mathrm{T}} + RS_e^{\mathrm{T}} - \epsilon_2 P - \epsilon_2 P + P_1, \\ \Xi_{23} &= \epsilon_2 P^{\mathrm{T}} A_d^{\mathrm{T}} - \epsilon_3 P + RS_d^{\mathrm{T}}, \\ \Xi_{33} &= -Q + \epsilon_3 P^{\mathrm{T}} A_d^{\mathrm{T}} + \epsilon_3 A_d P + \varepsilon MM^{\mathrm{T}}. \end{aligned}$$

Furthermore, if (2.34) is solvable, the desired controller gain is given in (2.28).

Remark 2.17. Note that Corollary 2.16 provides the delay-independent state feedback controller design method for the resultant closed-loop system to be admissible for all uncertainties. It can be found that the condition in Corollary 2.16 is a strict LMI which can be checked easily. [178, Theorem 9.6] and [184] also established the state feedback controller design methods for the same system. But it should be pointed out that the design method in [184], which is formulated by nonstrict LMIs, is invalid when the uncertain unforced systems are not admissible for all uncertainties. Moreover, solving the condition of [178, Theorem 9.6] is not a easy task because of the existing of the nonlinear term, and thus the design process is relatively more complicated and unreliable. Considering the above, the results in this chapter are more general and elegant than those in [178, Theorem 9.6] and [184].

Remark 2.18. In the case when $A_d = 0$, system (2.1) reduces to a SS with delay free. It is easy to find that Corollary 2.16 with $\epsilon_1 = \epsilon_2 = 1$ and $S_e = 0$ coincides with [65, Corollary 1]. In view of this, Corollary 2.16 can be viewed as an extension of existing results on discrete-time SSs to discrete-time SSs with time delay. The corresponding result on continuous-time SSs with time delay can be found in [178].

2.4 Numerical Examples

In this section, two numerical examples are introduced to demonstrate the effectiveness of the proposed methods.

Example 2.19. Consider the following uncertain system,

$$\begin{bmatrix} 2 & 0 \\ 0 & 0 \end{bmatrix} x(k+1) = \begin{bmatrix} 0.9977 + 0.1\alpha & 1.1972 \\ 0.1001 & A_{22} \end{bmatrix} x(k) + \begin{bmatrix} -1.1972 & 1.5772 \\ 0 & 0.9754 + 0.1\alpha \end{bmatrix} x(k-d),$$

where α is an uncertain parameter satisfying $|\alpha| \leqslant \bar{\alpha}$.

To compare our delay-dependent condition with that of [63], we assume the time delay $d = 1$ and $A_{22} = -1.9$. By the method in [63], the system is regular, causal and stable for any α satisfying $|\alpha| \leqslant 3.5055$. While by the result in Theorem 2.10, we find $\bar{\alpha} = 4.0038$, which is 14.21% larger than that in [63]. Table 2.1 also lists the comparisons of $\bar{\alpha}$, when the time delay $d = 2$, 3 and 4 with $A_{22} = -1.9$. Table 2.2 gives the comparison results on $\bar{\alpha}$ for various A_{22} and $d = 2$ by different methods. It can be seen the delay-dependent condition proposed in this chapter gives better results.

Table 2.1. Example 2.19: Comparisons of $\bar{\alpha}$ with different d

d	1	2	3	4
[63]	3.5055	1.9464	0.8033	0.1563
Theorem 2.10	4.0038	2.1359	1.0325	0.2853
Improvement rate	14.21%	9.74%	28.53%	82.53%

Table 2.2. Example 2.19: Comparisons of $\bar{\alpha}$ with different A_{d22}

A_{22}	-1.6	-1.8	-2	-2.2	-2.4	-2.6
[63]	1.5588	1.8384	2.0326	2.1404	2.1642	2.1567
Theorem 2.10	1.7046	2.0053	2.2533	2.4508	2.6040	2.7209
Improvement rate	9.35%	9.08%	10.86%	14.50%	20.33%	26.16%

Example 2.20. Consider system (2.1) with the following parameters,

$$E = \begin{bmatrix} 1 & 1 & 0 \\ 0 & 1 & 0 \\ 0 & 0 & 0 \end{bmatrix}, \ A = \begin{bmatrix} 1 & 0 & 0 \\ 0 & 2 & 1.5 \\ 1 & 2 & 0 \end{bmatrix}, \ A_d = \begin{bmatrix} 0.2 & 0.8 & 0 \\ 0.6 & 0 & 0.5 \\ 1 & 0.5 & 2 \end{bmatrix},$$

$$B = \begin{bmatrix} 1 & 2 \\ 1.2 & 1 \\ 1.5 & 2 \end{bmatrix}, \ M = \begin{bmatrix} 0.049 \\ 0.052 \\ 0.051 \end{bmatrix}, \ N_1 = N_2 = \begin{bmatrix} 0.02 \\ 0.02 \\ 0.2 \end{bmatrix}^{\mathrm{T}}, \ N_3 = \begin{bmatrix} 0.2 & 0.5 \end{bmatrix}.$$

In this example, we assume that the time delay $d = 5$. The purpose is the design of a state feedback controller such that, for all uncertainties, the resultant closed-loop system is regular, impulse free and stable. To this end, we choose $R = \begin{bmatrix} 0 & 0 & 1 \end{bmatrix}^{\mathrm{T}}$, $\epsilon_1 = \epsilon_2 = 1.02$ and $\epsilon_3 = -0.05$. Using Matlab LMI Control Toolbox to solve LMI (2.31), we get the state feedback controller has the following gain,

$$K = \begin{bmatrix} 0.2933 & -2.6839 & -3.3725 \\ -1.7937 & 1.3987 & -0.5579 \end{bmatrix}.$$

While the result of [184] fails to give the desired state feedback controller. Therefore, the result in this chapter improves the existing one.

2.5 Conclusion

The delay-dependent and delay-independent robust control problems for discrete-time SSs with time delay and norm bounded parametric uncertainties have been studied by introducing a finite sum inequality. The obtained results are all formulated by strict LMI and no decomposition of system matrices is involved, which makes the analysis and design relatively simple and reliable. When the LMI is feasible, the desired state feedback controller can be constructed directly. The effectiveness and reduced conservatism of the results proposed has been demonstrated via two numerical examples.

3

H_∞ Control for SSs with Time-Delay

3.1 Introduction

In the past decade, several delay-independent BRLs for singular time-delay systems have been provided and the delay-independent H_∞ control problem has been solved via the state feedback controller in [69, 186]. In terms of LMI approach, the problem of delay-dependent H_∞ control for singular time-delay systems has been solved in [3, 38, 193]. Several delay-dependent BRLs have been obtained and the design algorithms of desired controllers, including state feedback controller and output feedback controller, have also been given. It should be pointed out that, in [38, 193], the decomposition and transformation of original system coefficient matrices are required, which make the analysis and design procedures complex and unreliable. When norm-bounded parameter uncertainties arise, the delay-dependent robust H_∞ control problem for singular time-delay systems has been discussed in [200, 213], and some delay-dependent BRLs and sufficient conditions for the solvability of the problem have been obtained. The free-weighting matrix method has been used to deal with the delay-dependent H_∞ control problem for singular time-delay systems in [166, 187], and the obtained results have improved the conservatism of the results of [3, 38, 193, 200, 213] to a certain extent. It is should be pointed out that the proposed results of [3, 166, 187, 200, 213] are all formulated in terms of non-strict LMIs, whose solutions are difficult to calculate since equality constraints are often fragile and usually not met perfectly.

In this chapter, an augmented Lyapunov functional is proposed to discuss the delay-dependent H_∞ control problem for SSs with time-delay. Owing to the augmented Lyapunov functional, an improved delay-dependent BRL is derived, which guarantees the SSs with time-delay to be regular, impulse free and stable while satisfying a prescribed H_∞ performance level. Based on the BRL obtained, a strict LMI-based method is proposed to solve the delay-dependent H_∞ control problem and the desired state feedback controller can be constructed by solving a strict LMI. Numerical examples show that the proposed methods are much less conservative than the existing ones.

Z.-G. Wu et al.: *Anal. & Synth. of Singular Syst. with Time-Delays*, LNCIS 443, pp. 27–36.
DOI: 10.1007/978-3-642-37497-5_3

3.2 Problem Formulation

Consider the following SS with tim-delay:

$$\begin{cases} E\dot{x}(t) = Ax(t) + A_d x(t-d) + Bu(t) + B_\omega \omega(t), \\ z(t) = Cx(t) + Du(t), \\ x(t) = \phi(t),\ t \in [-d, 0], \end{cases} \tag{3.1}$$

where $x(t) \in \mathbb{R}^n$ is the state, $u(t) \in \mathbb{R}^m$ is the control input, $\omega(t) \in \mathbb{R}^p$ is the disturbance input that belongs to $\mathscr{L}_2[0,\infty)$, $z(t) \in \mathbb{R}^s$ is the controlled output, $d > 0$ is the constant time delay, and $\phi(t) \in C_{n,d}$ is a compatible vector valued initial function. The matrix $E \in \mathbb{R}^{n\times n}$ may be singular and it is assumed that $\operatorname{rank} E = r \leqslant n$. A, A_d, B, B_ω, C and D are known real constant matrices with appropriate dimensions.

Definition 3.1. *[178]*

1. System (1.1) is said to be regular and impulse free, if the pair (E, A) is regular and impulse free,
2. System (1.1) is said to be stable if for any scalar $\varepsilon > 0$, there exits a scalar $\delta(\varepsilon) > 0$ such that, for any compatible initial conditions $\phi(t)$ satisfying $\sup_{-d\leqslant t\leqslant 0} \|\phi(t)\| \leqslant \delta(\varepsilon)$, the solution $x(t)$ of system (1.1) satisfies $\|x(t)\| \leqslant \varepsilon$ for any $t \geqslant 0$, moreover $\lim\limits_{t\to\infty} x(t) = 0$,
3. System (1.1) is said to be admissible if it is regular, causal and stable.

In this chapter, we are interested in designing a state feedback controller

$$u(t) = Kx(t), \tag{3.2}$$

where $K \in \mathbb{R}^{m\times n}$ is the gain to be determined.

The aim of this chapter is, for a given scalar $\gamma > 0$, to develop a state feedback controller (3.2) such that the following requirements are satisfied:

1. The closed-loop system with $\omega(t) \equiv 0$ is admissible,
2. The closed-loop system possesses H_∞ performance γ, that is, under the zero initial condition, the closed-loop system satisfies

$$J_{z\omega} = \int_0^\infty \left(z(t)^{\mathrm{T}} z(t) - \gamma^2 \omega(t)^{\mathrm{T}} \omega(t)\right) \mathrm{d}t < 0, \tag{3.3}$$

for any nonzero $\omega(t) \in \mathscr{L}_2[0,\infty)$. In this case, the closed-loop system is said to be admissible with H_∞ performance γ.

3.3 Main Results

In this section, we will solve the delay-dependent H_∞ control problem for system (3.1) in terms of LMI approach.

3.3.1 BRL

A delay-dependent BRL is given which guarantees system (3.1) with $u(t) \equiv 0$ to be regular, impulse free, and stable while satisfying a prescribed H_∞ performance γ.

Theorem 3.2. *For a given scalar $\gamma > 0$, system (3.1) with $u(t) \equiv 0$ is admissible with H_∞ performance γ, if there exist matrices $Q > 0$,*

$$\begin{bmatrix} P_1 & P_2 \\ P_2^{\mathrm{T}} & P_3 \end{bmatrix} > 0, \begin{bmatrix} Z_1 & Z_2 \\ Z_2^{\mathrm{T}} & Z_3 \end{bmatrix} > 0,$$

S, T_1 and T_2 such that

$$\Xi = \begin{bmatrix} \Xi_{11} & \Xi_{12} & \Xi_{13} & \Xi_{14} & T_1^{\mathrm{T}} B_\omega & C^{\mathrm{T}} \\ * & \Xi_{22} & T_2^{\mathrm{T}} A_d & P_2 & T_2^{\mathrm{T}} B_\omega & 0 \\ * & * & \Xi_{33} & \Xi_{34} & 0 & 0 \\ * & * & * & -Z_1 & 0 & 0 \\ * & * & * & * & -\gamma^2 I & 0 \\ * & * & * & * & * & -I \end{bmatrix} < 0, \tag{3.4}$$

where $R \in \mathbb{R}^{n \times (n-r)}$ is any matrix with full column and satisfies $E^{\mathrm{T}} R = 0$, and

$$\begin{aligned} \Xi_{11} &= T_1^{\mathrm{T}} A + A^{\mathrm{T}} T_1 + E^{\mathrm{T}} P_2 + P_2^{\mathrm{T}} E + Q - E^{\mathrm{T}} Z_3 E + d^2 Z_1, \\ \Xi_{12} &= E^{\mathrm{T}} P_1 + S R^{\mathrm{T}} - T_1^{\mathrm{T}} + A^{\mathrm{T}} T_2 + d^2 Z_2, \\ \Xi_{13} &= T_1^{\mathrm{T}} A_d + E^{\mathrm{T}} Z_3 E - E^{\mathrm{T}} P_2, \\ \Xi_{14} &= P_3 - E^{\mathrm{T}} Z_2^{\mathrm{T}}, \\ \Xi_{22} &= -T_2 - T_2^{\mathrm{T}} + d^2 Z_3, \\ \Xi_{33} &= -Q - E^{\mathrm{T}} Z_3 E, \\ \Xi_{34} &= -P_3 + E^{\mathrm{T}} Z_2^{\mathrm{T}}. \end{aligned}$$

Proof. Firstly, we prove system (3.1) with $u(t) \equiv 0$ is regular, impulse free and stable. To this end, we consider system (1.1). It follows from (3.4) that

$$\begin{bmatrix} \Xi_{11} & \Xi_{12} & \Xi_{13} \\ * & \Xi_{22} & T_2^{\mathrm{T}} A_d \\ * & * & \Xi_{33} \end{bmatrix} - d^2 \begin{bmatrix} Z_1 & Z_2 & 0 \\ * & Z_3 & 0 \\ * & * & 0 \end{bmatrix} < 0. \tag{3.5}$$

Let

$$V = \begin{bmatrix} I & A^{\mathrm{T}} & 0 \\ 0 & A_d^{\mathrm{T}} & I \end{bmatrix}, \tag{3.6}$$

and pre- and post multiplying (3.5) by V and V^{T}, respectively, we get

$$\begin{bmatrix} \Upsilon_{11} & \Upsilon_{12} \\ * & \Xi_{33} \end{bmatrix} < 0, \tag{3.7}$$

where

$$\Upsilon_{11} = E^{\mathrm{T}}P_2 + P_2^{\mathrm{T}}E + Q - E^{\mathrm{T}}Z_3E + E^{\mathrm{T}}P_1A + SR^{\mathrm{T}}A + A^{\mathrm{T}}P_1E + A^{\mathrm{T}}RS^{\mathrm{T}},$$
$$\Upsilon_{12} = -E^{\mathrm{T}}P_2 + E^{\mathrm{T}}Z_3E + E^{\mathrm{T}}P_1A_d + SR^{\mathrm{T}}A_d.$$

Choose two nonsingular matrices M and N such that

$$MEN = \begin{bmatrix} I_r & 0 \\ 0 & 0 \end{bmatrix}. \tag{3.8}$$

Noting that $E^{\mathrm{T}}R = 0$ and $\operatorname{rank} R = n - r$, we can get

$$M^{-T}R = \begin{bmatrix} 0 \\ H \end{bmatrix}, \tag{3.9}$$

where $H \in \mathbb{R}^{(n-r)\times(n-r)}$ is any nonsingular matrix. Write

$$MAN = \begin{bmatrix} A_{11} & A_{12} \\ A_{21} & A_{22} \end{bmatrix}, \ N^{\mathrm{T}}S = \begin{bmatrix} S_1 \\ S_2 \end{bmatrix}. \tag{3.10}$$

Pre- and post multiplying $\Upsilon_{11} < 0$ by N^{T} and N, respectively, and then using the expressions in (3.8), (3.9) and (3.10), we have

$$S_2H^{\mathrm{T}}A_{22} + A_{22}^{\mathrm{T}}HS_2^{\mathrm{T}} < 0, \tag{3.11}$$

which implies A_{22} is nonsingular. Therefore, the pair (E, A) is regular and impulse free. Thus, according to Definition 3.1, system (1.1) is regular and impulse free.

Next, we shall show the stability of system (1.1). For any $t \geqslant d$, choose the following Lyapunov functional:

$$V(x_t) = V_1(x_t) + V_2(x_t) + V_3(x_t), \tag{3.12}$$

where

$$V_1(x_t) = \begin{bmatrix} Ex(t) \\ \int_{t-d}^{t} x(\alpha)\mathrm{d}\alpha \end{bmatrix}^{\mathrm{T}} \begin{bmatrix} P_1 & P_2 \\ P_2^{\mathrm{T}} & P_3 \end{bmatrix} \begin{bmatrix} Ex(t) \\ \int_{t-d}^{t} x(\alpha)\mathrm{d}\alpha \end{bmatrix},$$
$$V_2(x_t) = \int_{t-d}^{t} x(\alpha)^{\mathrm{T}}Qx(\alpha)\mathrm{d}\alpha,$$
$$V_3(x_t) = d\int_{-d}^{0}\int_{t+\beta}^{t} \begin{bmatrix} x(\alpha) \\ E\dot{x}(\alpha) \end{bmatrix}^{\mathrm{T}} \begin{bmatrix} Z_1 & Z_2 \\ * & Z_3 \end{bmatrix} \begin{bmatrix} x(\alpha) \\ E\dot{x}(\alpha) \end{bmatrix} \mathrm{d}\alpha\mathrm{d}\beta,$$

where $x_t = x(t+\theta)$, $-2d \leqslant \theta \leqslant 0$. Then, the time-derivative of $V(x_t)$ along the solution of system (1.1) gives

$$\dot{V}_1(x_t) = 2\begin{bmatrix} x(t) \\ \int_{t-d}^{t} x(\alpha)\,\mathrm{d}\alpha \end{bmatrix}^{\mathrm{T}} \begin{bmatrix} E^{\mathrm{T}}P_1 + SR^{\mathrm{T}} & E^{\mathrm{T}}P_2 \\ P_2^{\mathrm{T}} & P_3 \end{bmatrix} \begin{bmatrix} E\dot{x}(t) \\ x(t) - x(t-d) \end{bmatrix}, \tag{3.13}$$

$$\dot{V}_2(x_t) = x(t)^{\mathrm{T}}Qx(t) - x(t-d)^{\mathrm{T}}Qx(t-d), \tag{3.14}$$

$$\begin{aligned}\dot{V}_3(x_t) \leqslant{} & d^2 \begin{bmatrix} x(t) \\ E\dot{x}(t) \end{bmatrix}^{\mathrm{T}} \begin{bmatrix} Z_1 & Z_2 \\ * & Z_3 \end{bmatrix} \begin{bmatrix} x(t) \\ E\dot{x}(t) \end{bmatrix} \\ & - d \int_{t-d}^{t} \begin{bmatrix} x(\alpha) \\ E\dot{x}(\alpha) \end{bmatrix}^{\mathrm{T}} \begin{bmatrix} Z_1 & Z_2 \\ * & Z_3 \end{bmatrix} \begin{bmatrix} x(\alpha) \\ E\dot{x}(\alpha) \end{bmatrix} \mathrm{d}\alpha. \end{aligned} \tag{3.15}$$

On the other hand, for any appropriately dimensional matrices T_1 and T_2, the following equation is true:

$$\begin{aligned}\alpha(t) ={} & 2\left[x(t)^{\mathrm{T}}T_1^{\mathrm{T}} + (E\dot{x}(t))^{\mathrm{T}}T_2^{\mathrm{T}}\right]\left[E\dot{x}(t) + Ax(t) + A_d x(t-d)\right] \\ ={} & 0. \end{aligned} \tag{3.16}$$

Hence, taking into account to Lemma 1.7 and (3.4), we have that there exists a scalar $\lambda > 0$ such that

$$\begin{aligned}\dot{V}(x_t) ={} & \dot{V}_1(x_t) + \dot{V}_2(x_t) + \dot{V}_3(x_t) + \alpha(t) \\ \leqslant{} & \begin{bmatrix} x(t) \\ E\dot{x}(t) \\ x(t-d) \\ \int_{t-d}^{t} x(\alpha)\mathrm{d}\alpha \end{bmatrix}^{\mathrm{T}} \begin{bmatrix} \Xi_{11} & \Xi_{12} & \Xi_{13} & \Xi_{14} \\ * & \Xi_{22} & T_2^{\mathrm{T}}A_d & P_2 \\ * & * & \Xi_{33} & \Xi_{34} \\ * & * & * & -Z_1 \end{bmatrix} \begin{bmatrix} x(t) \\ E\dot{x}(t) \\ x(t-d) \\ \int_{t-d}^{t} x(\alpha)\mathrm{d}\alpha \end{bmatrix} \\ <{} & -\lambda\|x(t)\|^2. \end{aligned} \tag{3.17}$$

Note that the regularity and the absence of impulses of the pair (E, A) imply that there always exist two nonsingular matrices $\tilde{M}$ and $\tilde{N}$ such that

$$\tilde{M}E\tilde{N} = \begin{bmatrix} I_r & 0 \\ 0 & 0 \end{bmatrix}, \ \tilde{M}A\tilde{N} = \begin{bmatrix} A_1 & 0 \\ 0 & I_{n-r} \end{bmatrix}. \tag{3.18}$$

Write

$$\begin{aligned} & \tilde{M}A_d\tilde{N} = \begin{bmatrix} A_{d1} & A_{d2} \\ A_{d3} & A_{d4} \end{bmatrix}, \ \tilde{N}^{\mathrm{T}}Q\tilde{N} = \begin{bmatrix} Q_{11} & Q_{12} \\ * & Q_{22} \end{bmatrix}, \\ & \tilde{N}^{\mathrm{T}}S = \begin{bmatrix} S_{11} \\ S_{21} \end{bmatrix}, \ \tilde{M}^{-T}R = \begin{bmatrix} 0 \\ \tilde{H} \end{bmatrix}, \end{aligned} \tag{3.19}$$

where $\tilde{H} \in \mathbb{R}^{(n-r)\times(n-r)}$ is any nonsingular matrix. Pre- and post multiplying (3.7) by $\begin{bmatrix} \tilde{N} & 0 \\ 0 & \tilde{N} \end{bmatrix}^{\mathrm{T}}$ and $\begin{bmatrix} \tilde{N} & 0 \\ 0 & \tilde{N} \end{bmatrix}$, respectively, and then using the expressions in (3.18) and (3.19), we have

$$\begin{bmatrix} S_{21}\tilde{H}^{\mathrm{T}} + \tilde{H}S_{21}^{\mathrm{T}} + Q_{22} & S_{21}\tilde{H}^{\mathrm{T}}A_{d4} \\ * & -Q_{22} \end{bmatrix} < 0, \tag{3.20}$$

which implies [175]

$$\rho(A_{d4}) < 1. \tag{3.21}$$

Noting this, (3.12) and (3.17) and following a line similar to that in the proof of Theorem 1 of [175], we can deduce that system (1.1) is stable.

In the following, we will establish the H_∞ performance of system (3.1) with $u(t) \equiv 0$. Under zero initial condition, it can be shown that for any nonzero $\omega(t) \in \mathscr{L}_2[0, \infty)$,

$$\begin{aligned} J_{z\omega} &= \int_0^\infty \left(z(t)^{\mathrm{T}} z(t) - \gamma^2 \omega(t)^{\mathrm{T}} \omega(t)\right) \mathrm{d}t \\ &\leqslant \int_0^\infty \left(z(t)^{\mathrm{T}} z(t) - \gamma^2 \omega(t)^{\mathrm{T}} \omega(t) + \dot{V}(x_t)\right) \mathrm{d}t \\ &\leqslant \int_0^\infty \zeta(t)^{\mathrm{T}} \Omega \zeta(t) \, \mathrm{d}t, \end{aligned} \tag{3.22}$$

where

$$\zeta(t) = \begin{bmatrix} x(t) \\ E\dot{x}(t) \\ x(t-d) \\ \int_{t-d}^{t} x(\alpha)\mathrm{d}\alpha \\ \omega(t) \end{bmatrix}, \ \Omega = \begin{bmatrix} \Xi_{11} & \Xi_{12} & \Xi_{13} & \Xi_{14} & T_1^{\mathrm{T}} B_\omega \\ * & \Xi_{22} & T_2^{\mathrm{T}} A_d & P_2 & T_2^{\mathrm{T}} B_\omega \\ * & * & \Xi_{33} & \Xi_{34} & 0 \\ * & * & * & -Z_1 & 0 \\ * & * & * & * & -\gamma^2 I \end{bmatrix} + \begin{bmatrix} C \\ 0 \\ 0 \\ 0 \\ 0 \end{bmatrix}^{\mathrm{T}} \begin{bmatrix} C \\ 0 \\ 0 \\ 0 \\ 0 \end{bmatrix}.$$

by applying Schur complement to (3.4), we have $\Omega < 0$. Therefore, $J_{z\omega} < 0$ for any nonzero $\omega(t) \in \mathscr{L}_2[0, \infty)$. This completes the proof.

Remark 3.3. A new version of BRL for system (3.1) with $u(t) \equiv 0$ is proposed in Theorem 3.2, which is formulated by strict LMI with the coefficient matrices of the original system. Thus, the BRL is contrast to those of [3, 166, 187, 200, 213], where some nonstrict LMI conditions have been reported, and is also different from the conditions of [38, 193], where the decomposition of the given systems has been used. Testing such a strict LMI-based condition can avoid some numerical problems arising from equality constraints and decomposition of the original system. Thus, the BRL in this chapter is more elegant and has computational advantages from the mathematical point of view.

Remark 3.4. It is noted choosing $P_2 = P_3 = Z_1 = Z_2 = 0$, the Lyapunov functional (3.12) reduces that of [187], and thus the Lyapunov functional (3.12) is more generalized and includes more weighting matrices. Therefore, Theorem 3.2 has less conservatism than the result of [187], which will be demonstrated by numerical examples in Section 3.4.

3.3.2 H_∞ Controller Design

In the following theorem, we will apply Theorem 3.2 to design the state feedback controller (3.2) for system (3.1) such that the resultant closed-loop system is regular, impulse free and stable with H_∞ performance γ.

Theorem 3.5. *For a given scalar $\gamma > 0$, system (3.1) is admissible with H_∞ performance γ, if there exist matrices $Q > 0$,*

$$P = \begin{bmatrix} P_1 & P_2 \\ P_2^{\mathrm{T}} & P_3 \end{bmatrix} > 0,\ Z = \begin{bmatrix} Z_1 & Z_2 \\ Z_2^{\mathrm{T}} & Z_3 \end{bmatrix} > 0,$$

S, G and V such that

$$\Xi = \begin{bmatrix} \Xi_{11} & \Xi_{12} & \Xi_{13} & \Xi_{14} & \Xi_{15} & B_\omega \\ * & \Xi_{22} & G^{\mathrm{T}}A_d^{\mathrm{T}} & P_2 & \Xi_{15} & 0 \\ * & * & \Xi_{33} & \Xi_{34} & 0 & 0 \\ * & * & * & -Z_1 & 0 & 0 \\ * & * & * & * & -\gamma^2 I & 0 \\ * & * & * & * & * & -I \end{bmatrix} < 0, \tag{3.23}$$

where $R \in \mathbb{R}^{n\times(n-r)}$ is any matrix with full column and satisfies $ER = 0$ and

$$\begin{aligned}
\Xi_{11} &= G^{\mathrm{T}}A^{\mathrm{T}} + V^{\mathrm{T}}B^{\mathrm{T}} + AG + BV + EP_2 + P_2^{\mathrm{T}}E^{\mathrm{T}} + Q - EZ_1E^{\mathrm{T}} + d^2Z_1, \\
\Xi_{12} &= EP_1 + SR^{\mathrm{T}} - G^{\mathrm{T}} + AG + BV + d^2Z_2, \\
\Xi_{13} &= G^{\mathrm{T}}A_d^{\mathrm{T}} + EZ_3E^{\mathrm{T}} - EP_2, \\
\Xi_{14} &= P_3 - EZ_2^{\mathrm{T}}, \\
\Xi_{15} &= G^{\mathrm{T}}C^{\mathrm{T}} + V^{\mathrm{T}}D^{\mathrm{T}}, \\
\Xi_{22} &= -G - G^{\mathrm{T}} + d^2Z_3, \\
\Xi_{33} &= -Q - EZ_3E^{\mathrm{T}}, \\
\Xi_{34} &= -P_3 + EZ_2^{\mathrm{T}}.
\end{aligned}$$

Furthermore, if (3.23) is solvable, the desired controller gain is given as

$$K = VG^{-1}. \tag{3.24}$$

Proof. Substituting the state feedback controller $u(t) = Kx(t)$ to system (3.1) yields the following closed-loop system

$$\begin{cases} E\dot{x}(t) = (A + BK)x(t) + A_d x(t-d) + B_\omega \omega(t), \\ \quad z(t) = (C + DK)x(t). \end{cases} \tag{3.25}$$

Since $\det(sE - (A+BK)) = \det(sE^{\mathrm{T}} - (A+BK)^{\mathrm{T}})$, the pair $(E, (A+BK))$ is regular, impulse free if and only if the pair $(E^{\mathrm{T}}, (A+BK)^{\mathrm{T}})$ is regular and impulse free. Moreover, since the solution of $\det(sE - (A+BK) - e^{-ds}A_d) = 0$ is the same as that of $\det(sE^{\mathrm{T}} - (A+BK)^{\mathrm{T}} - e^{-ds}A_d^{\mathrm{T}}) = 0$ and the $\det(sE^{\mathrm{T}} - (A+BK)^{\mathrm{T}} - e^{-ds}A_d^{\mathrm{T}}) = 0$ and the

$$\|G(s)\|_\infty = \sup_{\omega\in[0,\infty)} \sigma_{\max}\{(C + DK)(j\omega E - (A + BK) - e^{-dj\omega}A_d)^{-1}B_\omega\} \tag{3.26}$$

is equal to

$$\begin{aligned}&\|H(s)\|_\infty\\ &= \sup_{\omega\in[0,\infty)} \sigma_{\max}\{B_\omega^{\mathrm{T}}(j\omega E^{\mathrm{T}} - (A+BK)^{\mathrm{T}} - e^{-dj\omega}A_d^{\mathrm{T}})^{-1}(C+DK)^{\mathrm{T}}\},\end{aligned} \tag{3.27}$$

as long as the regularity, absence of impulses and stability with H_∞ performance are the only concern, system (3.25) is equivalent to system

$$\begin{cases} E^{\mathrm{T}}\dot{x}(t) = (A+BK)^{\mathrm{T}}x(t) + A_d^{\mathrm{T}}x(t-d) + (C+DK)^{\mathrm{T}}\omega(t), \\ \quad z(t) = B_\omega^{\mathrm{T}}x(t). \end{cases} \tag{3.28}$$

Hence, applying Theorem 3.2 to the above system and setting $T_1 = T_2 = G$, and $V = KG$ yields (3.23) straightforwardly. This completes the proof.

Remark 3.6. Note that Theorem 3.5 provides a sufficient condition for the solvability of delay-dependent H_∞ control problem for system (3.1). The desired state feedback controller can be obtained by solving the strict LMI (3.23), which do not require any parameter tuning and decomposition or transformation of the original system, and thus can be solved numerically very efficiently by using LMI toolbox of Matlab. While the decomposition or transformation of system matrices is needed in [38, 193] and equality constraints appear in the state feedback controller design processes of [3, 166, 187, 200, 213]. Thus, Theorem 3.5 is much more general and elegant than the existing ones. Moreover, if (3.23) is feasible, it follows from $\Xi_{22} = -G - G^{\mathrm{T}} + d^2 Z_3 < 0$ that G is nonsingular and thus the desired state feedback gain K can be readily obtained.

3.4 Numerical Examples

In this section, some examples are used to demonstrate that the methods presented in this chapter are effective and are the improvement over the existing methods.

Example 3.7. Consider the following system:

$$\begin{bmatrix} 1 & 0 \\ 0 & 0 \end{bmatrix} \dot{x}(t) = \begin{bmatrix} 0.6341 & 0.5413 \\ -0.6121 & -1.1210 \end{bmatrix} x(t) + \begin{bmatrix} -0.4500 & 0 \\ 0 & -0.1210 \end{bmatrix} x(t-d)$$

By comparing the stability criterion in Theorem 3.2 with those of [5, 35, 38, 120, 193, 213] for the above system, we have Table 3.1. It is clear that, for this example, the stability criterion we derived is less conservative than those reported in the above-mentioned papers.

Table 3.1. Example 3.7: Comparisons of maximum allowed time delay d

[5]	[35]	[38, 193]	[120, 213]	Theorem 3.2
–	2.1328	2.1372	2.4841	2.4865

Example 3.8. To compare the delay-dependent BRL in Theorem 3.2 with the existing ones, we consider system (3.1) with $u(t) \equiv 0$ and

$$\begin{aligned} E &= \begin{bmatrix} 1 & 0 \\ 0 & 0 \end{bmatrix}, \ A = \begin{bmatrix} 0.6 & 0.5 \\ -0.6 & -1 \end{bmatrix}, \\ A_d &= \begin{bmatrix} -0.7 & 0 \\ 0 & -0.2 \end{bmatrix}, \ B_\omega = \begin{bmatrix} 0.5 \\ 2 \end{bmatrix}, \\ C &= \begin{bmatrix} 0.5 & 0.5 \end{bmatrix}. \end{aligned}$$

For a given $\gamma > 0$, the maximum allowed time delay d satisfying the LMI in (3.4) can be calculated by solving a quasi-convex optimization problem. Similarly, for a given $d > 0$, the minimum allowed γ satisfying the LMI in (3.4) can also be calculated by solving a quasi-convex optimization problem. Table 3.2 and Table 3.3 give the comparison results on the maximum allowed time delay d for given $\gamma > 0$ and the minimum allowed γ for given $d > 0$, respectively, via the methods in [3, 38, 166, 187, 193, 213] and Theorem 3.2 in our chapter. Additionally, the result of [3] can not deal with the above system. Thus, the BRL in Theorem 3.2 of this chapter is less conservative than those in [3, 38, 166, 187, 193, 213].

Table 3.2. Example 3.8: Comparisons of maximum allowed time delay d

γ	2.4	2.6	2.8	3.0	3.2
[213]	0.4760	0.5237	0.5607	0.5906	0.6156
[38, 193]	1.0533	1.1334	1.1864	1.2237	1.2512
[166, 187]	1.1102	1.2261	1.3061	1.3626	1.4034
Theorem 3.2	1.2865	1.3559	1.3973	1.4272	1.4525

Table 3.3. Example 3.8: Comparisons of minimum γ

d	1.00	1.05	1.10	1.15	1.20
[213]	6.3722	6.6540	6.9438	7.2440	7.5590
[38, 193]	2.3056	2.3933	2.5054	2.6547	2.8653
[166, 187]	2.2630	2.3205	2.3857	2.4604	2.5479
Theorem 3.2	2.0346	2.0751	2.1221	2.1774	2.2438

Example 3.9. To show the reduced conservatism of the H_∞ control result in Theorem 3.5 in this chapter, we now consider the following system[38]:

$$\begin{cases} E\dot{x}(t) = A_d x(t-d) + Bu(t) + B_\omega \omega(t), \\ z(t) = Cx(t) + Du(t), \end{cases}$$

where

$$E = \begin{bmatrix} 1 & 0 \\ 0 & 0 \end{bmatrix}, \; A_d = \begin{bmatrix} -1 & 0 \\ 1 & -1 \end{bmatrix},$$
$$B = \begin{bmatrix} -0.5 \\ 1 \end{bmatrix}, \; B_\omega = \begin{bmatrix} 1 \\ 1 \end{bmatrix},$$
$$C = \begin{bmatrix} 1 & 0.2 \end{bmatrix}, \; D = 0.1.$$

For a given time delay $d = 1.2$, the minimum γ is 21 and 15.0268 can be obtained using the methods of [38] and [187], respectively. However, by resorting to Theorem 3.5 in this chapter, for the same time delay, the minimum $\gamma = 9.6754$ by solving the strict LMI (3.23), which is 53.93% and 35.61% larger than those in [38] and [187], respectively. Furthermore, the state feedback controller achieving the minimum $\gamma = 9.6754$ can be obtained as

$$u(t) = \begin{bmatrix} 0.4834 & -2.3868 \end{bmatrix} x(t).$$

While the result of [3] can not deal with the H_∞ control problem for the above system. Therefore, Theorem 3.5 in this chapter is less conservative than [3, 38, 187].

3.5 Conclusion

The problem of delay-dependent H_∞ control for SSs with state delay has been solved by LMI approach and an augmented Lyapunov functional. A new version of delay-dependent BRL and the design method of the desired state feedback controller have been established. The obtained results are all formulated by strict LMIs involving no decomposition of system matrices, which can be tested easily by the LMI control toolbox and make the analysis and design relatively simple and reliable. Numerical examples have been given to demonstrate the reduced conservatism of the obtained stability, BRL as well as H_∞ control results in this chapter.

4

Reliable Passive Control for SSs with Time-Varying Delays

4.1 Introduction

It has been shown that the notion of passivity plays an important role in the analysis and design of linear and nonlinear systems, especially for high-order systems, and thus the passivity analysis approach has been used for a long time to deal with the control problems for some kinds of systems, see for instance, [12, 40, 75, 79, 155, 172, 190], and the references therein. The problem of robust passive control for uncertain singular time-delay systems has been considered in [76], where three types of controllers have been discussed, namely, state feedback controller, observer-based state feedback controller, and dynamic output feedback controller, and the controllers have been constructed to ensure the resultant closed-loop systems generalized quadratically stable and passive with dissipation η. However, the delay-independent approach has been applied in [76] to obtain the results, which have been proved to be more conservative than the ones with the delay-dependent approach. By the free-weighting matrix method, the problems of robust passivity analysis and passivity-based sliding mode control have been addressed for a kind of uncertain singular time-delay systems in [149], where a delay-dependent sufficient condition has been proposed in terms of LMI, which guarantees the sliding mode dynamics to be generalized quadratically stable and robustly passive, and the passification solvability condition has been also established. It should be mentioned that the involved time delays in [76, 149] are time-invariant, which limits the scope of applications of the given results.

On the other hand, much effort has been devoted to the reliable control since unexpected failures may result in substantial damage, and can even be hazardous to plant personnel and the environment. In the literature to date, several approaches in reliable control have been proposed, see for instance, [131, 135, 137, 195, 196] and the references therein. Up to now, to the best of our knowledge, the issue of reliable passive control for SSs with time-varying delays and actuator failures has not been fully investigated and remains to be important and challenging.

Z.-G. Wu et al.: *Anal. & Synth. of Singular Syst. with Time-Delays*, LNCIS 443, pp. 37–52.
DOI: 10.1007/978-3-642-37497-5_4

In this chapter, the problem of reliable passive control is discussed for SSs with time-varying delays and actuator failures by using LMI approach. The main objective is to design a state feedback controller such that, for all possible actuator failures, the resultant closed-loop system is delay-dependent exponentially admissible and passive. The delay-dependent conditions are proposed to guarantee the existence of the desired state feedback controller. Some numerical examples are exploited in order to illustrate the effectiveness of the proposed results.

4.2 Problem Formulation

Consider the following SS with time-varying delays:

$$\begin{cases} E\dot{x}(t) = Ax(t) + A_d x(t-d(t)) + Bu(t) + B_\omega \omega(t), \\ z(t) = Cx(t) + C_d x(t-d(t)) + D_\omega \omega(t), \\ x(t) = \phi(t),\ t \in [-d_2, 0], \end{cases} \tag{4.1}$$

where $x(t) \in \mathbb{R}^n$ is the state, $u(t) \in \mathbb{R}^m$ is the control input, $z(t) \in \mathbb{R}^s$ is the controlled output, and $\omega(t) \in \mathbb{R}^p$ is the disturbance input that belongs to $\mathscr{L}_2[0,\infty)$, and $\phi(t) \in C_{n,d_2}$ is a compatible vector valued initial function. The matrix $E \in \mathbb{R}^{n\times n}$ may be singular and it is assumed that $\operatorname{rank} E = r \leqslant n$. A, A_d, B, B_ω, C, C_d and D_ω are known real constant matrices with appropriate dimensions. $d(t)$ is a time-varying continuous function that satisfies $0 < d_1 \leqslant d(t) \leqslant d_2$ and $\dot{d}(t) \leqslant \mu$, where d_1 and d_2 are the lower and upper bounds of time-varying delay $d(t)$, respectively, and $0 \leqslant \mu < 1$ is the variation rate of time-varying delay $d(t)$.

When the actuators experience failures, we use $u^F(t)$ to describe the control signal sent from actuators. Consider the following actuator failure model with failure parameter α [195]:

$$u^F(t) = \alpha u(t), \tag{4.2}$$

where $\alpha = \operatorname{diag}\{\alpha_1, \alpha_2, \ldots, \alpha_m\}$ with

$$0 \leqslant \underline{\alpha}_i \leqslant \alpha_i \leqslant \bar{\alpha}_i \leqslant 1,\ i = 1, 2, \ldots, m. \tag{4.3}$$

Note that the parameters $\underline{\alpha}_i$ and $\bar{\alpha}_i$ characterize the admissible failures of the i-th actuator. Obviously, when $\underline{\alpha}_i = \bar{\alpha}_i = 0$, the failure model (4.2) corresponds to the case of the i-th actuator outage. When $0 < \underline{\alpha}_i < \bar{\alpha}_i < 1$, it corresponds to the case of partial failure of the i-th actuator. When $\underline{\alpha}_i = \bar{\alpha}_i = 1$, it implies that there is no failure in the i-th actuator. Denote [137]

$$\begin{aligned} \hat{\alpha} &= \operatorname{diag}\left\{\frac{\bar{\alpha}_1 + \underline{\alpha}_1}{2}, \frac{\bar{\alpha}_2 + \underline{\alpha}_2}{2}, \ldots, \frac{\bar{\alpha}_m + \underline{\alpha}_m}{2}\right\}, \\ \check{\alpha} &= \operatorname{diag}\left\{\frac{\bar{\alpha}_1 - \underline{\alpha}_1}{2}, \frac{\bar{\alpha}_2 - \underline{\alpha}_2}{2}, \ldots, \frac{\bar{\alpha}_m - \underline{\alpha}_m}{2}\right\}, \end{aligned} \tag{4.4}$$

and rewrite α as follows

$$\alpha = \hat{\alpha} + \Delta = \hat{\alpha} + \operatorname{diag}\{\delta_1, \delta_2, \ldots, \delta_m\}, \tag{4.5}$$

where

$$|\delta_i| \leqslant \frac{\bar{\alpha}_i - \underline{\alpha}_i}{2}, i = 1, 2, \ldots, m.$$

In this chapter, we consider the following controller

$$u(t) = Kx(t), \tag{4.6}$$

where $K \in \mathbb{R}^{m \times n}$ is the controller gain to be designed. Substituting $u^F(t)$ for $u(t)$ in (4.1), and considering (4.2) and (4.6), the resultant closed-loop system can be described by

$$\begin{cases} E\dot{x}(t) = (A + B\alpha K)x(t) + A_d x(t - d(t)) + B_\omega \omega(t), \\ \quad z(t) = Cx(t) + C_d x(t - d(t)) + D_\omega \omega(t). \end{cases} \tag{4.7}$$

Definition 4.1. *[125, 199]*

1. System

$$\begin{cases} E\dot{x}(t) = Ax(t) + A_d x(t - d(t)), \\ \quad x(t) = \phi(t), \, t \in [-d_2, 0], \end{cases} \tag{4.8}$$

is said to be exponentially stable, if there exist scalars $\alpha > 0$ and $\beta > 0$ such that $\|x(t)\| \leqslant \alpha e^{-\beta t} \|\phi(t)\|_{d_2}$, $t > 0$.
2. System (4.8) is said to be exponentially admissible, if it is regular, impulse free and exponentially stable.

In this chapter, we aim to design the state feedback controller (4.6) such that system (4.7) simultaneously satisfies the following two requirements:

(i) System (4.7) with $\omega(t) = 0$ is exponentially admissible,
(ii) Under zero initial condition, there exists a scalar $\gamma > 0$ such that

$$2 \int_0^{t^*} \omega(t)^{\mathrm{T}} z(t) \mathrm{d}t \geqslant -\gamma \int_0^{t^*} \omega(t)^{\mathrm{T}} \omega(t) \mathrm{d}t, \, \forall t^* > 0. \tag{4.9}$$

In this case, system (4.7) is said to be exponentially admissible and passive.

4.3 Main Results

4.3.1 Exponential Stability and Passivity Analysis

The following theorem provides a sufficient condition under which system (4.8) is exponentially admissible.

Theorem 4.2. *System (4.8) is exponentially admissible, if there exist matrices P, Y, $Z_1 > 0$, $Z_2 > 0$, $Z_3 > 0$, $S_1 > 0$ and $S_2 > 0$ such that*

$$E^{\mathrm{T}}P = P^{\mathrm{T}}E \geqslant 0, \tag{4.10}$$

$$\begin{bmatrix} \Xi_{11} & E^{\mathrm{T}}S_1E & P^{\mathrm{T}}A_d & 0 & d_1A^{\mathrm{T}}S_1 & d_{12}A^{\mathrm{T}}S_2 \\ * & \Xi_{22} & \Xi_{23} & E^{\mathrm{T}}YE & 0 & 0 \\ * & * & \Xi_{33} & \Xi_{34} & d_1A_d^{\mathrm{T}}S_1 & d_{12}A_d^{\mathrm{T}}S_2 \\ * & * & * & \Xi_{44} & 0 & 0 \\ * & * & * & * & -S_1 & 0 \\ * & * & * & * & * & -S_2 \end{bmatrix} < 0, \tag{4.11}$$

$$\begin{bmatrix} S_2 & Y \\ * & S_2 \end{bmatrix} > 0, \tag{4.12}$$

where $d_{12} = d_2 - d_1$ and

$$\begin{aligned}
\Xi_{11} &= P^{\mathrm{T}}A + A^{\mathrm{T}}P + Z_1 + Z_2 + Z_3 - E^{\mathrm{T}}S_1E, \\
\Xi_{22} &= -Z_1 - E^{\mathrm{T}}S_1E - E^{\mathrm{T}}S_2E, \\
\Xi_{23} &= E^{\mathrm{T}}S_2E - E^{\mathrm{T}}YE, \\
\Xi_{33} &= -(1-\mu)Z_2 - 2E^{\mathrm{T}}S_2E + E^{\mathrm{T}}YE + E^{\mathrm{T}}Y^{\mathrm{T}}E, \\
\Xi_{34} &= -E^{\mathrm{T}}YE + E^{\mathrm{T}}S_2E, \\
\Xi_{44} &= -Z_3 - E^{\mathrm{T}}S_2E.
\end{aligned}$$

Proof. Firstly, we prove the regularity and absence of impulses of system (4.8). Since $\operatorname{rank} E = r \leqslant n$, there exist two nonsingular matrices G and H such that

$$GEH = \begin{bmatrix} I_r & 0 \\ 0 & 0 \end{bmatrix}. \tag{4.13}$$

Denote

$$GAH = \begin{bmatrix} A_1 & A_2 \\ A_3 & A_4 \end{bmatrix}, \; G^{-\mathrm{T}}PH = \begin{bmatrix} \bar{P}_1 & \bar{P}_2 \\ \bar{P}_3 & \bar{P}_4 \end{bmatrix}. \tag{4.14}$$

From (4.10) and using the expressions in (4.13) and (4.14), it can be found that $\bar{P}_2 = 0$. Then, pre-multiplying and post-multiplying $\Xi_{11} < 0$ by H^{T} and H, respectively, we have

$$A_4^{\mathrm{T}}\bar{P}_4 + \bar{P}_4^{\mathrm{T}}A_4 < 0, \tag{4.15}$$

which implies A_4 is nonsingular and thus the pair (E, A) is regular and impulse free. Hence, by Definition 3.1, system (4.8) is regular and impulse free.

Next we will show that system (4.8) is stable. To the end, we consider the following Lyapunov functional for system (4.8):

$$V(x_t) = V_1(x_t) + V_2(x_t) + V_3(x_t), \tag{4.16}$$

where

$$
\begin{aligned}
V_1(x_t) &= x(t)^{\mathrm{T}} E^{\mathrm{T}} P x(t), \\
V_2(x_t) &= \int_{t-d_1}^{t} x(s)^{\mathrm{T}} Z_1 x(s) \mathrm{d}s + \int_{t-d(t)}^{t} x(s)^{\mathrm{T}} Z_2 x(s) \mathrm{d}s, \\
&\quad + \int_{t-d_2}^{t} x(s)^{\mathrm{T}} Z_3 x(s) \mathrm{d}s, \\
V_3(x_t) &= d_1 \int_{-d_1}^{0} \int_{t+\beta}^{t} \dot{x}(s)^{\mathrm{T}} E^{\mathrm{T}} S_1 E \dot{x}(s) \mathrm{d}s \mathrm{d}\beta \\
&\quad + d_{12} \int_{-d_2}^{-d_1} \int_{t+\beta}^{t} \dot{x}(s)^{\mathrm{T}} E^{\mathrm{T}} S_2 E \dot{x}(s) \mathrm{d}s \mathrm{d}\beta,
\end{aligned}
$$

where $\{x_t = x(t+\theta), -2d_2 \leqslant \theta \leqslant 0\}$. Calculating the time derivative of $V(x_t)$ along the solutions of system (4.8) yields

$$
\dot{V}_1(x_t) = 2x(t)^{\mathrm{T}} P^{\mathrm{T}} E \dot{x}(t), \tag{4.17}
$$

$$
\begin{aligned}
\dot{V}_2(x_t) \leqslant\ & x(t)^{\mathrm{T}} (Z_1 + Z_2 + Z_3) x(t) - x(t-d_1)^{\mathrm{T}} Z_1 x(t-d_1) \\
& - (1-\mu) x(t-d(t))^{\mathrm{T}} Z_2 x(t-d(t)) - x(t-d_2)^{\mathrm{T}} Z_3 x(t-d_2),
\end{aligned} \tag{4.18}
$$

$$
\begin{aligned}
\dot{V}_3(x_t) = \ & \dot{x}(t)^{\mathrm{T}} E^{\mathrm{T}} (d_1^2 S_1 + d_{12}^2 S_2) E \dot{x}(t) - d_1 \int_{t-d_1}^{t} \dot{x}(s)^{\mathrm{T}} E^{\mathrm{T}} S_1 E \dot{x}(s) \mathrm{d}s \\
& - d_{12} \int_{t-d_2}^{t-d_1} \dot{x}(s)^{\mathrm{T}} E^{\mathrm{T}} S_2 E \dot{x}(s) \mathrm{d}s.
\end{aligned} \tag{4.19}
$$

Applying Jensen inequality, we have that

$$
\begin{aligned}
-d_1 & \int_{t-d_1}^{t} \dot{x}(s)^{\mathrm{T}} E^{\mathrm{T}} S_1 E \dot{x}(s) \mathrm{d}s \\
& \leqslant - \int_{t-d_1}^{t} \dot{x}(s)^{\mathrm{T}} E^{\mathrm{T}} \mathrm{d}s S_1 \int_{t-d_1}^{t} \dot{x}(s)^{\mathrm{T}} E \dot{x}(s) \mathrm{d}s \\
& = \begin{bmatrix} x(t) \\ x(t-d_1) \end{bmatrix}^{\mathrm{T}} \begin{bmatrix} -E^{\mathrm{T}} S_1 E & E^{\mathrm{T}} S_1 E \\ * & -E^{\mathrm{T}} S_1 E \end{bmatrix} \begin{bmatrix} x(t) \\ x(t-d_1) \end{bmatrix}.
\end{aligned} \tag{4.20}
$$

On the other hand, when $d_1 < d(t) < d_2$, we have that

$$
\begin{aligned}
-d_{12} & \int_{t-d_2}^{t-d_1} \dot{x}(s)^{\mathrm{T}} E^{\mathrm{T}} S_2 E \dot{x}(s) \mathrm{d}s \\
= & - d_{12} \int_{t-d(t)}^{t-d_1} \dot{x}(s)^{\mathrm{T}} E^{\mathrm{T}} S_2 E \dot{x}(s) \mathrm{d}s - d_{12} \int_{t-d_2}^{t-d(t)} \dot{x}(s)^{\mathrm{T}} E^{\mathrm{T}} S_2 E \dot{x}(s) \mathrm{d}s
\end{aligned}
$$

$$
\begin{aligned}
&\leqslant -\frac{d_{12}}{d(t)-d_1}\int_{t-d(t)}^{t-d_1}\dot{x}(s)^{\mathrm{T}}E^{\mathrm{T}}\mathrm{d}sS_2\int_{t-d(t)}^{t-d_1}E\dot{x}(s)\mathrm{d}s \\
&\quad -\frac{d_{12}}{d_2-d(t)}\int_{t-d_2}^{t-d(t)}\dot{x}(s)^{\mathrm{T}}E^{\mathrm{T}}\mathrm{d}sS_2\int_{t-d_2}^{t-d(t)}E\dot{x}(s)\mathrm{d}s \\
&= -\int_{t-d(t)}^{t-d_1}\dot{x}(s)^{\mathrm{T}}E^{\mathrm{T}}\mathrm{d}sS_2\int_{t-d(t)}^{t-d_1}E\dot{x}(s)\mathrm{d}s \\
&\quad -\int_{t-d_2}^{t-d(t)}\dot{x}(s)^{\mathrm{T}}E^{\mathrm{T}}\mathrm{d}sS_2\int_{t-d_2}^{t-d(t)}E\dot{x}(s)\mathrm{d}s \\
&\quad -\frac{d_2-d(t)}{d(t)-d_1}\int_{t-d(t)}^{t-d_1}\dot{x}(s)^{\mathrm{T}}E^{\mathrm{T}}\mathrm{d}sS_2\int_{t-d(t)}^{t-d_1}E\dot{x}(s)\mathrm{d}s \\
&\quad -\frac{d(t)-d_1}{d_2-d(t)}\int_{t-d_2}^{t-d(t)}\dot{x}(s)^{\mathrm{T}}E^{\mathrm{T}}\mathrm{d}sS_2\int_{t-d_2}^{t-d(t)}E\dot{x}(s)\mathrm{d}s.
\end{aligned}
\tag{4.21}
$$

Based on the lower bounds lemma of [116], it can be found from (4.12) that

$$
\begin{bmatrix}\sqrt{\frac{d_2-d(t)}{d(t)-d_1}}\int_{t-d(t)}^{t-d_1}E\dot{x}(s)\mathrm{d}s\\ -\sqrt{\frac{d(t)-d_1}{d_2-d(t)}}\int_{t-d_2}^{t-d(t)}E\dot{x}(s)\mathrm{d}s\end{bmatrix}^{\mathrm{T}}\begin{bmatrix}S_2 & Y\\ * & S_2\end{bmatrix}\begin{bmatrix}\sqrt{\frac{d_2-d(t)}{d(t)-d_1}}\int_{t-d(t)}^{t-d_1}E\dot{x}(s)\mathrm{d}s\\ -\sqrt{\frac{d(t)-d_1}{d_2-d(t)}}\int_{t-d_2}^{t-d(t)}E\dot{x}(s)\mathrm{d}s\end{bmatrix}\geqslant 0,
\tag{4.22}
$$

which implies

$$
\begin{aligned}
&-\frac{d_2-d(t)}{d(t)-d_1}\int_{t-d(t)}^{t-d_1}\dot{x}(s)^{\mathrm{T}}E^{\mathrm{T}}\mathrm{d}sS_2\int_{t-d(t)}^{t-d_1}E\dot{x}(s)\mathrm{d}s \\
&-\frac{d(t)-d_1}{d_2-d(t)}\int_{t-d_2}^{t-d(t)}\dot{x}(s)^{\mathrm{T}}E^{\mathrm{T}}\mathrm{d}sS_2\int_{t-d_2}^{t-d(t)}E\dot{x}(s)\mathrm{d}s \\
\leqslant &-\int_{t-d(t)}^{t-d_1}\dot{x}(s)^{\mathrm{T}}E^{\mathrm{T}}\mathrm{d}sY\int_{t-d_2}^{t-d(t)}E\dot{x}(s)\mathrm{d}s \\
&-\int_{t-d_2}^{t-d(t)}\dot{x}(s)^{\mathrm{T}}E^{\mathrm{T}}\mathrm{d}sY^{\mathrm{T}}\int_{t-d(t)}^{t-d_1}E\dot{x}(s)\mathrm{d}s.
\end{aligned}
\tag{4.23}
$$

We can get from (4.21) and (4.23) that

$$
\begin{aligned}
&-d_{12}\int_{t-d_2}^{t-d_1}\dot{x}(s)^{\mathrm{T}}E^{\mathrm{T}}S_2E\dot{x}(s)\mathrm{d}s \\
&\qquad\leqslant -\begin{bmatrix}\int_{t-d(t)}^{t-d_1}E\dot{x}(s)\mathrm{d}s\\ \int_{t-d_2}^{t-d(t)}E\dot{x}(s)\mathrm{d}s\end{bmatrix}^{\mathrm{T}}\begin{bmatrix}S_2 & Y\\ * & S_2\end{bmatrix}\begin{bmatrix}\int_{t-d(t)}^{t-d_1}E\dot{x}(s)\mathrm{d}s\\ \int_{t-d_2}^{t-d(t)}E\dot{x}(s)\mathrm{d}s\end{bmatrix} \\
&\qquad= \begin{bmatrix}x(t-d_1)\\ x(t-d(t))\\ x(t-d_2)\end{bmatrix}^{\mathrm{T}}\Theta\begin{bmatrix}x(t-d_1)\\ x(t-d(t))\\ x(t-d_2)\end{bmatrix},
\end{aligned}
\tag{4.24}
$$

where

$$\Theta = \begin{bmatrix} -E^{\mathrm{T}}S_2E & E^{\mathrm{T}}S_2E - E^{\mathrm{T}}YE & E^{\mathrm{T}}YE \\ * & -2E^{\mathrm{T}}S_2E + E^{\mathrm{T}}YE + E^{\mathrm{T}}Y^{\mathrm{T}}E & -E^{\mathrm{T}}YE + E^{\mathrm{T}}S_2E \\ * & * & -E^{\mathrm{T}}S_2E \end{bmatrix}.$$

Note that when $d(t) = d_1$ or $d(t) = d_2$, we have $\int_{t-d(t)}^{t-d_1} E\dot{x}(s)\mathrm{d}s = 0$ or $\int_{t-d_2}^{t-d(t)} E\dot{x}(s)\mathrm{d}s = 0$, respectively. Thus, (4.24) still holds based on Jensen inequality. Then, we get from (4.17)-(4.20) and (4.24) that

$$\dot{V}(x_t) \leqslant \zeta(t)^{\mathrm{T}}\Sigma\zeta(t), \tag{4.25}$$

where

$$\zeta(t) = \begin{bmatrix} x(t) \\ x(t-d_1) \\ x(t-d(t)) \\ x(t-d_2) \end{bmatrix}, \Sigma = \begin{bmatrix} \Xi_{11} & E^{\mathrm{T}}S_1E & P^{\mathrm{T}}A_d & 0 \\ * & \Xi_{22} & \Xi_{23} & E^{\mathrm{T}}YE \\ * & * & \Xi_{33} & \Xi_{34} \\ * & * & * & \Xi_{44} \end{bmatrix} + \begin{bmatrix} A^{\mathrm{T}} \\ 0 \\ A_d^{\mathrm{T}} \\ 0 \end{bmatrix} S \begin{bmatrix} A^{\mathrm{T}} \\ 0 \\ A_d^{\mathrm{T}} \\ 0 \end{bmatrix}^{\mathrm{T}}.$$

and $S = d_1^2S_1 + d_{12}^2S_2$. According to Schur complement, we get from (4.11) that $\Sigma < 0$, which guarantees there exists a scalar $\lambda > 0$ such that

$$\dot{V}(t) < -\lambda||x(t)||^2. \tag{4.26}$$

Set

$$\hat{G} = \begin{bmatrix} I_r & -A_2A_4^{-1} \\ 0 & A_4^{-1} \end{bmatrix} G. \tag{4.27}$$

It is easy to get

$$\hat{G}EH = \begin{bmatrix} I_r & 0 \\ 0 & 0 \end{bmatrix}, \hat{G}AH = \begin{bmatrix} \hat{A}_1 & 0 \\ \hat{A}_3 & I \end{bmatrix}, \tag{4.28}$$

where $\hat{A}_1 = A_1 - A_2A_4^{-1}A_3$ and $\hat{A}_3 = A_4^{-1}A_3$. Denote

$$\hat{G}A_dH = \begin{bmatrix} A_{d1} & A_{d2} \\ A_{d3} & A_{d4} \end{bmatrix}, \hat{G}^{-\mathrm{T}}PH = \begin{bmatrix} P_1 & P_2 \\ P_3 & P_4 \end{bmatrix}, H^{\mathrm{T}}Z_2H = \begin{bmatrix} Z_{11} & Z_{21} \\ * & Z_{22} \end{bmatrix}. \tag{4.29}$$

It can be seen from (4.11) that

$$\begin{bmatrix} I \\ I \\ I \\ I \end{bmatrix}^{\mathrm{T}} \begin{bmatrix} \Xi_{11} & E^{\mathrm{T}}S_1E & P^{\mathrm{T}}A_d & 0 \\ * & \Xi_{22} & \Xi_{23} & E^{\mathrm{T}}YE \\ * & * & \Xi_{33} & \Xi_{34} \\ * & * & * & \Xi_{44} \end{bmatrix} \begin{bmatrix} I \\ I \\ I \\ I \end{bmatrix} < 0, \tag{4.30}$$

which implies

$$P^{\mathrm{T}}(A + A_d) + (A + A_d)^{\mathrm{T}}P < 0. \tag{4.31}$$

It is easy to find that matrix P satisfying the above inequality is nonsingular. Thus, considering (4.10), we can deduce that $P_1 > 0$ and $P_2 = 0$. Define

$$\hat{\zeta}(t) = \begin{bmatrix} \zeta_1(t) \\ \zeta_2(t) \end{bmatrix} = H^{-1}x(t), \tag{4.32}$$

then system (4.8) is equivalent to the following system:

$$\begin{cases} \dot{\zeta}_1(t) = \hat{A}_1\zeta_1(t) + A_{d1}\zeta_1(t-d(t)) + A_{d2}\zeta_2(t-d(t)), \\ -\zeta_2(t) = \hat{A}_3\zeta_1(t) + A_{d3}\zeta_1(t-d(t)) + A_{d4}\zeta_2(t-d(t)), \\ \hat{\zeta}(t) = \psi(t) = H^{-1}\phi(t),\ t \in [-d_2, 0]. \end{cases} \tag{4.33}$$

To prove the exponential stability of system (4.8), we define a function as

$$W(x_t, t) = e^{\varepsilon t}V(x_t, t), \tag{4.34}$$

where the scalar $\varepsilon > 0$. Taking its time derivative yields

$$\begin{aligned} \dot{W}(x_t, t) &= \varepsilon e^{\varepsilon t}V(x_t, t) + e^{\varepsilon t}\dot{V}(x_t, t) \\ &\leqslant \varepsilon e^{\varepsilon t}V(x_t, t) - \lambda e^{\varepsilon t}\|x(t)\|^2. \end{aligned} \tag{4.35}$$

Integrating both sides of (4.35) from 0 to t, we get that

$$W(x_t, t) \leqslant W(x_0, 0) + \int_0^t e^{\varepsilon s}\left[\varepsilon V(x_s, s) - \lambda\|x(s)\|^2\right] \mathrm{d}s. \tag{4.36}$$

By using the similar analysis method of [108], it can be seen from (4.16), (4.34) and (4.36) that, if the scalar ε is chosen small enough, a scalar $k > 0$ can be found such that for any $t > 0$,

$$V(x_t, t) \leqslant ke^{-\varepsilon t}\|\phi(t)\|_{d_2}^2. \tag{4.37}$$

Since $\lambda_{\min}(P_1)\|\zeta_1(t)\|^2 \leqslant x(t)^{\mathrm{T}}E^{\mathrm{T}}Px(t) \leqslant V(x_t, t)$, it can be shown from (4.37) that for any $t > 0$

$$\|\zeta_1(t)\|^2 \leqslant \alpha e^{-\varepsilon t}\|\phi(t)\|_{d_2}^2, \tag{4.38}$$

where $\alpha = \lambda_{\min}(P_1)^{-1}k$. Define

$$e(t) = \hat{A}_3\zeta_1(t) + A_{d3}\zeta_1(t-d(t)), \tag{4.39}$$

then, from (4.38), a scalar $m > 0$ can be found such that for any $t > 0$,

$$\|e(t)\|^2 \leqslant me^{-\varepsilon t}\|\phi(t)\|_{d_2}^2. \tag{4.40}$$

To study the exponential stability of $\zeta_2(t)$, we construct a function as

$$J(t) = \zeta_2(t)^{\mathrm{T}}Z_{22}\zeta_2(t) - \zeta_2(t-d(t))^{\mathrm{T}}Z_{22}\zeta_2(t-d(t)). \tag{4.41}$$

By pre-multiplying the second equation of (4.33) with $\zeta_2(t)^{\mathrm{T}}P_4^{\mathrm{T}}$, we obtain that

$$0 = 2\left[\zeta_2(t)^{\mathrm{T}}P_4^{\mathrm{T}}\zeta_2(t) + \zeta_2(t)^{\mathrm{T}}P_4^{\mathrm{T}}A_{d4}\zeta_2(t-d(t)) + \zeta_2(t)^{\mathrm{T}}P_4^{\mathrm{T}}e(t)\right]. \quad (4.42)$$

Adding (4.42) to (4.41) yields that

$$\begin{aligned} J(t) &= \zeta_2(t)^{\mathrm{T}}(P_4^{\mathrm{T}} + P_4 + Z_{22})\zeta_2(t) + 2\zeta_2(t)^{\mathrm{T}}P_4^{\mathrm{T}}A_{d4}\zeta_2(t-d(t)) \\ &\quad - \zeta_2(t-d(t))^{\mathrm{T}}Z_{22}\zeta_2(t-d(t)) + 2\zeta_2(t)^{\mathrm{T}}P_4^{\mathrm{T}}e(t) \\ &\leqslant \begin{bmatrix} \zeta_2(t) \\ \zeta_2(t-d(t)) \end{bmatrix}^{\mathrm{T}} \begin{bmatrix} P_4^{\mathrm{T}} + P_4 + Z_{22} & P_4^{\mathrm{T}}A_{d4} \\ * & -Z_{22} \end{bmatrix} \begin{bmatrix} \zeta_2(t) \\ \zeta_2(t-d(t)) \end{bmatrix} \\ &\quad + \eta_1\zeta_2(t)^{\mathrm{T}}\zeta_2(t) + \eta_1^{-1}e(t)^{\mathrm{T}}P_4P_4^{\mathrm{T}}e(t), \end{aligned} \quad (4.43)$$

where η_1 is any positive scalar. On the other hand, we can get from (4.11) that

$$\begin{bmatrix} Z_2 - E^{\mathrm{T}}S_1E + P^{\mathrm{T}}A + A^{\mathrm{T}}P & P^{\mathrm{T}}A_d \\ * & -Z_2 - 2E^{\mathrm{T}}S_2E + E^{\mathrm{T}}YE + E^{\mathrm{T}}Y^{\mathrm{T}}E \end{bmatrix} < 0. \quad (4.44)$$

Pre-multiplying and post-multiplying (4.44) by $\begin{bmatrix} H & 0 \\ 0 & H \end{bmatrix}^{\mathrm{T}}$ and $\begin{bmatrix} H & 0 \\ 0 & H \end{bmatrix}$, respectively, a scalar $\eta_2 > 0$ can be found such that

$$\begin{bmatrix} P_4^{\mathrm{T}} + P_4 + Z_{22} & P_4^{\mathrm{T}}A_{d4} \\ * & -Z_{22} \end{bmatrix} \leqslant -\begin{bmatrix} \eta_2 I & 0 \\ 0 & 0 \end{bmatrix}. \quad (4.45)$$

On the other hand, since η_1 can be chosen arbitrarily, η_1 can be chosen small enough such that $\eta_2 - \eta_1 > 0$. Then a scalar $\eta_3 > 1$ can always be found such that

$$Z_{22} - (\eta_1 - \eta_2)I \geqslant \eta_3 Z_{22}. \quad (4.46)$$

It follows from (4.41), (4.43), (4.45) and (4.46) that

$$\zeta_2(t)^{\mathrm{T}}Z_{22}\zeta_2(t) \leqslant \eta_3^{-1}\zeta_2(t-d(t))^{\mathrm{T}}Z_{22}\zeta_2(t-d(t)) + (\eta_1\eta_3)^{-1}e(t)^{\mathrm{T}}P_4P_4^{\mathrm{T}}e(t), \quad (4.47)$$

which infers

$$f(t) \leqslant \eta_3^{-1} \sup_{t-d_2 \leqslant s \leqslant t} f(s) + \xi e^{-\delta t}, \quad (4.48)$$

where $0 < \delta < \min\{\varepsilon, d_2^{-1}\ln\eta_3\}$, $\xi = (\eta_1\eta_3)^{-1}m\|P_4\|^2\|\phi(t)\|_{d_2}^2$ and $f(t) = \zeta_2(t)^{\mathrm{T}}Z_{22}\zeta_2(t)$. Therefore, applying Lemma 1.10 to (4.48) yields that

$$\|\zeta_2(s)\|^2 \leqslant \lambda_{\min}^{-1}(Z_{22})\lambda_{\max}(Z_{22})e^{-\delta t}\|\zeta_2(s)\|_{d_2}^2 + \frac{\lambda_{\min}^{-1}(Z_{22})\xi e^{-\delta t}}{1 - \eta_3^{-1}e^{\delta d_2}}. \quad (4.49)$$

We can find from (4.38) and (4.49) that system (4.8) is exponentially stable. This completes the proof.

Remark 4.3. It is noted that a sufficient condition is established in Theorem 4.2 to guarantee the exponential admissibility of SSs with time-varying delays. Based on the lower bounds lemma of [116], a matrix Y is introduced to deal with the terms $-\frac{d_2-d(t)}{d(t)-d_1}\int_{t-d(t)}^{t-d_1}\dot{x}(s)^{\mathrm{T}}E^{\mathrm{T}}\mathrm{d}sS_2\int_{t-d(t)}^{t-d_1}E\dot{x}(s)\mathrm{d}s$ and $-\frac{d(t)-d_1}{d_2-d(t)}\int_{t-d_2}^{t-d(t)}\dot{x}(s)^{\mathrm{T}}E^{\mathrm{T}}\mathrm{d}sS_2\int_{t-d_2}^{t-d(t)}E\dot{x}(s)\mathrm{d}s$ in (4.21). However, the above two terms are ignored by [50, 163] and not fully considered in [219], which may lead to conservatism to some extent. Thus, our result derived here is expected to have less conservatism than those of [50, 163, 219].

Next, let us deal with the problem of passivity analysis for the following system,

$$\begin{cases} E\dot{x}(t) = Ax(t) + A_d x(t-d(t)) + B_\omega \omega(t), \\ z(t) = Cx(t) + C_d x(t-d(t)) + D_\omega \omega(t), \\ x(t) = \phi(t),\ t \in [-d_2, 0]. \end{cases} \tag{4.50}$$

Theorem 4.4. *System (4.50) is exponentially admissible and passive, if there exist matrices P, Y, $Z_1 > 0$, $Z_2 > 0$, $Z_3 > 0$, $S_1 > 0$ and $S_2 > 0$, and a scalar $\gamma > 0$ such that*

$$E^{\mathrm{T}}P = P^{\mathrm{T}}E \geqslant 0, \tag{4.51}$$

$$\begin{bmatrix} \Xi_{11} & E^{\mathrm{T}}S_1E & P^{\mathrm{T}}A_d & 0 & P^{\mathrm{T}}B_\omega - C^{\mathrm{T}} & d_1A^{\mathrm{T}}S_1 & d_{12}A^{\mathrm{T}}S_2 \\ * & \Xi_{22} & \Xi_{23} & E^{\mathrm{T}}YE & 0 & 0 & 0 \\ * & * & \Xi_{33} & \Xi_{34} & -C_d^{\mathrm{T}} & d_1A_d^{\mathrm{T}}S_1 & d_{12}A_d^{\mathrm{T}}S_2 \\ * & * & * & \Xi_{44} & 0 & 0 & 0 \\ * & * & * & * & -\gamma I - D_\omega - D_\omega^{\mathrm{T}} & d_1B_\omega^{\mathrm{T}}S_1 & d_{12}B_\omega^{\mathrm{T}}S_2 \\ * & * & * & * & * & -S_1 & 0 \\ * & * & * & * & * & * & -S_2 \end{bmatrix} < 0, \tag{4.52}$$

$$\begin{bmatrix} S_2 & Y \\ * & S_2 \end{bmatrix} > 0, \tag{4.53}$$

where d_{12}, Ξ_{11}, Ξ_{22}, Ξ_{23}, Ξ_{33}, Ξ_{34} and Ξ_{44} are given in Theorem 4.2.

Proof. It is obvious that (4.51)-(4.53) implies (4.10)-(4.12) hold which, according to Theorem 4.2, guarantees the exponential admissibility of system (4.50) with $\omega(t) = 0$. For the passivity analysis, we consider Lyapunov functional (4.16) for system (4.50). Following the similar method employed in Theorem 4.2, we can get that

$$\dot{V}(x_t) - 2\omega(t)^{\mathrm{T}}z(t) - \gamma\omega(t)^{\mathrm{T}}\omega(t) \leqslant \bar{\zeta}(t)^{\mathrm{T}}\hat{\Sigma}\bar{\zeta}(t), \tag{4.54}$$

where

$$\bar{\zeta}(t) = \begin{bmatrix} x(t) \\ x(t-d_1) \\ x(t-d(t)) \\ x(t-d_2) \\ \omega(t) \end{bmatrix},$$

$$\hat{\Sigma} = \begin{bmatrix} \Xi_{11} & E^{\mathrm{T}}S_1E & P^{\mathrm{T}}A_d & 0 & P^{\mathrm{T}}B_\omega - C^{\mathrm{T}} \\ * & \Xi_{22} & \Xi_{23} & E^{\mathrm{T}}YE & 0 \\ * & * & \Xi_{33} & \Xi_{34} & -C_d^{\mathrm{T}} \\ * & * & * & \Xi_{44} & 0 \\ * & * & * & * & -\gamma I - D_\omega - D_\omega^{\mathrm{T}} \end{bmatrix} + \begin{bmatrix} A^{\mathrm{T}} \\ 0 \\ A_d^{\mathrm{T}} \\ 0 \\ B_\omega^{\mathrm{T}} \end{bmatrix} S \begin{bmatrix} A^{\mathrm{T}} \\ 0 \\ A_d^{\mathrm{T}} \\ 0 \\ B_\omega^{\mathrm{T}} \end{bmatrix}^{\mathrm{T}}.$$

and $S = d_1^2 S_1 + d_{12}^2 S_2$. According to Schur complement, we get from (4.52) that $\hat{\Sigma} < 0$, which implies

$$\dot{V}(x_t) - 2\omega(t)^{\mathrm{T}} z(t) - \gamma\omega(t)^{\mathrm{T}}\omega(t) \leqslant 0. \tag{4.55}$$

Integrating both sides of the inequality (4.55) with respect to t over the time period $[0, t^*]$ gives rise to

$$V(x_{t^*}) - V(x_0) - 2\int_0^{t^*} \omega(t)^{\mathrm{T}} z(t)\mathrm{d}t - \gamma\int_0^{t^*} \omega(t)^{\mathrm{T}}\omega(t)\mathrm{d}t \leqslant 0. \tag{4.56}$$

Under the zero initial condition, we have that $V(x_{t^*}) \geqslant 0$ and $V(x_0) = 0$, and thus the inequality (4.56) guarantees (4.9), which means system (4.50) is passive. This completes the proof.

Remark 4.5. Based on Theorem 4.2, Theorem 4.4 provides a delay-dependent condition on passivity analysis for SSs with time-varying delays. It is noted that the involved matrix E of [149] has been assumed to be $\mathrm{diag}\{I_r, 0\}$, which sometimes makes the analysis procedure relatively intricate and complicated, though this assumption is without loss of generality. Moreover, the considered time delay in [149] is time-invariant, which limits the scope of applications of the proposed results. Thus, the condition given in this chapter is more desirable and elegant than that of [149].

4.3.2 Reliable Controller Design

According to the passivity analysis condition given in Theorem 4.4, we are now in the position to provide a solution to the reliable passive control problem for system (4.1). The design method of the desired state feedback controller is established in the following theorem.

Theorem 4.6. *System (4.7) is exponentially admissible and passive, if there exist matrices L, $\hat{Y}$, V, $\hat{Z}_1 > 0$, $\hat{Z}_2 > 0$, $\hat{Z}_3 > 0$, $\hat{S}_1 > 0$, $\hat{S}_2 > 0$ and $M = \mathrm{diag}\{m_1, m_2, \cdots, m_i\} > 0$, and a scalar $\gamma > 0$ such that*

$$L^{\mathrm{T}}E^{\mathrm{T}} = EL \geqslant 0, \tag{4.57}$$

$$\begin{bmatrix} \hat{\Xi}_{11} & E\hat{S}_1E^{\mathrm{T}} & A_dL & 0 & \hat{\Xi}_{15} & d_1\mathcal{F}^{\mathrm{T}} & d_{12}\mathcal{F}^{\mathrm{T}} & BM\breve{\alpha} & V^{\mathrm{T}} \\ * & \hat{\Xi}_{22} & \hat{\Xi}_{23} & E\hat{Y}_1E^{\mathrm{T}} & 0 & 0 & 0 & 0 & 0 \\ * & * & \hat{\Xi}_{33} & \hat{\Xi}_{34} & -L^{\mathrm{T}}C_d^{\mathrm{T}} & d_1L^{\mathrm{T}}A_d^{\mathrm{T}} & d_{12}L^{\mathrm{T}}A_d^{\mathrm{T}} & 0 & 0 \\ * & * & * & \hat{\Xi}_{44} & 0 & 0 & 0 & 0 & 0 \\ * & * & * & * & \hat{\Xi}_{55} & d_1B_\omega^{\mathrm{T}} & d_{12}B_\omega^{\mathrm{T}} & 0 & 0 \\ * & * & * & * & * & \hat{\Xi}_{66} & 0 & d_1BM\breve{\alpha} & 0 \\ * & * & * & * & * & * & \hat{\Xi}_{77} & d_{12}BM\breve{\alpha} & 0 \\ * & * & * & * & * & * & * & -M & 0 \\ * & * & * & * & * & * & * & * & -M \end{bmatrix} < 0, \tag{4.58}$$

$$\begin{bmatrix} \hat{S}_2 & \hat{Y} \\ * & \hat{S}_2 \end{bmatrix} > 0, \tag{4.59}$$

where d_{12} *is given in Theorem 4.2 and*

$$\begin{aligned}
\hat{\Xi}_{11} &= AL + B\hat{\alpha}V + (AL + B\hat{\alpha}V)^{\mathrm{T}} + \hat{Z}_1 + \hat{Z}_2 + \hat{Z}_3 - E\hat{S}_1E^{\mathrm{T}}, \\
\hat{\Xi}_{15} &= B_\omega - L^{\mathrm{T}}C^{\mathrm{T}}, \\
\hat{\Xi}_{22} &= -\hat{Z}_1 - E\hat{S}_1E^{\mathrm{T}} - E\hat{S}_2E^{\mathrm{T}}, \\
\hat{\Xi}_{23} &= E\hat{S}_2E^{\mathrm{T}} - E\hat{Y}E^{\mathrm{T}} \\
\hat{\Xi}_{33} &= -(1-\mu)\hat{Z}_2 - 2E\hat{S}_2E^{\mathrm{T}} + E\hat{Y}E^{\mathrm{T}} + E\hat{Y}^{\mathrm{T}}E^{\mathrm{T}}, \\
\hat{\Xi}_{34} &= -E\hat{Y}E^{\mathrm{T}} + E\hat{S}_2E^{\mathrm{T}}, \\
\hat{\Xi}_{44} &= -\hat{Z}_3 - E\hat{S}_2E^{\mathrm{T}}, \\
\hat{\Xi}_{55} &= -\gamma I - D_\omega - D_\omega^{\mathrm{T}}, \\
\hat{\Xi}_{66} &= \hat{S}_1 - L^{\mathrm{T}} - L, \\
\hat{\Xi}_{77} &= \hat{S}_2 - L^{\mathrm{T}} - L, \\
\mathcal{F} &= AL + B\hat{\alpha}V.
\end{aligned}$$

Furthermore, if (4.57)-(4.59) *are solvable, the desired controller gain is given as*

$$K = VL^{-1}. \tag{4.60}$$

Proof. Setting $L = P^{-1}$, and pre- and post-multiplying (4.51) by L^{T} and L, respectively, we have (4.57). Furthermore, replacing A in (4.52) with $A + B\alpha K$, we have

$$\begin{bmatrix} \bar{\Xi}_{11} & E^{\mathrm{T}}S_1E & P^{\mathrm{T}}A_d & 0 & P^{\mathrm{T}}B_{\omega}-C^{\mathrm{T}} & \bar{\Xi}_{16} & \bar{\Xi}_{17} \\ * & \Xi_{22} & \Xi_{23} & E^{\mathrm{T}}YE & 0 & 0 & 0 \\ * & * & \Xi_{33} & \Xi_{34} & -C_d^{\mathrm{T}} & d_1A_d^{\mathrm{T}}S_1 & d_{12}A_d^{\mathrm{T}}S_2 \\ * & * & * & \Xi_{44} & 0 & 0 & 0 \\ * & * & * & * & \hat{\Xi}_{55} & d_1B_{\omega}^{\mathrm{T}}S_1 & d_{12}B_{\omega}^{\mathrm{T}}S_2 \\ * & * & * & * & * & -S_1 & 0 \\ * & * & * & * & * & * & -S_2 \end{bmatrix} < 0, \tag{4.61}$$

where

$$\begin{aligned}\bar{\Xi}_{11} &= P^{\mathrm{T}}(A+B\alpha K)+(A+B\alpha K)^{\mathrm{T}}P \\ &\quad + Z_1+Z_2+Z_3-E^{\mathrm{T}}S_1E, \\ \bar{\Xi}_{16} &= d_1(A+B\alpha K)^{\mathrm{T}}S_1, \\ \bar{\Xi}_{17} &= d_{12}(A+B\alpha K)^{\mathrm{T}}S_2.\end{aligned}$$

On the other hand, define $J=\mathrm{diag}\{L,L,L,L,I,S_1^{-1},S_2^{-1}\}$, $\hat{Z}_1=L^{\mathrm{T}}Z_1L$, $\hat{Z}_2=L^{\mathrm{T}}Z_2L$, $\hat{Z}_3=L^{\mathrm{T}}Z_3L$, $\hat{S}_1=LS_1L^{\mathrm{T}}$, $\hat{S}_2=LS_2L^{\mathrm{T}}$, $\hat{Y}=LYL^{\mathrm{T}}$, and $V=KL$. Then, pre- and post-multiplying the left-hand side of (4.61) by J^{T} and J, respectively, we can find that (4.61) is equivalent to

$$\hat{\Sigma}+\Lambda_1\Delta\Lambda_2^{\mathrm{T}}+\Lambda_2\Delta^{\mathrm{T}}\Lambda_1^{\mathrm{T}}<0, \tag{4.62}$$

where

$$\hat{\Sigma}=\begin{bmatrix} \hat{\Xi}_{11} & E\hat{S}_1E^{\mathrm{T}} & A_dL & 0 & B_{\omega}-L^{\mathrm{T}}C^{\mathrm{T}} & d_1\mathcal{F}^{\mathrm{T}} & d_{12}\mathcal{F}^{\mathrm{T}} \\ * & \hat{\Xi}_{22} & \hat{\Xi}_{23} & E\hat{Y}E^{\mathrm{T}} & 0 & 0 & 0 \\ * & * & \hat{\Xi}_{33} & \hat{\Xi}_{34} & -L^{\mathrm{T}}C_d^{\mathrm{T}} & d_1L^{\mathrm{T}}A_d^{\mathrm{T}} & d_{12}L^{\mathrm{T}}A_d^{\mathrm{T}} \\ * & * & * & \hat{\Xi}_{44} & 0 & 0 & 0 \\ * & * & * & * & -\gamma I-D_{\omega}-D_{\omega}^{\mathrm{T}} & d_1B_{\omega}^{\mathrm{T}} & d_{12}B_{\omega}^{\mathrm{T}} \\ * & * & * & * & * & -L^{\mathrm{T}}\hat{S}_1^{-1}L & 0 \\ * & * & * & * & * & * & -L^{\mathrm{T}}\hat{S}_2^{-1}L \end{bmatrix},$$

$$\Lambda_1=\begin{bmatrix} B \\ 0 \\ 0 \\ 0 \\ 0 \\ d_1B \\ d_{12}B \end{bmatrix}, \Lambda_2=\begin{bmatrix} V^{\mathrm{T}} \\ 0 \\ 0 \\ 0 \\ 0 \\ 0 \\ 0 \end{bmatrix}.$$

Noting $\hat{S}_1>0$ and $\hat{S}_2>0$, we have

$$\begin{aligned}-L^{\mathrm{T}}\hat{S}_1^{-1}L &\leqslant \hat{S}_1-L^{\mathrm{T}}-L, \\ -L^{\mathrm{T}}\hat{S}_2^{-1}L &\leqslant \hat{S}_2-L^{\mathrm{T}}-L.\end{aligned} \tag{4.63}$$

By Schur complement, and combining (4.58) and (4.63), we can obtain

$$\hat{\Sigma} + \Lambda_1 M \breve{\alpha}^2 \Lambda_1^{\mathrm{T}} + \Lambda_2 M^{-1} \Lambda_2^{\mathrm{T}} < 0. \tag{4.64}$$

From (4.4), (4.5) and the elementary inequality $x^{\mathrm{T}}y + y^{\mathrm{T}}x \leqslant \varepsilon x^{\mathrm{T}}x + \varepsilon^{-1}y^{\mathrm{T}}y$, we have that (4.64) holds implies (4.62) holds. On the other hand, pre- and post-multiplying the left side of (4.53) by $\mathrm{diag}\{L, L\}$ and $\mathrm{diag}\{L^{\mathrm{T}}, L^{\mathrm{T}}\}$, respectively, we can find that (4.53) is equivalent to (4.59). This completes the proof.

It is noted that the equality constraint is involved in Theorem 4.6, which may lead to some numerical problems when checking such nonstrict LMI condition since equality constraint is often fragile and usually not met perfectly. Applying the similar methods of [175, 178, 200], we introduce the matrix $G \in \mathbb{R}^{n\times(n-r)}$ satisfying $EG = 0$ and $\mathrm{rank}\, G = n - r$, and define $L = L_1 E^{\mathrm{T}} + GW^{\mathrm{T}}$, where $L_1 > 0$ and $W \in \mathbb{R}^{n\times(n-r)}$. Then, we have the following theorem.

Theorem 4.7. *System* (4.7) *is exponentially admissible and passive, if there exist matrices* $L_1 > 0$, W, $\hat{Y}$, V, $\hat{Z}_1 > 0$, $\hat{Z}_2 > 0$, $\hat{Z}_3 > 0$, $\hat{S}_1 > 0$, $\hat{S}_2 > 0$ *and* $M = \mathrm{diag}\{m_1, m_2, \cdots, m_i\} > 0$, *and a scalar* $\gamma > 0$ *such that* $(4.58)|_{L=L_1E^{\mathrm{T}}+GW^{\mathrm{T}}}$ *and* (4.59) *hold. Furthermore, if* $(4.58)|_{L=L_1E^{\mathrm{T}}+GW^{\mathrm{T}}}$ *and* (4.59) *are solvable, the desired controller gain is given as*

$$K = V(L_1 E^{\mathrm{T}} + GW^{\mathrm{T}})^{-1}. \tag{4.65}$$

Remark 4.8. Notice that Theorem 4.7 provides a sufficient condition for the solvability of the reliable passive control problem for SS with time-varying delays. It is noted that the minimum passivity performance γ can be readily found by solving the following convex optimization problem using the LMI toolbox in MATLAB: Minimize γ subject to $(4.58)|_{L=L_1E^{\mathrm{T}}+GW^{\mathrm{T}}}$ and (4.59) over $L_1 > 0$, W, $\hat{Y}$, V, $\hat{Z}_1 > 0$, $\hat{Z}_2 > 0$, $\hat{Z}_3 > 0$, $\hat{S}_1 > 0$, $\hat{S}_2 > 0$ and $M = \mathrm{diag}\{m_1, m_2, \cdots, m_i\} > 0$.

Remark 4.9. It is worth pointing out that by applying the similar method, the other conditions of this chapter can also be transformed to strict LMI conditions. Thus, all the developed results in the chapter can be checked easily by using the existing powerful tools like the LMI toolbox of Matlab or any equivalent tool.

4.4 Numerical Examples

In this section, two examples are presented to demonstrate the effectiveness of the results developed in this chapter.

Example 4.10. Consider system (4.8) with [163]

$$E = \begin{bmatrix} 9 & 3 \\ 6 & 2 \end{bmatrix}, A = \begin{bmatrix} -13.1 & -13.7 \\ -15.4 & -23.8 \end{bmatrix}, A_d = \begin{bmatrix} -18.6 & -10.4 \\ -25.2 & -16.8 \end{bmatrix}.$$

In this example, we choose $\mu = 0.5$. Table 4.1 lists the allowable upper bound d_2 for various d_1 by different methods. It is seen from Tables 4.1 that the stability criterion proposed here gives less conservative results than those in [50, 163, 219].

Table 4.1. Example 4.10: Comparisons of the allowed d_2 for various d_1

d_1	1.4	1.6	1.8	2.0	2.2
[163]	2.1121	2.1450	2.2841	2.4328	2.5852
[50] ($\alpha \to 0$)	2.1121	2.1450	2.2841	2.4328	2.5852
[219]	2.2314	2.2761	2.4041	2.5383	2.6777
Theorem 4.2	2.3372	2.3730	2.4923	2.6181	2.7494

It is assumed that $d(t) = 1.85 + 0.45\sin((0.5/0.45)t)$. A straightforward calculation gives $d_2 = 2.3$, $d_1 = 1.4$ and $\mu = 0.5$. Fig. 4.1 shows the the trajectories of the state responses of system (4.8), from which we find that the corresponding state responses converge to zeros.

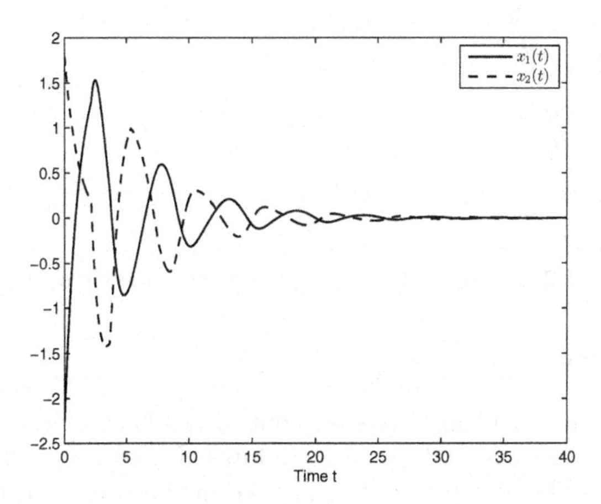

Fig. 4.1. Example 4.10: State responses of system (4.8)

Example 4.11. Consider system (4.1) with

$$E = \begin{bmatrix} 1 & 2 \\ 0 & 0 \end{bmatrix}, \; A = \begin{bmatrix} 0 & 1 \\ 0 & 0 \end{bmatrix}, \; A_d = \begin{bmatrix} 0 & 0.5 \\ 0.2 & -1 \end{bmatrix}, \; B = \begin{bmatrix} 2 \\ 5 \end{bmatrix},$$

$$B_\omega = \begin{bmatrix} 1 \\ 0.5 \end{bmatrix}, \; C = \begin{bmatrix} 1 & 2 \end{bmatrix}, \; C_d = \begin{bmatrix} 3 & 1 \end{bmatrix}, \; D_\omega = 3.$$

In this example, it is assumed that $d_1 = 0.2$, $d_2 = 0.8$, $\mu = 0.5$, $\underline{\alpha}_1 = 0.4$ and $\bar{\alpha}_1 = 0.7$. Our purpose is to design a state feedback controller (4.6) such that the closed-loop system is exponentially admissible and passive. Applying Theorem 4.7, the gain of the state feedback controller (4.6) is given as

$$K = \begin{bmatrix} -0.1310 & -1.6874 \end{bmatrix},$$

and the obtained minimum passivity performance $\gamma = 0.7180$. Under the above given gain, Fig. 4.2 shows the trajectories of the state responses of system (4.7), from which we find that the corresponding state responses converge to zeros.

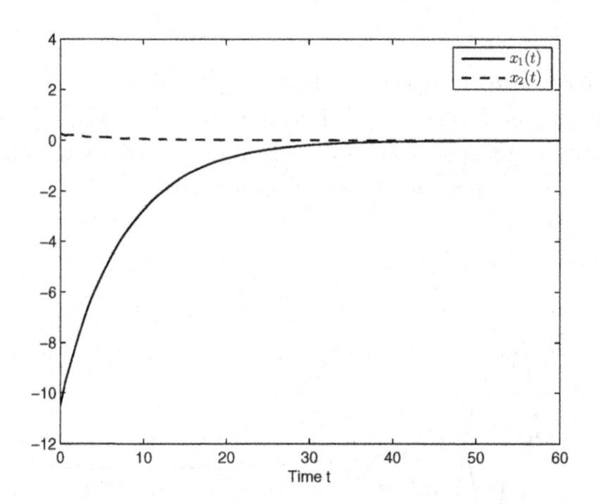

Fig. 4.2. Example 4.11: State responses of system (4.7)

4.5 Conclusion

In this chapter, the problem of passive control has been investigated for SSs with time-varying delays and actuator failures. Based on the lower bounds lemma of [116], a delay-dependent condition has been established to ensure the considered system to be regular, impulse-free, exponentially stable and passive. The state feedback controller has been designed to ensure, for all possible actuator failures, the resultant closed-loop system is exponentially admissible and passive. Numerical examples have been carried out to demonstrate the effectiveness of the proposed methods.

5

Dissipativity Analysis for SSs with Time-Varying Delays

5.1 Introduction

It has been shown that the theory of dissipative systems plays an important role in system and control areas, and thus has been attracting a great deal of attention. The dissipative theory gives a framework for the analysis and design of control systems using an input-output description based on energy-related considerations [11]. It serves as a powerful or even indispensable tool in characterizing important system behaviors such as stability and passivity, and has close connections with passivity theorem, the bounded real lemma, the Kalman-Yakubovich lemma, and the circle criterion. With the advancement of LMI approach, many interesting and important results on dissipativity analysis and synthesis have been reported for different kinds of dynamic systems. For example, the problem of reliable dissipative control has been investigated in [202] for a type of stochastic hybrid systems in terms of LMI approach, and linear state feedback controllers and impulsive controllers are designed such that, for all uncertainties as well as actuator failure occurring among a prespecified subset of actuators, the stochastic hybrid system is stochastically robustly stable and strictly dissipative. The problem of delay-dependent dissipativity analysis has been investigated for neural networks in [32, 153, 157], and some sufficient conditions have been given to guarantee the neural network is dissipative. Recently, some necessary and sufficient conditions of dissipativity of SSs have been proposed in [111, 112], and the synthesis of control gains to attain dissipativity of feedback systems has also been discussed. The results on dissipativity analysis of discrete-time SSs can be found in [28]. As for singular time-delay systems, some results have been developed for the dissipative analysis and state-feedback synthesis in [106] via LMI approach. Very recently, some improved results have been given in [34]. However, it is noted that the considered time-delays of [34, 106] are all time invariant, which limits the scope of applications of the proposed results.

Addressing continuous-time/discrete-time SSs with time-varying delays, the problem of dissipativity analysis is investigated using LMI approach and

Z.-G. Wu et al.: *Anal. & Synth. of Singular Syst. with Time-Delays*, LNCIS 443, pp. 53–69.
DOI: 10.1007/978-3-642-37497-5_5

the delay partitioning method. Some delay-dependent conditions have been established to guarantee the dissipativity of SSs. All the results given in this chapter are delay-dependent as well as partition dependent. Some examples are given to show the validness and reduced conservatism of the proposed methods.

5.2 Continuous-Time Systems

5.2.1 *Preliminaries*

Consider the following continuous-time SS with time-varying delays:

$$\begin{cases} E\dot{x}(t) = Ax(t) + A_d x(t - d(t)) + B_\omega \omega(t), \\ z(t) = Cx(t) + C_d x(t - d(t)) + D_\omega \omega(t), \\ x(t) = \phi(t), t \in [-d_2, 0], \end{cases} \tag{5.1}$$

where $x(t) \in \mathbb{R}^n$ is the state, $u(t) \in \mathbb{R}^m$ is the control input, $z(t) \in \mathbb{R}^s$ is the controlled output, $\omega(t) \in \mathbb{R}^p$ is the disturbance input that belongs to $\mathscr{L}_2[0, \infty)$, and $\phi(t) \in C_{n,d_2}$ is a compatible vector valued initial function. The matrix $E \in \mathbb{R}^{n \times n}$ may be singular and it is assumed that $\operatorname{rank} E = r \leqslant n$. A, A_d, B_ω, C, C_d and D_ω are known real constant matrices with appropriate dimensions. $d(t)$ is a time-varying continuous function that satisfies $0 < d_1 \leqslant d(t) \leqslant d_2$ and $\dot{d}(t) \leqslant \mu$, where d_1 and d_2 are the lower and upper bounds of time-varying delay $d(t)$, respectively, and $0 \leqslant \mu < 1$ is the variation rate of time-varying delay $d(t)$.

We are now in a position to introduce the property of dissipativity. Let the energy supply function of system (5.1) be defined by

$$G(\omega, z, \tau) = \langle z, \mathcal{Q}z\rangle_\tau + 2\langle z, \mathcal{S}\omega\rangle_\tau + \langle \omega, \mathcal{R}\omega\rangle_\tau, \quad \forall \tau \geqslant 0, \tag{5.2}$$

where $\mathcal{Q}$, $\mathcal{S}$ and $\mathcal{R}$ are real matrices with $\mathcal{Q}$, $\mathcal{R}$ symmetric, and $\langle a, b\rangle_\tau = \int_0^\tau a^{\mathrm{T}} b \, dt$. Without loss of generality, it is assumed that $\mathcal{Q} \leq 0$ and denoted that $-\mathcal{Q} = \mathcal{Q}_-^{\mathrm{T}} \mathcal{Q}_-$ for some $\mathcal{Q}_-$.

Definition 5.1. *System* (5.1) *is said to be strictly* $(\mathcal{Q}, \mathcal{S}, \mathcal{R})$-$\gamma$-*dissipative if, for some scalar* $\gamma > 0$, *the following inequality*

$$G(\omega, z, \tau) \geqslant \gamma \langle \omega, \omega\rangle_\tau, \quad \forall \tau \geqslant 0 \tag{5.3}$$

holds under zero initial condition.

5.2.2 *Main Results*

In this subsection, the delay partitioning technique will be developed to investigate the dissipativity of system (5.1). For the sake's of vector and matrix representation, the followings are denoted

$$\begin{aligned}
\Upsilon(t) &= \left[x(t)^{\mathrm{T}}\ x(t-\tfrac{1}{m}d_1)^{\mathrm{T}}\ x(t-\tfrac{2}{m}d_1)^{\mathrm{T}}\ \ldots\ x(t-\tfrac{m-1}{m}d_1)^{\mathrm{T}}\right]^{\mathrm{T}},\\
\eta(t) &= \left[\Upsilon(t)^{\mathrm{T}}\ x(t-d_1)^{\mathrm{T}}\right]^{\mathrm{T}},\\
\zeta(t) &= \left[\eta(t)^{\mathrm{T}}\ x(t-d(t))^{\mathrm{T}}\ x(t-d_2)^{\mathrm{T}}\ \omega(t)^{\mathrm{T}}\right]^{\mathrm{T}},\\
W_1 &= \left[I_{mn}\ 0_{mn\times n}\right],\\
W_2 &= \left[0_{mn\times n}\ I_{mn}\right],\\
g_l &= \left[0_{n\times(l-1)n}\ I_n\ 0_{n\times(m-l+1)n}\right],\ l=1,2,\cdots,m+1,\\
d_{12} &= d_2-d_1,\\
\mathcal{D} &= \left(\frac{d_1}{m}\right)^2\sum_{i=1}^{m}S_i + d_{12}^2 S_{m+1}.
\end{aligned}$$

This way, system (5.1) can be rewritten as

$$\begin{cases} E\dot{x}(t) = Ag_1\eta(t) + A_d x(t-d(t)) + B_\omega\omega(t),\\ z(t) = Cg_1\eta(t) + C_d x(t-d(t)) + D_\omega\omega(t),\\ x(t) = \phi(t),\ t\in[-d_2,0]. \end{cases} \tag{5.4}$$

Theorem 5.2. *For a given integer $m>0$ and a scalar $\gamma>0$, system (5.1) is exponentially admissible and strictly $(\mathcal{Q},\mathcal{S},\mathcal{R})$-$\gamma$-dissipative, if there exist matrices P, $Q>0$, $Z_1>0$, $Z_2>0$ and $S_i>0$ $(i=1,2,\cdots,m+1)$ such that*

$$E^{\mathrm{T}}P = P^{\mathrm{T}}E \geqslant 0, \tag{5.5}$$

$$\begin{bmatrix} \Xi_{11} & \Xi_{12} & 0 & \Xi_{14} & g_1^{\mathrm{T}}A^{\mathrm{T}}\mathcal{D} & g_1^{\mathrm{T}}C^{\mathrm{T}}\mathcal{Q}_-^{\mathrm{T}}\\ * & \Xi_{22} & \Xi_{23} & -C_d^{\mathrm{T}}\mathcal{S} & A_d^{\mathrm{T}}\mathcal{D} & C_d^{\mathrm{T}}\mathcal{Q}_-^{\mathrm{T}}\\ * & * & \Xi_{33} & 0 & 0 & 0\\ * & * & * & \Xi_{44} & B_\omega^{\mathrm{T}}\mathcal{D} & D_\omega^{\mathrm{T}}\mathcal{Q}_-^{\mathrm{T}}\\ * & * & * & * & -\mathcal{D} & 0\\ * & * & * & * & * & -I \end{bmatrix} < 0, \tag{5.6}$$

where

$$\begin{aligned}
\Xi_{11} =&\ g_1^{\mathrm{T}}P^{\mathrm{T}}Ag_1 + g_1^{\mathrm{T}}A^{\mathrm{T}}Pg_1 + W_1^{\mathrm{T}}QW_1 - W_2^{\mathrm{T}}QW_2\\
&-\sum_{i=1}^{m}(g_i-g_{i+1})^{\mathrm{T}}E^{\mathrm{T}}S_iE(g_i-g_{i+1})\\
&+g_{m+1}^{\mathrm{T}}Z_1g_{m+1} + g_1^{\mathrm{T}}Z_2g_1 - g_{m+1}^{\mathrm{T}}E^{\mathrm{T}}S_{m+1}Eg_{m+1},\\
\Xi_{12} =&\ g_1^{\mathrm{T}}P^{\mathrm{T}}A_d + g_{m+1}^{\mathrm{T}}E^{\mathrm{T}}S_{m+1}E,\\
\Xi_{14} =&\ g_1^{\mathrm{T}}P^{\mathrm{T}}B_\omega - g_1^{\mathrm{T}}C^{\mathrm{T}}\mathcal{S},\\
\Xi_{22} =&\ -(1-\mu)Z_2 - 2E^{\mathrm{T}}S_{m+1}E,\\
\Xi_{23} =&\ E^{\mathrm{T}}S_{m+1}E,
\end{aligned}$$

$$\Xi_{33} = -Z_1 - E^{\mathrm{T}}S_{m+1}E,$$
$$\Xi_{44} = -\mathcal{R} + \gamma I - \mathcal{S}^{\mathrm{T}}D_\omega - D_\omega^{\mathrm{T}}\mathcal{S}.$$

Proof. Firstly, we prove the regularity and absence of impulses of system (5.1) with $\omega(t) = 0$. Since $\mathrm{rank}\, E = r \leqslant n$, there exist two nonsingular matrices G and H such that

$$GEH = \begin{bmatrix} I_r & 0 \\ 0 & 0 \end{bmatrix}. \tag{5.7}$$

Denote

$$GAH = \begin{bmatrix} A_1 & A_2 \\ A_3 & A_4 \end{bmatrix}, \; G^{-\mathrm{T}}PH = \begin{bmatrix} \bar{P}_1 & \bar{P}_2 \\ \bar{P}_3 & \bar{P}_4 \end{bmatrix}. \tag{5.8}$$

From (5.5) and using the expressions in (5.7) and (5.8), it is straightforward that $\bar{P}_2 = 0$. It can be found from $\Xi_{11} < 0$ that

$$A^{\mathrm{T}}P + P^{\mathrm{T}}A - E^{\mathrm{T}}S_1E < 0. \tag{5.9}$$

Then, pre-multiplying and post-multiplying (5.9) by H^{T} and H, respectively, we have

$$A_4^{\mathrm{T}}\bar{P}_4 + \bar{P}_4^{\mathrm{T}}A_4 < 0, \tag{5.10}$$

which implies A_4 is nonsingular and thus the pair (E, A) is regular and impulse free. Hence, by Definition 3.1, system (5.1) with $\omega(t) = 0$ is regular and impulse free.

Next, we will show that system (5.1) is stable and strictly $(\mathcal{Q}, \mathcal{S}, \mathcal{R})$-$\gamma$-dissipative. To the end, the following Lyapunov functional is considered for system (5.1):

$$V(x_t) = \sum_{l=1}^{4} V_l(x_t), \tag{5.11}$$

where

$$V_1(x_t) = x(t)^{\mathrm{T}}E^{\mathrm{T}}Px(t),$$
$$V_2(x_t) = \int_{t-\frac{d_1}{m}}^{t} \Upsilon(s)^{\mathrm{T}}Q\Upsilon(s)\mathrm{d}s,$$
$$V_3(x_t) = \int_{t-d_2}^{t-d_1} x(s)^{\mathrm{T}}Z_1x(s)\mathrm{d}s + \int_{t-d(t)}^{t} x(s)^{\mathrm{T}}Z_2x(s)\mathrm{d}s,$$
$$V_4(x_t) = \frac{d_1}{m}\sum_{i=1}^{m}\int_{-\frac{i}{m}d_1}^{-\frac{i-1}{m}d_1}\int_{t+\beta}^{t} \dot{x}(s)^{\mathrm{T}}E^{\mathrm{T}}S_iE\dot{x}(s)\mathrm{d}s\mathrm{d}\beta$$
$$+ d_{12}\int_{-d_2}^{-d_1}\int_{t+\beta}^{t} \dot{x}(s)^{\mathrm{T}}E^{\mathrm{T}}S_{m+1}E\dot{x}(s)\mathrm{d}s\mathrm{d}\beta,$$

where $\{x_t = x(t+\theta), -2d_2 \leqslant \theta \leqslant 0\}$. Calculating the time derivative of $V(x_t)$ along the solution of system (5.1) yields

$$\dot{V}_1(x_t) = 2x(t)^{\mathrm{T}}P^{\mathrm{T}}E\dot{x}(t) = 2\eta(t)^{\mathrm{T}}g_1^{\mathrm{T}}P^{\mathrm{T}}(Ag_1\eta(t) + A_dx(t-d(t)) + B_\omega\omega(t)), \tag{5.12}$$

$$\begin{aligned}\dot{V}_2(x_t) &= \Upsilon(t)^{\mathrm{T}}Q\Upsilon(t) - \Upsilon(t-\frac{d_1}{m})^{\mathrm{T}}Q\Upsilon(t-\frac{d_1}{m}) \\ &= \eta(t)^{\mathrm{T}}W_1^{\mathrm{T}}QW_1\eta(t) - \eta(t)^{\mathrm{T}}W_2^{\mathrm{T}}QW_2\eta(t),\end{aligned} \tag{5.13}$$

$$\begin{aligned}\dot{V}_3(x_t) \leqslant\ & x(t-d_1)^{\mathrm{T}}Z_1x(t-d_1) - x(t-d_2)^{\mathrm{T}}Z_1x(t-d_2) \\ & + x(t)^{\mathrm{T}}Z_2x(t) - (1-\mu)x(t-d(t))^{\mathrm{T}}Z_2x(t-d(t)) \\ =\ & \eta(t)^{\mathrm{T}}g_{m+1}^{\mathrm{T}}Z_1g_{m+1}\eta(t) - x(t-d_2)^{\mathrm{T}}Z_1x(t-d_2) \\ & + \eta(t)^{\mathrm{T}}g_1^{\mathrm{T}}Z_2g_1\eta(t) - (1-\mu)x(t-d(t))^{\mathrm{T}}Z_2x(t-d(t)),\end{aligned} \tag{5.14}$$

$$\begin{aligned}\dot{V}_4(x_t) =\ & \left(\frac{d_1}{m}\right)^2\sum_{i=1}^{m}\dot{x}(t)^{\mathrm{T}}E^{\mathrm{T}}S_iE\dot{x}(t) - \frac{d_1}{m}\sum_{i=1}^{m}\int_{t-\frac{i}{m}d_1}^{t-\frac{i-1}{m}d_1}\dot{x}(s)^{\mathrm{T}}E^{\mathrm{T}}S_iE\dot{x}(s)\mathrm{d}s \\ & + d_{12}^2\dot{x}(t)^{\mathrm{T}}E^{\mathrm{T}}S_{m+1}E\dot{x}(t) - d_{12}\int_{t-d_2}^{t-d_1}\dot{x}(s)^{\mathrm{T}}E^{\mathrm{T}}S_{m+1}E\dot{x}(s)\mathrm{d}s \\ =\ & \dot{x}(t)^{\mathrm{T}}E^{\mathrm{T}}\mathcal{D}E\dot{x}(t) - \frac{d_1}{m}\sum_{i=1}^{m}\int_{t-\frac{i}{m}d_1}^{t-\frac{i-1}{m}d_1}\dot{x}(s)^{\mathrm{T}}E^{\mathrm{T}}S_iE\dot{x}(s)\mathrm{d}s \\ & - d_{12}\int_{t-d(t)}^{t-d_1}\dot{x}(s)^{\mathrm{T}}E^{\mathrm{T}}S_{m+1}E\dot{x}(s)\mathrm{d}s \\ & - d_{12}\int_{t-d_2}^{t-d(t)}\dot{x}(s)^{\mathrm{T}}E^{\mathrm{T}}S_{m+1}E\dot{x}(s)\mathrm{d}s \\ \leqslant\ & \dot{x}(t)^{\mathrm{T}}E^{\mathrm{T}}\mathcal{D}E\dot{x}(t) - \sum_{i=1}^{m}\eta(t)^{\mathrm{T}}(g_i - g_{i+1})^{\mathrm{T}}E^{\mathrm{T}}S_iE(g_i - g_{i+1})\eta(t) \\ & - (g_{m+1}\eta(t) - x(t-d(t)))^{\mathrm{T}}E^{\mathrm{T}}S_{m+1}E(g_{m+1}\eta(t) - x(t-d(t))) \\ & - (x(t-d(t)) - x(t-d_2))^{\mathrm{T}}E^{\mathrm{T}}S_{m+1}E(x(t-d(t)) - x(t-d_2)),\end{aligned} \tag{5.15}$$

where Jensen inequality is applied. Then, we get from (5.12)-(5.15) that

$$\dot{V}(x_t) - z(t)^{\mathrm{T}}\mathcal{Q}z(t) - 2z(t)^{\mathrm{T}}\mathcal{S}\omega(t) - \omega(t)^{\mathrm{T}}(\mathcal{R} - \gamma I)\omega(t) \leqslant \zeta(t)^{\mathrm{T}}\Sigma\zeta(t), \tag{5.16}$$

where

$$\Sigma = \begin{bmatrix} \Xi_{11} & \Xi_{12} & \Xi_{13} & \Xi_{14} \\ * & \Xi_{22} & \Xi_{23} & -C_d^{\mathrm{T}}\mathcal{S} \\ * & * & \Xi_{33} & 0 \\ * & * & * & \Xi_{44} \end{bmatrix} + \begin{bmatrix} g_1^{\mathrm{T}} A^{\mathrm{T}} \\ A_d^{\mathrm{T}} \\ 0 \\ B_\omega^{\mathrm{T}} \end{bmatrix} \mathcal{D} \begin{bmatrix} g_1^{\mathrm{T}} A^{\mathrm{T}} \\ A_d^{\mathrm{T}} \\ 0 \\ B_\omega^{\mathrm{T}} \end{bmatrix}^{\mathrm{T}} - \begin{bmatrix} g_1^{\mathrm{T}} C^{\mathrm{T}} \\ C_d^{\mathrm{T}} \\ 0 \\ D_\omega^{\mathrm{T}} \end{bmatrix} \mathcal{Q} \begin{bmatrix} g_1^{\mathrm{T}} C^{\mathrm{T}} \\ C_d^{\mathrm{T}} \\ 0 \\ D_\omega^{\mathrm{T}} \end{bmatrix}^{\mathrm{T}}.$$

According to Schur complement, we get from (5.6) that $\Sigma < 0$, and thus,

$$\dot{V}(x_t) - z(t)^{\mathrm{T}}\mathcal{Q}z(t) - 2z(t)^{\mathrm{T}}\mathcal{S}\omega(t) - \omega(t)^{\mathrm{T}}(\mathcal{R} - \gamma I)\omega(t) \leqslant 0. \tag{5.17}$$

Integrating both sides of the inequality (5.17) with respect to t over the time period $[0, t^*]$ gives rise to

$$\begin{aligned} &V(x_{t^*}) - V(x_0) - \int_0^{t^*} z(t)^{\mathrm{T}}\mathcal{Q}z(t)\mathrm{d}t \\ &- 2\int_0^{t^*} z(t)^{\mathrm{T}}\mathcal{S}\omega(t)\mathrm{d}t - \int_0^{t^*} \omega(t)^{\mathrm{T}}(\mathcal{R} - \gamma I)\omega(t)\mathrm{d}t \leqslant 0. \end{aligned} \tag{5.18}$$

Under zero initial condition, we have that $V(x_{t^*}) \geqslant 0$ and $V(x_0) = 0$, and thus the inequality (5.18) guarantees (5.3) holds, which means system (5.1) is strictly $(\mathcal{Q}, \mathcal{S}, \mathcal{R})$-$\gamma$-dissipative. On the other hand, by application of Schur complement and (5.6), we obtain that

$$\begin{bmatrix} \Xi_{11} & \Xi_{12} & 0 \\ * & \Xi_{22} & \Xi_{23} \\ * & * & \Xi_{33} \end{bmatrix} + \begin{bmatrix} g_1^{\mathrm{T}} A^{\mathrm{T}} \\ A_d^{\mathrm{T}} \\ 0 \end{bmatrix} \mathcal{D} \begin{bmatrix} g_1^{\mathrm{T}} A^{\mathrm{T}} \\ A_d^{\mathrm{T}} \\ 0 \end{bmatrix}^{\mathrm{T}} < 0, \tag{5.19}$$

which guarantees there exists a scalar $\lambda > 0$ such that

$$\dot{V}(t) < -\lambda ||x(t)||^2 \tag{5.20}$$

in case of $\omega(t) \equiv 0$. Set

$$\hat{G} = \begin{bmatrix} I_r & -A_2 A_4^{-1} \\ 0 & A_4^{-1} \end{bmatrix} G. \tag{5.21}$$

It is easy to get

$$\hat{G}EH = \begin{bmatrix} I_r & 0 \\ 0 & 0 \end{bmatrix}, \hat{G}AH = \begin{bmatrix} \hat{A}_1 & 0 \\ \hat{A}_3 & I \end{bmatrix}, \tag{5.22}$$

where $\hat{A}_1 = A_1 - A_2 A_4^{-1} A_3$ and $\hat{A}_3 = A_4^{-1} A_3$. Denote

$$\hat{G}A_d H = \begin{bmatrix} A_{d1} & A_{d2} \\ A_{d3} & A_{d4} \end{bmatrix}, \hat{G}^{-\mathrm{T}}PH = \begin{bmatrix} P_1 & P_2 \\ P_3 & P_4 \end{bmatrix}, H^{\mathrm{T}}Z_2 H = \begin{bmatrix} Z_{11} & Z_{21} \\ * & Z_{22} \end{bmatrix}. \tag{5.23}$$

It can be seen from (5.6) that

$$W_m \begin{bmatrix} \Xi_{11} & \Xi_{12} & 0 \\ * & \Xi_{22} & \Xi_{23} \\ * & * & \Xi_{33} \end{bmatrix} W_m^{\mathrm{T}} < 0, \tag{5.24}$$

where $W_m = \begin{bmatrix} I_n & I_n & \cdots & I_n \end{bmatrix} \in \mathbb{R}^{n\times(m+3)n}$. We can get from (5.24) that

$$P^{\mathrm{T}}(A + A_d) + (A + A_d)^{\mathrm{T}} P < 0. \tag{5.25}$$

It is easy to find the matrix P satisfying the above inequality is nonsingular. Thus, considering (5.5), we can deduce that $P_1 > 0$ and $P_2 = 0$. By using the similar analysis method applied in Theorem 4.2, it can be concluded that system (5.1) with $\omega(t) = 0$ is exponentially stable. This completes the proof.

Remark 5.3. It is noted that by the delay partitioning technique, Theorem 5.2 provides a sufficient condition to ensure SSs with time-varying delays to be exponentially admissible and strictly $(\mathcal{Q}, \mathcal{S}, \mathcal{R})$-$\gamma$-dissipative. It should be pointed out that the conservatism of Theorem 5.2 lies in the parameter m, which refers to the number of delay partitioning, that is, the conservatism is reduced as the partitions grow. On the other hand, the computational complexity is also dependent on the partition number m, that is, the computational complexity is increased as the partitioning becomes thinner. Therefore, there is a tradeoff between the computational complexity and the dissipativity of system.

The admissibility criterion of system (4.8) can be easily obtained from Theorem 5.2.

Corollary 5.4. *For a given integer $m > 0$, system (4.8) is exponentially admissible, if there exist matrices P, $Q > 0$, $Z_1 > 0$, $Z_2 > 0$ and $S_i > 0$ $(i = 1, 2, \cdots, m+1)$ such that (5.5) and the following LMI hold,*

$$\begin{bmatrix} \Xi_{11} & \Xi_{12} & 0 & g_1^{\mathrm{T}} A^{\mathrm{T}} \mathcal{D} \\ * & \Xi_{22} & \Xi_{23} & A_d^{\mathrm{T}} \mathcal{D} \\ * & * & \Xi_{33} & 0 \\ * & * & * & -\mathcal{D} \end{bmatrix} < 0, \tag{5.26}$$

where Ξ_{11}, Ξ_{12}, Ξ_{22}, Ξ_{23} and Ξ_{33} are given in Theorem 5.2.

5.2.3 Numerical Examples

Two numerical examples are presented to illustrate the usefulness and flexibility of the results developed in the above subsection.

Example 5.5. Consider system (4.8) with [163]

$$E = \begin{bmatrix} 9 & 3 \\ 6 & 2 \end{bmatrix}, A = \begin{bmatrix} -13.1 & -13.7 \\ -15.4 & -23.8 \end{bmatrix}, A_d = \begin{bmatrix} -18.6 & -10.4 \\ -25.2 & -16.8 \end{bmatrix}.$$

For $\mu = 0.2$, the admissible upper bound d_2 is given with different lower bound d_1 in Table 5.1. From Table 5.1, it can be easily seen that even for $m = 1$, our criterion is less conservative than the methods in [50, 163].

Table 5.1. Example 5.5: Comparisons of the allowed upper bound d_2 for various d_1

d_1	0.5	1.0	1.5
[50, 163]	3.0401	3.0416	3.0441
Corollary 5.4 ($m = 1$)	3.1091	3.2331	3.3173
Corollary 5.4 ($m = 2$)	3.1132	3.2524	3.3812
Corollary 5.4 ($m = 4$)	3.1143	3.2573	3.3961
Corollary 5.4 ($m = 6$)	3.1145	3.2583	3.3988

Example 5.6. Consider system (5.1) with

$$E = \begin{bmatrix} 1 & 1 \\ 0 & 0 \end{bmatrix}, A = \begin{bmatrix} -2 & 0 \\ 0 & -3 \end{bmatrix}, A_d = \begin{bmatrix} -1.5 & 0.5 \\ 0.4 & -0.3 \end{bmatrix},$$

$$B_\omega = \begin{bmatrix} 0.1 & 0 \\ 2.5 & 0 \end{bmatrix}, C = \begin{bmatrix} -1 & 0.4 \\ 1 & 1 \end{bmatrix}, C_d = 0, D_\omega = \begin{bmatrix} 1 & 1.5 \\ 3 & 2 \end{bmatrix}.$$

The purpose of this example is to find the optimal dissipativity performance γ such that the considered system is exponentially admissible and strictly $(\mathcal{Q}, \mathcal{S}, \mathcal{R})$-$\gamma$-dissipative. To this end, we choose $d_1 = 0.8$, $d_2 = 1.5$, $\mu = 0.3$, and

$$\mathcal{S} = \begin{bmatrix} 1.1 & 0.5 \\ 3 & 2 \end{bmatrix}, \mathcal{Q} = \begin{bmatrix} -0.04 & 0 \\ 0 & -1 \end{bmatrix}, \mathcal{R} = \begin{bmatrix} 3 & 0 \\ 0 & 1 \end{bmatrix}.$$

By application of Theorem 5.2, the optimal dissipativity performance γ for different partitioning size is given in Table 5.2. From Table 5.2, it can be found that the larger partition number m corresponds to the larger dissipativity performance γ. In addition, the results of [34, 106] are invalid in this example.

Table 5.2. Example 5.6: Optimal dissipativity performance γ for various m

m	1	2	4	6	8	10
Theorem 5.2	0.3514	0.4052	0.4173	0.4195	0.4202	0.4206

5.3 Discrete-Time Systems

5.3.1 Preliminaries

Consider the following discrete-time SS with time-varying delays:

$$\begin{cases} Ex(k+1) = Ax(k) + A_d x(k-d(k)) + A_\tau \sum_{v=1}^{\tau(k)} x(k-v) + B_\omega \omega(k), \\ z(k) = Cx(k) + C_d x(k-d(k)) + C_\tau \sum_{v=1}^{\tau(k)} x(k-v) + D_\omega \omega(k), \\ x(k) = \phi(k),\ k \in \mathbb{N}[-\max\{d_2, \tau_2\}, 0], \end{cases} \tag{5.27}$$

where $x(k) \in \mathbb{R}^n$ is the state vector, $z(k) \in \mathbb{R}^s$ is the output vector, $\omega(k) \in \mathbb{R}^p$ is the disturbance input that belongs to $l_2[0, \infty)$, and $\phi(k)$ is a compatible vector valued initial function. The matrix $E \in \mathbb{R}^{n\times n}$ may be singular and it is assumed that $\operatorname{rank} E = r \leqslant n$. A, A_d, A_τ, B_ω, C, C_d, C_τ and D_ω are known real constant matrices with appropriate dimensions. $d(k)$ and $\tau(k)$ denote discrete delay and distributed delay, respectively, and satisfy $0 < d_1 \leqslant d(k) \leqslant d_2$, and $0 < \tau_1 \leqslant \tau(k) \leqslant \tau_2$, where d_1, d_2, τ_1 and τ_2 are known integers.

Remark 5.7. It should be pointed out that in system (5.27), the term $\sum_{v=1}^{\tau(k)} x(k-v)$ originated from [92] is named as finite-distributed delay in the discrete-time setting, which can be regarded as the discrete analog of the finite-distributed delay $\int_{t-\tau(t)}^{t} x(s)\mathrm{d}s$ in continuous-time systems. For the continuous-time case, SSs with finite-distributed delay has been studied extensively [125, 199, 200]. However, few results have been given for discrete-time SSs with finite-distributed delay.

Let the energy supply function of system (5.27) be defined by (5.2), where τ is an integer, and $\langle a, b\rangle_\tau = \sum_{k=0}^{\tau} a(k)^{\mathrm{T}} b(k)$ for discrete-time systems.

Definition 5.8. *System* (5.27) *is said to be strictly* $(\mathcal{Q}, \mathcal{S}, \mathcal{R})$-$\gamma$-*dissipative if, for some scalar* $\gamma > 0$, (5.3) *holds under zero initial condition.*

5.3.2 Main Results

In this subsection, we will investigate the problem of dissipativity for discrete-time SSs with discrete and distributed time-varying delays. Before giving our main results, for the sake's of vector and matrix representation, the followings are denoted

$$\Upsilon(t) = \left[x(k)^{\mathrm{T}}\ x(k-\tfrac{1}{m}d_1)^{\mathrm{T}}\ x(k-\tfrac{2}{m}d_1)^{\mathrm{T}} \ldots x(k-\tfrac{m-1}{m}d_1)^{\mathrm{T}}\right]^{\mathrm{T}},$$
$$\eta(k) = \left[\Upsilon(k)^{\mathrm{T}}\ x(k-d_1)^{\mathrm{T}}\right]^{\mathrm{T}},$$
$$\hat{\varsigma}(k) = \left[\eta(k)^{\mathrm{T}}\ x(k-d(k))^{\mathrm{T}}\ x(k-d_2)^{\mathrm{T}} \sum_{v=1}^{\tau(k)} x(k-v)^{\mathrm{T}}\ \omega(k)^{\mathrm{T}}\right]^{\mathrm{T}},$$
$$\vartheta = \frac{\tau_2(\tau_2+\tau_1)(\tau_2-\tau_1+1)}{2}.$$

This way, system (5.27) can be rewritten as

$$\begin{cases} Ex(k+1) = Ag_1\eta(k) + A_d x(k-d(k)) + A_\tau \sum_{v=1}^{\tau(k)} x(k-v) + B_\omega \omega(k), \\ z(k) = Cg_1\eta(k) + C_d x(k-d(k)) + C_\tau \sum_{v=1}^{\tau(k)} x(k-v) + D_\omega \omega(k), \\ x(k) = \phi(k), k \in \mathbb{N}[-\max\{d_2, \tau_2\}, 0], \end{cases} \tag{5.28}$$

where g_1 is given in subsection 5.2.2.

Theorem 5.9. *For a given integer $m > 0$, system (5.27) is admissible and strictly $(\mathcal{Q}, \mathcal{S}, \mathcal{R})$-$\gamma$-dissipative, if there exist matrices $P > 0$, $Q > 0$, $Z_1 > 0$, $Z_2 > 0$, $U > 0$, $S_i > 0$ $(i = 1, 2, \cdots, m)$, $\begin{bmatrix} S_{m+1} & \mathcal{Y} \\ * & S_{m+1} \end{bmatrix} \geqslant 0$ and W, and a scalar $\gamma > 0$ such that*

$$\begin{bmatrix} \Xi_{11} & \Xi_{12} & \Xi_{13} & \Xi_{14} & \Delta_1 & \Xi_{15} & g_1^{\mathrm{T}}A^{\mathrm{T}}P & g_1^{\mathrm{T}}C^{\mathrm{T}}\mathcal{Q}_-^{\mathrm{T}} \\ * & \Xi_{22} & \Xi_{23} & 0 & -C_d^{\mathrm{T}}\mathcal{S} & A_d^{\mathrm{T}}\mathcal{D} & A_d^{\mathrm{T}}P & C_d^{\mathrm{T}}\mathcal{Q}_-^{\mathrm{T}} \\ * & * & \Xi_{33} & 0 & 0 & 0 & 0 & 0 \\ * & * & * & -U & -C_\tau^{\mathrm{T}}\mathcal{S} & A_\tau^{\mathrm{T}}\mathcal{D} & A_\tau^{\mathrm{T}}P & C_\tau^{\mathrm{T}}\mathcal{Q}_-^{\mathrm{T}} \\ * & * & * & * & \Delta_2 & B_\omega^{\mathrm{T}}\mathcal{D} & B_\omega^{\mathrm{T}}P & D_\omega^{\mathrm{T}}\mathcal{Q}_-^{\mathrm{T}} \\ * & * & * & * & * & -\mathcal{D} & 0 & 0 \\ * & * & * & * & * & * & -P & 0 \\ * & * & * & * & * & * & * & -I \end{bmatrix} < 0, \tag{5.29}$$

where $R \in \mathbb{R}^{n\times(n-r)}$ is any matrix with full column and satisfies $E^{\mathrm{T}}R = 0$, g_l, W_1, W_2, d_{12} and $\mathcal{D}$ are given in subsection 5.2.2, and

$$\begin{aligned} \Xi_{11} = &-g_1^{\mathrm{T}}E^{\mathrm{T}}PEg_1 + W_1^{\mathrm{T}}QW_1 - W_2^{\mathrm{T}}QW_2 \\ &- \sum_{i=1}^{m}(g_i - g_{i+1})^{\mathrm{T}}E^{\mathrm{T}}S_iE(g_i - g_{i+1}) \\ &+ g_{m+1}^{\mathrm{T}}Z_1g_{m+1} + (d_{12}+1)g_{m+1}^{\mathrm{T}}Z_2g_{m+1} + \vartheta g_1^{\mathrm{T}}Ug_1 \\ &- g_{m+1}^{\mathrm{T}}E^{\mathrm{T}}S_{m+1}Eg_{m+1} + g_1^{\mathrm{T}}WR^{\mathrm{T}}Ag_1 + g_1^{\mathrm{T}}A^{\mathrm{T}}RW^{\mathrm{T}}g_1, \end{aligned}$$

$$\Xi_{12} = g_{m+1}^{\mathrm{T}} E^{\mathrm{T}} S_{m+1} E - g_{m+1}^{\mathrm{T}} E^{\mathrm{T}} \mathcal{Y} E + g_1^{\mathrm{T}} W R^{\mathrm{T}} A_d,$$
$$\Xi_{13} = g_{m+1}^{\mathrm{T}} E^{\mathrm{T}} \mathcal{Y} E,$$
$$\Xi_{14} = g_1^{\mathrm{T}} W R^{\mathrm{T}} A_\tau,$$
$$\Xi_{15} = g_1^{\mathrm{T}} (A - E)^{\mathrm{T}} \mathcal{D},$$
$$\Xi_{22} = -Z_2 - 2E^{\mathrm{T}} S_{m+1} E + E^{\mathrm{T}} \mathcal{Y} E + E^{\mathrm{T}} \mathcal{Y}^{\mathrm{T}} E,$$
$$\Xi_{23} = -E^{\mathrm{T}} \mathcal{Y} E + E^{\mathrm{T}} S_{m+1} E,$$
$$\Xi_{33} = -Z_1 - E^{\mathrm{T}} S_{m+1} E,$$
$$\Delta_1 = g_1^{\mathrm{T}} W R^{\mathrm{T}} B_\omega - g_1^{\mathrm{T}} C^{\mathrm{T}} \mathcal{S},$$
$$\Delta_2 = -\mathcal{R} + \gamma I - \mathcal{S}^{\mathrm{T}} D_\omega - D_\omega^{\mathrm{T}} \mathcal{S}.$$

Proof. Under the given condition, we first show that system (5.27) with $\omega(k) = 0$, that is, system

$$\begin{cases} Ex(k+1) = Ax(k) + A_d x(k - d(k)) + A_\tau \sum_{v=1}^{\tau(k)} x(k - v), \\ \qquad x(k) = \phi(k), \; k \in \mathbb{N}[-\max\{d_2, \tau_2\}, 0] \end{cases} \tag{5.30}$$

is regular and causal. Since $\operatorname{rank} E = r$, we choose two nonsingular matrices M and G such that

$$MEG = \begin{bmatrix} I_r & 0 \\ 0 & 0 \end{bmatrix}. \tag{5.31}$$

Set

$$MAG = \begin{bmatrix} A_1 & A_2 \\ A_3 & A_4 \end{bmatrix}, \; G^{\mathrm{T}} W = \begin{bmatrix} W_1 \\ W_2 \end{bmatrix}, \; M^{-\mathrm{T}} R = \begin{bmatrix} 0 \\ I \end{bmatrix} F, \tag{5.32}$$

where $F \in \mathbb{R}^{(n-r)\times(n-r)}$ is any nonsingular matrix. It can be seen that $\Xi_{11} < 0$ implies

$$-E^{\mathrm{T}} P E + W R^{\mathrm{T}} A + A^{\mathrm{T}} R W^{\mathrm{T}} - E^{\mathrm{T}} S_1 E < 0. \tag{5.33}$$

Pre-multiplying and post-multiplying (5.33) by G^{T} and G, respectively, we have $W_2 F^{\mathrm{T}} A_4 + A_4^{\mathrm{T}} F S_2^{\mathrm{T}} < 0$, which implies A_4 is nonsingular. Thus, the pair (E, A) is regular and causal. According to Definition 2.2, system (5.30) is regular and causal.

Next we will show that system (5.30) is stable. To the end, we define $\delta(k) = x(k+1) - x(k)$ and consider the following Lyapunov functional for system (5.30):

$$V(x(k), k) = \sum_{l=1}^{5} V_l(x(k), k), \tag{5.34}$$

where

$$V_1(x(k),k) = x(k)^{\mathrm{T}}E^{\mathrm{T}}PEx(k),$$

$$V_2(x(k),k) = \sum_{s=k-\frac{d_1}{m}}^{k-1} \Upsilon(s)^{\mathrm{T}}Q\Upsilon(s),$$

$$V_3(x(k),k) = \sum_{s=k-d_2}^{k-d_1-1} x(s)^{\mathrm{T}}Z_1x(s) + \sum_{j=-d_2+1}^{-d_1+1}\sum_{s=k-1+j}^{k-d_1-1} x(s)^{\mathrm{T}}Z_2x(s),$$

$$V_4(x(k),k) = \frac{d_1}{m}\sum_{i=1}^{m}\sum_{g=-\frac{i}{m}d_1}^{-\frac{i-1}{m}d_1-1}\sum_{s=k+g}^{k-1} \delta(s)^{\mathrm{T}}E^{\mathrm{T}}S_iE\delta(s)$$
$$+ d_{12}\sum_{g=-d_2}^{-d_1-1}\sum_{s=k+g}^{k-1} \delta(s)^{\mathrm{T}}E^{\mathrm{T}}S_{m+1}E\delta(s),$$

$$V_5(x(k),k) = \tau_2\sum_{\beta=\tau_1}^{\tau_2}\sum_{v=1}^{\beta}\sum_{\alpha=k-v}^{k-1} x(\alpha)^{\mathrm{T}}Ux(\alpha).$$

Then, along the solution of system (5.30), we have that

$$\begin{aligned}\Delta V_1(k) &= x(k+1)^{\mathrm{T}}E^{\mathrm{T}}PEx(k+1) - x(k)^{\mathrm{T}}E^{\mathrm{T}}PEx(k)\\ &= x(k+1)^{\mathrm{T}}E^{\mathrm{T}}PEx(k+1) - \eta(k)^{\mathrm{T}}g_1^{\mathrm{T}}E^{\mathrm{T}}PEg_1\eta(k),\end{aligned} \tag{5.35}$$

$$\begin{aligned}\Delta V_2(k) &= \Upsilon(k)^{\mathrm{T}}Q\Upsilon(k) - \Upsilon\left(k-\frac{d_1}{m}\right)^{\mathrm{T}}Q\Upsilon\left(k-\frac{d_1}{m}\right)\\ &= \eta(k)^{\mathrm{T}}W_1^{\mathrm{T}}QW_1\eta(k) - \eta(k)^{\mathrm{T}}W_2^{\mathrm{T}}QW_2\eta(k),\end{aligned} \tag{5.36}$$

$$\begin{aligned}\Delta V_3(k) &= x(k-d_1)^{\mathrm{T}}Z_1x(k-d_1) - x(k-d_2)^{\mathrm{T}}Z_1x(k-d_2)\\ &\quad + (d_{12}+1)x(k-d_1)^{\mathrm{T}}Z_2x(k-d_1) - \sum_{s=k-d_2}^{k-d_1} x(s)^{\mathrm{T}}Z_2x(s)\\ &\leqslant \eta(k)^{\mathrm{T}}g_{m+1}^{\mathrm{T}}Z_1g_{m+1}\eta(k) - x(k-d_2)^{\mathrm{T}}Z_1x(k-d_2)\\ &\quad + (d_{12}+1)\eta(k)^{\mathrm{T}}g_{m+1}^{\mathrm{T}}Z_2g_{m+1}\eta(k) - x(k-d(k))^{\mathrm{T}}Z_2x(k-d(k)),\end{aligned} \tag{5.37}$$

$$\begin{aligned}\Delta V_4(k) &= \left(\frac{d_1}{m}\right)^2\sum_{i=1}^{m}\delta(k)^{\mathrm{T}}E^{\mathrm{T}}S_iE\delta(k) - \frac{d_1}{m}\sum_{i=1}^{m}\sum_{s=k-\frac{i}{m}d_1}^{k-\frac{i-1}{m}d_1-1} \delta(s)^{\mathrm{T}}E^{\mathrm{T}}S_iE\delta(s)\\ &\quad + d_{12}^2\delta(k)^{\mathrm{T}}E^{\mathrm{T}}S_{m+1}E\delta(k) - d_{12}\sum_{s=k-d_2}^{k-d_1-1} \delta(s)^{\mathrm{T}}E^{\mathrm{T}}S_{m+1}E\delta(s),\end{aligned} \tag{5.38}$$

$$\begin{aligned}\Delta V_5(k) &= \vartheta x(k)^{\mathrm{T}} U x(k) - \tau_2 \sum_{\beta=\tau_1}^{\tau_2} \sum_{v=1}^{\beta} x(k-v)^{\mathrm{T}} U x(k-v) \\ &\leqslant \vartheta \eta(k)^{\mathrm{T}} g_1^{\mathrm{T}} U g_1 \eta(k) - \tau_2 \sum_{v=1}^{\tau(k)} x(k-v)^{\mathrm{T}} U x(k-v). \end{aligned} \tag{5.39}$$

By use of Lemma 1.8, we get

$$-d_{12} \sum_{s=k-d_2}^{k-d_1-1} \delta(s)^{\mathrm{T}} E^{\mathrm{T}} S_{m+1} E \delta(s) \leqslant \begin{bmatrix} \eta(k) \\ x(k-d(k)) \\ x(k-d_2) \end{bmatrix}^{\mathrm{T}} \Gamma \begin{bmatrix} \eta(k) \\ x(k-d(k)) \\ x(k-d_2) \end{bmatrix}, \tag{5.40}$$

where

$$\Gamma = \begin{bmatrix} -g_{m+1}^{\mathrm{T}} E^{\mathrm{T}} S_{m+1} E g_{m+1} & g_{m+1}^{\mathrm{T}} E^{\mathrm{T}} S_{m+1} E - g_{m+1}^{\mathrm{T}} E^{\mathrm{T}} \mathcal{Y} E & g_{m+1}^{\mathrm{T}} E^{\mathrm{T}} \mathcal{Y} E \\ * & \bar{\Delta}_1 & \bar{\Delta}_2 \\ * & * & -E^{\mathrm{T}} S_{m+1} E \end{bmatrix},$$

$$\begin{aligned}\bar{\Delta}_1 &= -2E^{\mathrm{T}} S_{m+1} E + E^{\mathrm{T}} \mathcal{Y} E + E^{\mathrm{T}} \mathcal{Y}^{\mathrm{T}} E, \\ \bar{\Delta}_2 &= -E^{\mathrm{T}} \mathcal{Y} E + E^{\mathrm{T}} S_{m+1} E.\end{aligned}$$

Furthermore, applying discretized Jensen inequality, we obtain

$$\begin{aligned} & -\frac{d_1}{m} \sum_{i=1}^{m} \sum_{s=k-\frac{i}{m}d_1}^{k-\frac{i-1}{m}d_1-1} \delta(s)^{\mathrm{T}} E^{\mathrm{T}} S_i E \delta(s) \\ & \leqslant -\sum_{i=1}^{m} \sum_{s=k-\frac{i}{m}d_1}^{k-\frac{i-1}{m}d_1-1} \delta(s)^{\mathrm{T}} E^{\mathrm{T}} S_i E \sum_{s=k-\frac{i}{m}d_1}^{k-\frac{i-1}{m}d_1-1} \delta(s) \\ & = -\sum_{i=1}^{m} \eta(k)^{\mathrm{T}} (g_i - g_{i+1})^{\mathrm{T}} E^{\mathrm{T}} S_i E (g_i - g_{i+1}) \eta(k), \end{aligned} \tag{5.41}$$

and

$$-\tau_2 \sum_{v=1}^{\tau(k)} x(k-v)^{\mathrm{T}} U x(k-v) \leqslant -\sum_{v=1}^{\tau(k)} x(k-v)^{\mathrm{T}} U \sum_{v=1}^{\tau(k)} x(k-v). \tag{5.42}$$

On the other hand, it is clear that for any appropriately dimensioned matrix W, the following equation holds

$$\begin{aligned} f(k) &= 2x(k)^{\mathrm{T}} W R^{\mathrm{T}} E x(k+1) \\ &= 2\eta(k)^{\mathrm{T}} g_1^{\mathrm{T}} W R^{\mathrm{T}} E x(k+1) = 0. \end{aligned} \tag{5.43}$$

Thus,

$$\Delta V(k) = \sum_{l=1}^{5} \Delta V_l(k) + f(k) \leqslant \xi(k)^T \Theta \xi(k), \tag{5.44}$$

where

$$\xi(k) = \left[\eta(k)^{\mathrm{T}} \; x(k-d(k))^{\mathrm{T}} \; x(k-d_2)^{\mathrm{T}} \sum_{v=1}^{\tau(k)} x(k-v)^{\mathrm{T}} \right]^{\mathrm{T}},$$

$$\Theta = \begin{bmatrix} \Xi_{11} & \Xi_{12} & \Xi_{13} & \Xi_{14} \\ * & \Xi_{22} & \Xi_{23} & 0 \\ * & * & \Xi_{33} & 0 \\ * & * & * & -U \end{bmatrix} + \begin{bmatrix} g_1^{\mathrm{T}} A^{\mathrm{T}} \\ A_d^{\mathrm{T}} \\ 0 \\ A_\tau^{\mathrm{T}} \end{bmatrix} P \begin{bmatrix} g_1^{\mathrm{T}} A^{\mathrm{T}} \\ A_d^{\mathrm{T}} \\ 0 \\ A_\tau^{\mathrm{T}} \end{bmatrix}^{\mathrm{T}}$$
$$+ \begin{bmatrix} g_1^{\mathrm{T}} (A-E)^{\mathrm{T}} \\ A_d^{\mathrm{T}} \\ 0 \\ A_\tau^{\mathrm{T}} \end{bmatrix} \mathcal{D} \begin{bmatrix} g_1^{\mathrm{T}} (A-E)^{\mathrm{T}} \\ A_d^{\mathrm{T}} \\ 0 \\ A_\tau^{\mathrm{T}} \end{bmatrix}^{\mathrm{T}}.$$

According to Schur complement, we get from (5.29) that $\Theta < 0$. Therefore,

$$\Delta V(k) \leqslant -\alpha ||x(k)||^2, \tag{5.45}$$

where $\alpha = -\lambda_{\max}(\Theta) > 0$. Thus, we have

$$\sum_{i=0}^{k} ||x(i)||^2 \leqslant \frac{1}{\alpha} V(x(0), 0) < \infty, \tag{5.46}$$

that is, series $\sum_{i=0}^{\infty} ||x(i)||^2$ converge, which implies that $\lim_{k \to \infty} x(k) = 0$. According to Definition 2.2, system (5.30) is stable.

To prove the dissipativity of system (5.27), we consider Lyapunov functional (5.34) and the following index for system (5.27):

$$J_{\mathcal{T},\omega} = \sum_{k=0}^{\mathcal{T}} [z(k)^{\mathrm{T}} \mathcal{Q} z(k) + 2z(k)^{\mathrm{T}} \mathcal{S} \omega(k) + \omega(k)^{\mathrm{T}} (\mathcal{R} - \gamma I) \omega(k)]. \tag{5.47}$$

It can be seen that

$$\sum_{k=0}^{\mathcal{T}} \Delta V(k) - J_{\mathcal{T},\omega} \leqslant \sum_{k=0}^{\mathcal{T}} \hat{\varsigma}(k)^{\mathrm{T}} \Xi \hat{\varsigma}(k), \tag{5.48}$$

where

$$\Xi = \begin{bmatrix} \Xi_{11} & \Xi_{12} & \Xi_{13} & \Xi_{14} & \Delta_1 \\ * & \Xi_{22} & \Xi_{23} & 0 & -C_d^{\mathrm{T}}\mathcal{S} \\ * & * & \Xi_{33} & 0 & 0 \\ * & * & * & -U & -C_\tau^{\mathrm{T}}\mathcal{S} \\ * & * & * & * & \Delta_2 \end{bmatrix} + \begin{bmatrix} g_1^{\mathrm{T}}A^{\mathrm{T}} \\ A_d^{\mathrm{T}} \\ 0 \\ A_\tau^{\mathrm{T}} \\ B_\omega^{\mathrm{T}} \end{bmatrix} P \begin{bmatrix} g_1^{\mathrm{T}}A^{\mathrm{T}} \\ A_d^{\mathrm{T}} \\ 0 \\ A_\tau^{\mathrm{T}} \\ B_\omega^{\mathrm{T}} \end{bmatrix}^{\mathrm{T}}$$
$$+ \begin{bmatrix} g_1^{\mathrm{T}}(A-E)^{\mathrm{T}} \\ A_d^{\mathrm{T}} \\ 0 \\ A_\tau^{\mathrm{T}} \\ B_\omega^{\mathrm{T}} \end{bmatrix} \mathcal{D} \begin{bmatrix} g_1^{\mathrm{T}}(A-E)^{\mathrm{T}} \\ A_d^{\mathrm{T}} \\ 0 \\ A_\tau^{\mathrm{T}} \\ B_\omega^{\mathrm{T}} \end{bmatrix}^{\mathrm{T}} - \begin{bmatrix} g_1^{\mathrm{T}}C^{\mathrm{T}} \\ C_d^{\mathrm{T}} \\ 0 \\ C_\tau^{\mathrm{T}} \\ D_\omega^{\mathrm{T}} \end{bmatrix} \mathcal{Q} \begin{bmatrix} g_1^{\mathrm{T}}C^{\mathrm{T}} \\ C_d^{\mathrm{T}} \\ 0 \\ C_\tau^{\mathrm{T}} \\ D_\omega^{\mathrm{T}} \end{bmatrix}^{\mathrm{T}}.$$

According to Schur complement, we get from (5.29) that $\Xi < 0$. Therefore,

$$\sum_{k=0}^{\tau} \Delta V(k) \leqslant J_{\tau,\omega}, \tag{5.49}$$

which implies

$$V(x(\tau+1)) - V(x(0)) \leqslant J_{\tau,\omega}. \tag{5.50}$$

Thus, (5.3) holds under zero initial condition. Therefore, according to Definition 5.8, system (5.27) is strictly $(\mathcal{Q},\mathcal{S},\mathcal{R})$-$\gamma$-dissipative. This completes the proof.

Remark 5.10. It is noted that Theorem 5.9 provides a delay-dependent condition on the dissipativity of system (5.27). It should be pointed out that by setting $\delta = -\gamma$ and minimizing δ subject to (5.29), we can obtain the optimal dissipativity performance γ^* (by $\gamma^* = -\delta$).

The admissibility criterion of system (5.30) can be easily obtained based on Theorem 5.9.

Corollary 5.11. *For a given integer $m > 0$, system (5.30) is admissible, if there exist matrices $P > 0$, $Q > 0$, $Z_1 > 0$, $Z_2 > 0$, $U > 0$, $S_i > 0$ $(i = 1, 2, \cdots, m)$, $\begin{bmatrix} S_{m+1} & \mathcal{Y} \\ * & S_{m+1} \end{bmatrix} \geqslant 0$ and W such that*

$$\begin{bmatrix} \Xi_{11} & \Xi_{12} & \Xi_{13} & \Xi_{14} & \Xi_{15} & g_1^{\mathrm{T}}A^{\mathrm{T}}P \\ * & \Xi_{22} & \Xi_{23} & 0 & A_d^{\mathrm{T}}\mathcal{D} & A_d^{\mathrm{T}}P \\ * & * & \Xi_{33} & 0 & 0 & 0 \\ * & * & * & -U & A_\tau^{\mathrm{T}}\mathcal{D} & A_\tau^{\mathrm{T}}P \\ * & * & * & * & -\mathcal{D} & 0 \\ * & * & * & * & * & -P \end{bmatrix} < 0, \tag{5.51}$$

where Ξ_{11}, Ξ_{12}, Ξ_{13}, Ξ_{14}, Ξ_{15}, Ξ_{22}, Ξ_{23} and Ξ_{33} are given in Theorem 5.9.

5.3.3 Numerical Examples

Three numerical examples are introduced to demonstrate the effectiveness of the proposed methods.

Example 5.12. Consider system (5.30) with

$$E = \begin{bmatrix} 1 & 0 \\ 0 & 0 \end{bmatrix}, \ A = \begin{bmatrix} 0.8 & 0 \\ 0.05 & 0.9 \end{bmatrix}, \ A_d = \begin{bmatrix} -0.1 & 0 \\ -0.2 & -0.1 \end{bmatrix}, \ A_\tau = \begin{bmatrix} 0 & 0 \\ 0 & 0 \end{bmatrix}.$$

In this example, we choose $d_1 = 3$ and

$$R = \begin{bmatrix} 0 \\ 1 \end{bmatrix}.$$

Applying the methods of [30, 103] and Corollary 5.11 of this chapter, the allowable maximum values of d_2 ensuring the admissibility of the considered system are listed in Table 5.3, which shows that our condition gives better results than those in [30, 103] even for the case of $m = 1$.

Table 5.3. Example 5.12: Comparisons of the allowed upper bound d_2

[103]	[30]	Corollary 5.11 ($m = 1$)	Corollary 5.11 ($m = 3$)
8	16	18	19

Example 5.13. Consider system (5.30) with

$$E = \begin{bmatrix} 2 & 1 \\ 6 & 3 \end{bmatrix}, \ A = \begin{bmatrix} 5.36 & 9.88 \\ 6.68 & 6.94 \end{bmatrix}, \ A_d = \begin{bmatrix} -1.50 & -1.55 \\ -1.50 & -1.15 \end{bmatrix}, \ A_\tau = \begin{bmatrix} 3.68 & 2.00 \\ 1.84 & 1.00 \end{bmatrix}.$$

In this example, we choose $d_1 = 6$, $d_2 = 14$, $\tau_1 = 5$, $\tau_2 = 12$, and

$$R = \begin{bmatrix} 3 \\ -1 \end{bmatrix}.$$

By using the Matlab LMI toolbox to solve the LMI in Corollary 5.11 with $m = 2$, it can be checked that considered system is admissible.

Example 5.14. Consider system (5.27) with

$$E = \begin{bmatrix} 56 & 35 \\ 64 & 40 \end{bmatrix}, \ A = \begin{bmatrix} 132.48 & 90 \\ 182.72 & 125 \end{bmatrix},$$

$$A_d = \begin{bmatrix} -24.4 & -16.05 \\ -33.6 & -22.2 \end{bmatrix}, \ A_\tau = \begin{bmatrix} 30.72 & 19.36 \\ 46.08 & 29.04 \end{bmatrix},$$

$$B_\omega = \begin{bmatrix} 0.03 & 0.01 \\ 0.04 & 0.02 \end{bmatrix}, \ C = \begin{bmatrix} -0.4 & 0.2 \\ 6 & 4 \end{bmatrix}, \ C_d = \begin{bmatrix} 1.8 & 1.3 \\ 0 & 0 \end{bmatrix},$$

$$C_\tau = \begin{bmatrix} 2 & 1.4 \\ 0.2 & 0.1 \end{bmatrix}, \ D_\omega = \begin{bmatrix} 0.1 & 0.01 \\ 0.2 & 0 \end{bmatrix}.$$

In this example, we choose $d_1 = 6$, $d_2 = 9$, $\tau_1 = 3$, $\tau_2 = 8$ and

$$\mathcal{S} = \begin{bmatrix} 1 & 0.6 \\ 1 & 1 \end{bmatrix}, \ \mathcal{Q} = \begin{bmatrix} -0.04 & 0 \\ 0 & -1 \end{bmatrix}, \ \mathcal{R} = \begin{bmatrix} 1 & 0 \\ 0 & 1 \end{bmatrix}.$$

The purpose of this example is to find the optimal dissipativity performance γ such that the considered system is admissible and strictly $(\mathcal{Q}, \mathcal{S}, \mathcal{R})$-$\gamma$-dissipative. By application of Theorem 5.9 with

$$R = \begin{bmatrix} 8 \\ -7 \end{bmatrix},$$

the optimal dissipativity performance γ for different partitioning size is given in Table 5.4, from which we can find that the larger partition number m corresponds to the larger dissipativity performance γ.

Table 5.4. Example 5.14: Optimal dissipativity performance γ for various m

m	1	2	3	6
Theorem 5.9	0.2884	0.3557	0.3652	0.3705

5.4 Conclusions

The problem of dissipativity has been studied for continuous-time and discrete-time SSs with time-varying delay. By taking advantage of the delay partitioning technique, several delay-dependent dissipativity conditions have been derived. Some delay-dependent criteria have also been given to ensure SSs admissible. All the results presented depend upon not only the time-delay, but also the partitioning size. Numerical examples have been proposed to demonstrate the effectiveness of the proposed methods.

6

$\mathscr{L}_2 - \mathscr{L}_\infty$ Filtering for SSs with Constant Delays

6.1 Introduction

Over the past decades, the filtering problem has been extensively investigated due to the fact that filtering is of both theoretical and practical importance in signal processing. When external noises are not precisely known, the Kalman filtering scheme is no longer applicable. In this case, the H_∞ filtering and $\mathscr{L}_2 - \mathscr{L}_\infty$ filtering are more involved and so far, various methodologies have been developed for the filter designs in different contexts. For H_∞ filter designs, there have been fruitful results by LMI approach [82, 183, 214]. While the results on $\mathscr{L}_2 - \mathscr{L}_\infty$ filtering problem have also been presented in the literature [42, 44, 168]. Recently, the filtering problem for SSs has been studied by LMI approach. For example, the H_∞ filtering issues have been investigated in [179, 185] for SSs. While the $\mathscr{L}_2 - \mathscr{L}_\infty$ filtering results for SSs have been presented in [217]. When time delays appear, the H_∞ filtering problem for singular time-delay systems has been discussed in [38, 199, 201], where some sufficient conditions have been proposed for the existence of linear H_∞ filter.

This chapter is concerned with the problem of delay-dependent $\mathscr{L}_2 - \mathscr{L}_\infty$ filtering problem for SSs with time delays. The purpose is to design a filter such that the error system is delay-dependent exponentially admissible with a prescribed $\mathscr{L}_2 - \mathscr{L}_\infty$ performance index. By dividing the whole interval of time delay into two equal subintervals and defining a Lyapunov functional for each subinterval, two delay-dependent conditions are proposed, which guarantee the considered SSs to be exponentially admissible with a prescribed $\mathscr{L}_2 - \mathscr{L}_\infty$ performance level. A method for designing a linear full-order filter is also given. Some numerical examples are given to show the effectiveness of the methods.

Z.-G. Wu et al.: *Anal. & Synth. of Singular Syst. with Time-Delays*, LNCIS 443, pp. 71–87.
DOI: 10.1007/978-3-642-37497-5_6

6.2 Problem Formulation

Consider the following SS with time-delays:

$$\begin{cases} E\dot{x}(t) = Ax(t) + A_d x(t-d) + B_\omega \omega(t), \\ y(t) = Cx(t) + C_d x(t-d) + D_\omega \omega(t), \\ z(t) = Lx(t), \\ x(t) = \phi(t),\ t \in [-d, 0], \end{cases} \tag{6.1}$$

where $x(t) \in \mathbb{R}^n$ is the state, $y(t) \in \mathbb{R}^s$ is the measurement, $z(t) \in \mathbb{R}^q$ is the signal to be estimated, $\omega(t) \in \mathbb{R}^p$ is the disturbance input that belongs to $\mathscr{L}_2[0,\infty)$, d is the constant time delay, and $\phi(t) \in C_{n,d}$ is a compatible vector valued initial function. The matrix $E \in \mathbb{R}^{n\times n}$ may be singular and it is assumed that $\operatorname{rank} E = r \leqslant n$. A, A_d, B_ω, C, C_d, D_ω and L are known real constant matrices with appropriate dimensions.

In this chapter, we consider a full-order filter with the following form:

$$\begin{cases} E_f\dot{\hat{x}}(t) = A_f\hat{x}(t) + B_f y(t), \\ \hat{z}(t) = C_f\hat{x}(t), \end{cases} \tag{6.2}$$

where $\hat{x} \in \mathbb{R}^n$, $\hat{z}(t) \in \mathbb{R}^q$, and the constant matrices E_f, A_f, B_f and C_f are the filter matrices with appropriate dimensions, which are to be designed. Augmenting (6.1) to include the states of the filter, we obtain the following filtering error system

$$\begin{cases} \bar{E}\dot{\bar{x}}(t) = \bar{A}\bar{x}(t) + \bar{A}_d\bar{x}(t-d) + \bar{B}_\omega\omega(t), \\ \bar{z}(t) = \bar{L}\bar{x}(t), \end{cases} \tag{6.3}$$

where

$$\bar{z}(t) = z(t) - \hat{z}(t),$$
$$\bar{x}(t) = \left[x(t)^{\mathrm{T}}\ \hat{x}(t)^{\mathrm{T}}\right]^{\mathrm{T}},$$

and

$$\bar{E} = \begin{bmatrix} E & 0 \\ 0 & E_f \end{bmatrix},\ \bar{A} = \begin{bmatrix} A & 0 \\ B_f C & A_f \end{bmatrix},$$
$$\bar{A}_d = \begin{bmatrix} A_d & 0 \\ B_f C_d & 0 \end{bmatrix},\ \bar{B}_\omega = \begin{bmatrix} B_\omega \\ B_f D_\omega \end{bmatrix},$$
$$\bar{L} = \begin{bmatrix} L & -C_f \end{bmatrix}.$$

The problem to be addressed in this chapter is as follows: given a scalar $\gamma > 0$ and system (6.1), design a full-order filter of the form (6.2) such that the filtering error system (6.3) with $\omega(t) = 0$ is exponentially admissible, and under the zero-initial condition,

$$\sup_t \sqrt{z(t)^{\mathrm{T}} z(t)} < \gamma \sqrt{\int_0^\infty \omega(t)^{\mathrm{T}} \omega(t) \mathrm{d}t} \tag{6.4}$$

is guaranteed for all nonzero $\omega(t) \in \mathscr{L}_2[0, \infty)$.

Before ending this section, we give the following lemma that will be used in the proof of our main results in the next section.

Lemma 6.1. *Suppose system (1.1) is regular and impulse free, then there exists a scalar $\kappa > 0$ such that*

$$\sup_{-d \leqslant s \leqslant d} \|x(s)\|^2 \leqslant \kappa \|\phi(t)\|_d^2. \tag{6.5}$$

Proof. Note that the regularity and the absence of impulses of the pair (E, A) imply that there always exist two nonsingular matrices M and N such that

$$MEN = \begin{bmatrix} I_r & 0 \\ 0 & 0 \end{bmatrix}, \; MAN = \begin{bmatrix} A_1 & 0 \\ 0 & I_{n-r} \end{bmatrix}. \tag{6.6}$$

Write

$$MA_dN = \begin{bmatrix} A_{d1} & A_{d2} \\ A_{d3} & A_{d4} \end{bmatrix}, \; \xi(t) = \begin{bmatrix} \xi_1(t) \\ \xi_2(t) \end{bmatrix} = N^{-1} x(t). \tag{6.7}$$

Then, system (1.1) can be written as

$$\begin{cases} \dot{\xi}_1(t) = A_1 \xi_1(t) + A_{d1} \xi_1(t-d) + A_{d2} \xi_2(t-d), \\ -\xi_2(t) = A_{d3} \xi_1(t-d) + A_{d4} \xi_2(t-d), \\ \xi(t) = \psi(t) = N^{-1} \phi(t), \; t \in [-d, 0]. \end{cases} \tag{6.8}$$

Then, for any $0 \leqslant t \leqslant d$, we have

$$\|\xi_1(t)\| \leqslant (2k_1 d + 1) \|\psi\|_d + k_1 \int_0^t \|\xi_1(\alpha)\| \mathrm{d}\alpha. \tag{6.9}$$

where $k_1 = \max\{\|A_1\|, \|A_{d1}\|, \|A_{d2}\|\}$. Applying the well known Gronwall Lemma, we obtain from (6.9) that for any $0 \leqslant t \leqslant d$,

$$\|\xi_1(t)\| \leqslant (2k_1 d + 1) \|\psi\|_d e^{k_1 d}. \tag{6.10}$$

Thus,

$$\sup_{0 \leqslant s \leqslant d} \|\xi_1(s)\|^2 \leqslant (2k_1 d + 1)^2 \|\psi\|_d^2 e^{2k_1 d}. \tag{6.11}$$

It is easy to get from (6.8) that

$$\sup_{0 \leqslant s \leqslant d} \|\xi_2(s)\|^2 \leqslant 4k_2^2 \|\psi\|_d^2, \tag{6.12}$$

where $k_2 = \max\{\|A_{d3}\|, \|A_{d4}\|\}$. Hence, there exists a scalar $\kappa > 0$ such that (6.5) holds.

6.3 Main Results

In this section, a delay-dependent approach to the design of $\mathscr{L}_2 - \mathscr{L}_\infty$ filter will be developed.

6.3.1 $\mathscr{L}_2 - \mathscr{L}_\infty$ *Performance Analysis*

We shall make use of the Lyapunov functional approach combined with LMI approach to investigate the exponential admissibility and $\mathscr{L}_2 - \mathscr{L}_\infty$ performance of the following system:

$$\begin{cases} E\dot{x}(t) = Ax(t) + A_d x(t-d) + B_\omega \omega(t), \\ z(t) = Lx(t), \\ x(t) = \phi(t), \ t \in [-d, 0]. \end{cases} \tag{6.13}$$

Theorem 6.2. *For a given scalar $\gamma > 0$, system (6.13) is exponentially admissible with $\mathscr{L}_2 - \mathscr{L}_\infty$ performance γ, if there exist matrix P, $Q_1 > 0$, $Q_2 > 0$, $Z_1 > 0$ and $Z_2 > 0$ such that*

$$E^{\mathrm{T}} P = P^{\mathrm{T}} E \geqslant 0, \tag{6.14}$$

$$\begin{bmatrix} \Xi_{11} & P^{\mathrm{T}} A_d & E^{\mathrm{T}} Z_1 E & P^{\mathrm{T}} B_\omega & d^2 A^{\mathrm{T}}(Z_1 + Z_2) \\ * & \Xi_{22} & E^{\mathrm{T}} Z_2 E & 0 & d^2 A_d^{\mathrm{T}}(Z_1 + Z_2) \\ * & * & \Xi_{33} & 0 & 0 \\ * & * & * & -I & d^2 B_\omega^{\mathrm{T}}(Z_1 + Z_2) \\ * & * & * & * & -4d^2(Z_1 + Z_2) \end{bmatrix} < 0, \tag{6.15}$$

$$\begin{bmatrix} E^{\mathrm{T}} P & L^{\mathrm{T}} \\ * & \gamma^2 I \end{bmatrix} \geqslant 0, \tag{6.16}$$

where

$$\begin{aligned} \Xi_{11} &= P^{\mathrm{T}} A + A^{\mathrm{T}} P + Q_1 - E^{\mathrm{T}} Z_1 E, \\ \Xi_{22} &= -Q_2 - E^{\mathrm{T}} Z_2 E, \\ \Xi_{33} &= -Q_1 + Q_2 - E^{\mathrm{T}}(Z_1 + Z_2) E. \end{aligned}$$

Proof. Under the given condition, we first show the exponentially admissibility of system (6.13) with $\omega(t) \equiv 0$. Noting that $\operatorname{rank} E = r \leqslant n$, we can always find two nonsingular matrices G and H such that

$$GEH = \begin{bmatrix} I_r & 0 \\ 0 & 0 \end{bmatrix}. \tag{6.17}$$

Denote

$$GAH = \begin{bmatrix} A_1 & A_2 \\ A_3 & A_4 \end{bmatrix}, \ G^{-T} P H = \begin{bmatrix} \bar{P}_1 & \bar{P}_2 \\ \bar{P}_3 & \bar{P}_4 \end{bmatrix}. \tag{6.18}$$

From (6.14) and using the expressions in (6.17) and (6.18), it is easy to obtain that $\bar{P}_2 = 0$. Then, pre-multiplying and post-multiplying $\Xi_{11} < 0$ by H^{T} and H, respectively, we have

$$A_4^{\mathrm{T}}\bar{P}_4 + \bar{P}_4^{\mathrm{T}}A_4 < 0, \tag{6.19}$$

which implies A_4 is nonsingular and thus the pair (E, A) is regular and impulse free. Thus, according to Definition 3.1, system (6.13) with $\omega(t) \equiv 0$ is regular and impulse free.

Next, we will show the exponential stability of system (6.13). To this end, define the following Lyapunov functional for system (6.13) with $\omega(t) \equiv 0$,

$$\begin{aligned} V(x_t, t) = {} & x(t)^{\mathrm{T}}E^{\mathrm{T}}Px(t) + \int_{t-\frac{d}{2}}^{t} x(s)^{\mathrm{T}}Q_1x(s)\,\mathrm{d}s \\ & + \int_{t-d}^{t-\frac{d}{2}} x(s)^{\mathrm{T}}Q_2x(s)\,\mathrm{d}s \\ & + \frac{d}{2}\int_{-\frac{d}{2}}^{0}\int_{t+\beta}^{t} \dot{x}(\alpha)^{\mathrm{T}}E^{\mathrm{T}}Z_1E\dot{x}(\alpha)\,\mathrm{d}\alpha\mathrm{d}\beta \\ & + \frac{d}{2}\int_{-d}^{-\frac{d}{2}}\int_{t+\beta}^{t} \dot{x}(\alpha)^{\mathrm{T}}E^{\mathrm{T}}Z_2E\dot{x}(\alpha)\,\mathrm{d}\alpha\mathrm{d}\beta, \end{aligned} \tag{6.20}$$

where $x_t = x(t+\theta)$, $-2d \leqslant \theta \leqslant 0$. Calculating the time derivative of $V(x_t, t)$ along the solutions of system (6.13) with $\omega(t) \equiv 0$ yields for any $t \geqslant d$,

$$\begin{aligned} \dot{V}(x_t, t) = {} & 2x(t)^{\mathrm{T}}E^{\mathrm{T}}P\dot{x}(t) + x(t)^{\mathrm{T}}Q_1x(t) - x(t-d)^{\mathrm{T}}Q_2x(t-d) \\ & + x(t-\frac{d}{2})^{\mathrm{T}}(Q_2 - Q_1)x(t-\frac{d}{2}) + \frac{d^2}{4}\dot{x}(t)^{\mathrm{T}}E^{\mathrm{T}}(Z_1+Z_2)E\dot{x}(t) \\ & - \frac{d}{2}\int_{t-\frac{d}{2}}^{t} \dot{x}(\alpha)^{\mathrm{T}}E^{\mathrm{T}}Z_1E\dot{x}(\alpha)\,\mathrm{d}\alpha - \frac{d}{2}\int_{t-d}^{t-\frac{d}{2}} \dot{x}(\alpha)^{\mathrm{T}}E^{\mathrm{T}}Z_2E\dot{x}(\alpha)\,\mathrm{d}\alpha. \end{aligned} \tag{6.21}$$

According to Jensen inequality, the following equations are true:

$$\begin{aligned} & -\frac{d}{2}\int_{t-\frac{d}{2}}^{t} \dot{x}(\alpha)^{\mathrm{T}}E^{\mathrm{T}}Z_1E\dot{x}(\alpha)\,\mathrm{d}\alpha \\ & \qquad \leqslant -\left(\int_{t-\frac{d}{2}}^{t} E\dot{x}(\alpha)\,\mathrm{d}\alpha\right)^{\mathrm{T}} Z_1 \int_{t-\frac{d}{2}}^{t} E\dot{x}(\alpha)\,\mathrm{d}\alpha \\ & \qquad = \begin{bmatrix} x(t) \\ x(t-\frac{d}{2}) \end{bmatrix}^{\mathrm{T}} \begin{bmatrix} -E^{\mathrm{T}}Z_1E & E^{\mathrm{T}}Z_1E \\ * & -E^{\mathrm{T}}Z_1E \end{bmatrix} \begin{bmatrix} x(t) \\ x(t-\frac{d}{2}) \end{bmatrix} \end{aligned} \tag{6.22}$$

and

$$-\frac{d}{2}\int_{t-d}^{t-\frac{d}{2}} \dot{x}(\alpha)^{\mathrm{T}}E^{\mathrm{T}}Z_2E\dot{x}(\alpha)\,\mathrm{d}\alpha$$

$$\leqslant -\left(\int_{t-d}^{t-\frac{d}{2}} E\dot{x}(\alpha)\,\mathrm{d}\alpha\right)^{\mathrm{T}} Z_2 \int_{t-d}^{t-\frac{d}{2}} E\dot{x}(\alpha)\,\mathrm{d}\alpha$$

$$= \begin{bmatrix} x(t-d) \\ x(t-\frac{d}{2}) \end{bmatrix}^{\mathrm{T}} \begin{bmatrix} -E^{\mathrm{T}}Z_2E & E^{\mathrm{T}}Z_2E \\ * & -E^{\mathrm{T}}Z_2E \end{bmatrix} \begin{bmatrix} x(t-d) \\ x(t-\frac{d}{2}) \end{bmatrix}. \tag{6.23}$$

Thus, through algebraic manipulations, we have that when $t \geqslant d$,

$$\dot{V}(x_t, t) \leqslant \eta(t)^{\mathrm{T}}\Psi\eta(t), \tag{6.24}$$

where

$$\Psi = \begin{bmatrix} \Xi_{11} & P^{\mathrm{T}}A_d & E^{\mathrm{T}}Z_1E \\ * & \Xi_{22} & E^{\mathrm{T}}Z_2E \\ * & * & \Xi_{33} \end{bmatrix} + \frac{d^2}{4}\begin{bmatrix} A^{\mathrm{T}} \\ A_d^{\mathrm{T}} \\ 0 \end{bmatrix}(Z_1+Z_2)\begin{bmatrix} A^{\mathrm{T}} \\ A_d^{\mathrm{T}} \\ 0 \end{bmatrix}^{\mathrm{T}},$$

$$\eta(t) = \begin{bmatrix} x(t) \\ x(t-d) \\ x(t-\frac{d}{2}) \end{bmatrix}.$$

Now, applying Schur complements, it is easy to see from (6.15) that $\Psi < 0$, which implies that there exits a scalar $\lambda_0 > 0$ such that

$$\dot{V}(x_t, t) \leqslant -\lambda_0\|x(t)\|^2. \tag{6.25}$$

Moreover, by the definition of $V(x_t, t)$, there exist positive scalars, there exist scalars λ_1, λ_2 and λ_3 such that for any $t \geqslant d$,

$$V(x_t, t) \leqslant \lambda_1\|x(t)\|^2 + \lambda_2\int_{t-d}^{t}\|x(s)\|^2\,\mathrm{d}s + \lambda_3\int_{t-d}^{t}\|x(s-d)\|^2\,\mathrm{d}s. \tag{6.26}$$

Then, according to Lemma 6.1, it is easy to get there exists a scalar $\lambda_4 > 0$ such that

$$V(x_d, d) \leqslant \lambda_4\|\phi(t)\|_d^2. \tag{6.27}$$

Set

$$\hat{G} = \begin{bmatrix} I_r & -A_2A_4^{-1} \\ 0 & A_4^{-1} \end{bmatrix} G. \tag{6.28}$$

It is easy to get

$$\hat{G}EH = \begin{bmatrix} I_r & 0 \\ 0 & 0 \end{bmatrix}, \hat{G}AH = \begin{bmatrix} \hat{A}_1 & 0 \\ \hat{A}_3 & I \end{bmatrix}, \tag{6.29}$$

where $\hat{A}_1 = A_1 - A_2A_4^{-1}A_3$ and $\hat{A}_3 = A_4^{-1}A_3$. Denote

$$\hat{G}A_dH = \begin{bmatrix} A_{d1} & A_{d2} \\ A_{d3} & A_{d4} \end{bmatrix}, \hat{G}^{-\mathrm{T}}PH = \begin{bmatrix} P_1 & P_2 \\ P_3 & P_4 \end{bmatrix}, H^{\mathrm{T}}Q_1H = \begin{bmatrix} Q_{11} & Q_{21} \\ * & Q_{22} \end{bmatrix}. \tag{6.30}$$

Considering (6.14) and nonsingularity of P, we can deduce that $P_1 > 0$ and $P_2 = 0$. Define

$$\zeta(t) = \begin{bmatrix} \zeta_1(t) \\ \zeta_2(t) \end{bmatrix} = H^{-1}x(t), \tag{6.31}$$

then system (6.13) with $\omega(t) \equiv 0$ is equivalent to

$$\begin{cases} \dot{\zeta}_1(t) = \hat{A}_1\zeta_1(t) + A_{d1}\zeta_1(t-d) + A_{d2}\zeta_2(t-d), \\ -\zeta_2(t) = \hat{A}_{13}\zeta_1(t) + A_{d3}\zeta_1(t-d) + A_{d4}\zeta_2(t-d), \\ \zeta(t) = \psi(t) = H^{-1}\phi(t), \ t \in [-d, 0]. \end{cases} \tag{6.32}$$

To prove the exponential stability, we define a new function as

$$W(x_t, t) = e^{\varepsilon t}V(x_t, t), \ t \geqslant d, \tag{6.33}$$

where the scalar $\varepsilon > 0$. Then, we find that for any $t \geqslant d$,

$$\begin{aligned} W(x_t, t) - W(x_d, d) &\leqslant \int_d^t e^{\varepsilon s}\left[\varepsilon V(x_s, s) - \lambda_0\|x(s)\|^2\right] \mathrm{d}s \\ &\leqslant \int_d^t e^{\varepsilon\alpha}\left[\varepsilon\lambda_1\|x(\alpha)\|^2 + \varepsilon\lambda_2\int_{\alpha-d}^{\alpha}\|x(s)\|^2\,\mathrm{d}s \right. \\ &\quad \left. +\varepsilon\lambda_3\int_{\alpha-d}^{\alpha}\|x(s-d)\|^2\,\mathrm{d}s - \lambda_0\|x(\alpha)\|^2\right] \mathrm{d}\alpha. \end{aligned} \tag{6.34}$$

By Lemma 6.1 and interchanging the integration sequence, we get for any $t \geqslant d$,

$$\begin{aligned} \int_d^t e^{\varepsilon\alpha}\,\mathrm{d}\alpha\int_{\alpha-d}^{\alpha}\|x(s)\|^2\,\mathrm{d}s &\leqslant de^{\varepsilon d}\int_0^t e^{\varepsilon\alpha}\|x(\alpha)\|^2\,\mathrm{d}\alpha \\ &\leqslant de^{\varepsilon d}\int_d^t e^{\varepsilon\alpha}\|x(\alpha)\|^2\,\mathrm{d}\alpha + d^2e^{2\varepsilon d}\kappa\|\phi(t)\|_d^2 \end{aligned} \tag{6.35}$$

and

$$\begin{aligned} \int_d^t e^{\varepsilon\alpha}\,\mathrm{d}\alpha\int_{\alpha-d}^{\alpha}\|x(s-d)\|^2\,\mathrm{d}s &\leqslant de^{\varepsilon d}\int_0^t e^{\varepsilon\alpha}\|x(\alpha-d)\|^2\,\mathrm{d}\alpha \\ &\leqslant de^{2\varepsilon d}\int_d^t e^{\varepsilon\alpha}\|x(\alpha)\|^2\,\mathrm{d}\alpha + 2d^2e^{3\varepsilon d}\kappa\|\phi(t)\|_d^2. \end{aligned} \tag{6.36}$$

Let the scalar ε small enough such that

$$\varepsilon\lambda_1 + \varepsilon de^{\varepsilon d}\lambda_2 + \varepsilon de^{2\varepsilon d}\lambda_3 - \lambda_0 \leqslant 0. \tag{6.37}$$

Then, substituting (6.35) and (6.36) into (6.34) gives that there exists a scalar $k > 0$ such that for any $t \geqslant d$,

$$\lambda_{\min}(P_1)\|\zeta_1(t)\|^2 \leqslant x(t)^{\mathrm{T}} E^{\mathrm{T}} P x(t) \leqslant V(x_t, t) \leqslant k e^{-\varepsilon t}\|\phi(t)\|_d^2. \tag{6.38}$$

Hence, for any $t \geqslant d$,

$$\|\zeta_1(t)\|^2 \leqslant \lambda_{\min}(P_1)^{-1} k e^{-\varepsilon t}\|\phi(t)\|_d^2, \tag{6.39}$$

Combining Lemma 6.1 yields for any $t > 0$,

$$\|\zeta_1(t)\|^2 \leqslant a e^{-\varepsilon t}\|\phi(t)\|_d^2, \tag{6.40}$$

where $a = \max\{\lambda_{\min}(P_1)^{-1}k, \kappa e^{\varepsilon d}\}$. Define

$$e(t) = \hat{A}_{13}\zeta_1(t) + A_{d3}\zeta_1(t-d), \tag{6.41}$$

then, from (6.40), a scalar $m > 0$ can be found such that for any $t > 0$,

$$\|e(t)\|^2 \leqslant m e^{-\varepsilon t}\|\phi(t)\|_d^2. \tag{6.42}$$

To study the exponential stability of $\zeta_2(t)$, we construct a function as

$$J(t) = \zeta_2(t)^{\mathrm{T}} Q_{22}\zeta_2(t) - \zeta_2(t-d)^{\mathrm{T}} Q_{22}\zeta_2(t-d). \tag{6.43}$$

By pre-multiplying the second equation of (6.32) with $\zeta_2(t)^{\mathrm{T}} P_4^{\mathrm{T}}$, we obtain that

$$0 = \zeta_2(t)^{\mathrm{T}} P_4^{\mathrm{T}}\zeta_2(t) + \zeta_2(t)^{\mathrm{T}} P_4^{\mathrm{T}} A_{d4}\zeta_2(t-d) + \zeta_2(t)^{\mathrm{T}} P_4^{\mathrm{T}} e(t). \tag{6.44}$$

Adding (6.44) to (6.43) yields that

$$\begin{aligned} J(t) &= \zeta_2(t)^{\mathrm{T}}(P_4^{\mathrm{T}} + P_4 + Q_{22})\zeta_2(t) + 2\zeta_2(t)^{\mathrm{T}} P_4^{\mathrm{T}} A_{d4}\zeta_2(t-d) \\ &\quad - \zeta_2(t-d)^{\mathrm{T}} Q_{22}\zeta_2(t-d) + 2\zeta_2(t)^{\mathrm{T}} P_4^{\mathrm{T}} e(t) \\ &\leqslant \begin{bmatrix} \zeta_2(t) \\ \zeta_2(t-d) \end{bmatrix}^{\mathrm{T}} \begin{bmatrix} P_4^{\mathrm{T}} + P_4 + Q_{22} & P_4^{\mathrm{T}} A_{d4} \\ * & -Q_{22} \end{bmatrix} \begin{bmatrix} \zeta_2(t) \\ \zeta_2(t-d) \end{bmatrix} \\ &\quad + \eta_1\zeta_2(t)^{\mathrm{T}}\zeta_2(t) + \eta_1^{-1} e(t)^{\mathrm{T}} P_4 P_4^{\mathrm{T}} e(t), \end{aligned} \tag{6.45}$$

where η_1 is any positive scalar. Now, pre-multiplying and post-multiplying

$$\begin{bmatrix} \Xi_{11} & P^{\mathrm{T}} A_d & E^{\mathrm{T}} Z_1 E \\ * & \Xi_{22} & E^{\mathrm{T}} Z_2 E \\ * & * & \Xi_{33} \end{bmatrix} < 0, \tag{6.46}$$

by $\Pi = \begin{bmatrix} I & 0 & 0 \\ 0 & I & I \end{bmatrix}$ and Π^{T}, respectively, we get

$$\begin{bmatrix} P^{\mathrm{T}} A + A^{\mathrm{T}} P + Q_1 - E^{\mathrm{T}} Z_1 E & P^{\mathrm{T}} A_d + E^{\mathrm{T}} Z_1 E \\ * & -Q_1 - E^{\mathrm{T}} Z_1 E \end{bmatrix} < 0. \tag{6.47}$$

Pre-multiplying and post-multiplying (6.47) by $\begin{bmatrix} H & 0 \\ 0 & H \end{bmatrix}^{\mathrm{T}}$ and $\begin{bmatrix} H & 0 \\ 0 & H \end{bmatrix}$, respectively, a scalar $\eta_2 > 0$ can be found such that

$$\begin{bmatrix} P_4^{\mathrm{T}} + P_4 + Q_{22} & P_4^{\mathrm{T}} A_{d4} \\ * & -Q_{22} \end{bmatrix} \leqslant - \begin{bmatrix} \eta_2 I & 0 \\ 0 & 0 \end{bmatrix}. \tag{6.48}$$

On the other hand, since η_1 can be chosen arbitrarily, η_1 can be chosen small enough such that $\eta_2 - \eta_1 > 0$. Then a scalar $\eta_3 > 1$ can always be found such that

$$Q_{22} - (\eta_1 - \eta_2) I \geqslant \eta_3 Q_{22}. \tag{6.49}$$

It follows from (6.43), (6.45) and (6.49) that

$$\zeta_2(t)^{\mathrm{T}} Q_{22} \zeta_2(t) \leqslant \eta_3^{-1} \zeta_2(t-d)^{\mathrm{T}} Q_{22} \zeta_2(t-d) + (\eta_1 \eta_3)^{-1} e(t)^{\mathrm{T}} P_4 P_4^{\mathrm{T}} e(t), \tag{6.50}$$

which infers

$$f(t) \leqslant \eta_3^{-1} \sup_{t-d \leqslant s \leqslant t} f(s) + \xi e^{-\delta t}, \tag{6.51}$$

where $\delta = \min\{\varepsilon, d^{-1} \ln \eta_3\}$, $f(t) = \zeta_2(t)^{\mathrm{T}} Q_{22} \zeta_2(t)$, and

$$\xi = (\eta_1 \eta_3)^{-1} m \|P_4\|^2 \|\phi(t)\|_d^2.$$

Therefore, applying Lemma 1.10 to the above inequality yields

$$\|\zeta_2(s)\|^2 \leqslant \lambda_{\min}^{-1}(Q_{22}) \lambda_{\max}(Q_{22}) e^{-\delta t} \|\zeta_2(s)\|_d^2 + \frac{\lambda_{\min}^{-1}(Q_{22}) \xi e^{-\delta t}}{1 - \eta_3^{-1} e^{\delta d}}, \tag{6.52}$$

which means, combining (6.40), that system (6.13) is exponentially stable.

Next, we will establish the $\mathscr{L}_2 - \mathscr{L}_\infty$ performance. Choose the Lyapunov functional as in (6.20) and the following index for system (6.13):

$$W(t) = V(x_t, t) - \int_0^t \omega(s)^{\mathrm{T}} \omega(s) \, \mathrm{d}s. \tag{6.53}$$

Under zero initial condition, it easy to see that

$$\begin{aligned} W(t) &= \int_0^t \left[\dot{V}(x_s, s) - \omega(s)^{\mathrm{T}} \omega(s) \right] \mathrm{d}s \\ &\leqslant \int_0^t \begin{bmatrix} \eta(s) \\ \omega(s) \end{bmatrix}^{\mathrm{T}} \Theta \begin{bmatrix} \eta(s) \\ \omega(s) \end{bmatrix} \mathrm{d}s, \end{aligned} \tag{6.54}$$

where

$$\Theta = \begin{bmatrix} \Xi_{11} & P^{\mathrm{T}} A_d & E^{\mathrm{T}} Z_1 E & P^{\mathrm{T}} B_\omega \\ * & \Xi_{22} & E^{\mathrm{T}} Z_2 E & 0 \\ * & * & \Xi_{33} & 0 \\ * & * & * & -I \end{bmatrix} + \frac{d^2}{4} \begin{bmatrix} A^{\mathrm{T}} \\ A_d^{\mathrm{T}} \\ 0 \\ B_\omega^{\mathrm{T}} \end{bmatrix} (Z_1 + Z_2) \begin{bmatrix} A^{\mathrm{T}} \\ A_d^{\mathrm{T}} \\ 0 \\ B_\omega^{\mathrm{T}} \end{bmatrix}^{\mathrm{T}}.$$

Hence, by the Schur complements, (6.15) implies that for any $t > 0$, $W(t) < 0$ for any non-zero $\omega(t) \in \mathscr{L}_2[0, \infty)$. Thus,

$$V(x_t, t) < \int_0^t \omega(s)^{\mathrm{T}} \omega(s) \, \mathrm{d}s. \tag{6.55}$$

On the other hand, (6.16) implies

$$L^{\mathrm{T}} L \leqslant \gamma^2 E^{\mathrm{T}} P. \tag{6.56}$$

Therefore, for any $t > 0$,

$$\begin{aligned} z(t)^{\mathrm{T}} z(t) &= x(t)^{\mathrm{T}} L^{\mathrm{T}} L x(t) \\ &\leqslant \gamma^2 x(t)^{\mathrm{T}} E^{\mathrm{T}} P x(t) \\ &< \gamma^2 \int_0^t \omega(s)^{\mathrm{T}} \omega(s) \, \mathrm{d}s \leqslant \gamma^2 \int_0^\infty \omega(s)^{\mathrm{T}} \omega(s) \, \mathrm{d}s. \end{aligned} \tag{6.57}$$

Thus, (6.4) is guaranteed. Therefore, system (6.13) is exponentially admissible with $\mathscr{L}_2 - \mathscr{L}_\infty$ performance γ. This completes the proof.

During the proof of Theorem 6.2, the well known Jensen inequality is applied to deal with the integral terms $-\frac{d}{2} \int_{t-\frac{d}{2}}^{t} \dot{x}(\alpha)^{\mathrm{T}} E^{\mathrm{T}} Z_1 E \dot{x}(\alpha) \, \mathrm{d}\alpha$ and $-\frac{d}{2} \int_{t-d}^{t-\frac{d}{2}} \dot{x}(\alpha)^{\mathrm{T}} E^{\mathrm{T}} Z_2 E \dot{x}(\alpha) \, \mathrm{d}\alpha$. The free-weighting matrix method, which is based on the well known Newton-Leibniz formula, is regarded as an effective method to get less conservative results for all sorts of time-delay systems. Now, we would like to point out that the free-weighting matrix method can also be used to treat with the above two integral terms. Introducing the following null terms

$$\begin{aligned} &\left[x(t)^{\mathrm{T}} T_1 + x(t-d)^{\mathrm{T}} T_2 + x(t-\frac{d}{2})^{\mathrm{T}} T_3 \right] \\ &\times \left[E x(t) - E x(t-\frac{d}{2}) - \int_{t-\frac{d}{2}}^{t} E \dot{x}(s) \, \mathrm{d}s \right] = 0 \end{aligned} \tag{6.58}$$

and

$$\begin{aligned} &\left[x(t)^{\mathrm{T}} F_1 + x(t-d)^{\mathrm{T}} F_2 + x(t-\frac{d}{2})^{\mathrm{T}} F_3 \right] \\ &\times \left[E x(t-\frac{d}{2}) - E x(t-d) - \int_{t-d}^{t-\frac{d}{2}} E \dot{x}(s) \, \mathrm{d}s \right] = 0 \end{aligned} \tag{6.59}$$

to deal with the related integral terms and following the same method of Theorem 6.2, we can get another condition to ensure system (6.13) is exponentially admissible with $\mathscr{L}_2 - \mathscr{L}_\infty$ performance γ as follows.

Theorem 6.3. *For a given scalar $\gamma > 0$, system (6.13) is exponentially admissible with $\mathscr{L}_2-\mathscr{L}_\infty$ performance γ, if there exist matrices $Q_1 > 0$, $Q_2 > 0$, $Z_1 > 0$, $Z_2 > 0$, P, T_1, T_2, T_3, F_1, F_2 and F_3 such that (6.14), (6.16) and (6.60) hold,*

$$\begin{bmatrix} \Psi_{11} & \Psi_{12} & \Psi_{13} & P^{\mathrm{T}}B_\omega & d^2A^{\mathrm{T}}(Z_1+Z_2) & -T_1 & -F_1 \\ * & \Psi_{22} & \Psi_{23} & 0 & d^2A_d^{\mathrm{T}}(Z_1+Z_2) & -T_2 & -F_2 \\ * & * & \Psi_{33} & 0 & 0 & -T_3 & -F_3 \\ * & * & * & -I & d^2B_\omega^{\mathrm{T}}(Z_1+Z_2) & 0 & 0 \\ * & * & * & * & -4d^2(Z_1+Z_2) & 0 & 0 \\ * & * & * & * & * & -Z_1 & 0 \\ * & * & * & * & * & 0 & -Z_2 \end{bmatrix} < 0, \tag{6.60}$$

where

$$\begin{aligned} \Psi_{11} &= P^{\mathrm{T}}A + A^{\mathrm{T}}P + Q_1 + T_1E + E^{\mathrm{T}}T_1^{\mathrm{T}}, \\ \Psi_{12} &= P^{\mathrm{T}}A_d - F_1E + E^{\mathrm{T}}T_2^{\mathrm{T}}, \\ \Psi_{13} &= -T_1E + F_1E + E^{\mathrm{T}}T_3^{\mathrm{T}}, \\ \Psi_{22} &= -Q_2 - F_2E - E^{\mathrm{T}}F_2^{\mathrm{T}}, \\ \Psi_{23} &= -T_2E + F_2E - E^{\mathrm{T}}F_3^{\mathrm{T}}, \\ \Psi_{33} &= -Q_1 + Q_2 - T_3E - E^{\mathrm{T}}T_3^{\mathrm{T}} + F_3E + E^{\mathrm{T}}F_3^{\mathrm{T}}. \end{aligned}$$

Now, we are in a position to show the relationship between Theorem 6.2 and Theorem 6.3.

Theorem 6.4. *There exist matrices $Q_1 > 0$, $Q_2 > 0$, $Z_1 > 0$, $Z_2 > 0$, P, T_1, T_2, T_3, F_1, F_2 and F_3 such that (6.14), (6.16) and (6.60) hold if and only if there exist matrices P, $Q_1 > 0$, $Q_2 > 0$, $Z_1 > 0$ and $Z_2 > 0$ such that (6.14)-(6.16) hold.*

Proof. (*Sufficiency*) If there exist matrices P, $Q_1 > 0$, $Q_2 > 0$, $Z_1 > 0$ and $Z_2 > 0$ such that (6.14)-(6.16) hold, choosing $T_1 = -T_3 = -E^{\mathrm{T}}Z_1$, $F_2 = -F_3 = E^{\mathrm{T}}Z_2$ and $T_2 = F_1 = 0$, it is easy to get (6.60).

(*Necessity*) If there exist matrices $Q_1 > 0$, $Q_2 > 0$, $Z_1 > 0$, $Z_2 > 0$, P, T_1, T_2, T_3, F_1, F_2 and F_3 such that (6.14), (6.16) and (6.60) hold. Set

$$\Pi = \begin{bmatrix} I & 0 & 0 & 0 & 0 \\ 0 & I & 0 & 0 & 0 \\ 0 & 0 & I & 0 & 0 \\ 0 & 0 & 0 & I & 0 \\ 0 & 0 & 0 & 0 & I \\ E & 0 & -E & 0 & 0 \\ 0 & -E & E & 0 & 0 \end{bmatrix}. \tag{6.61}$$

Pre-multiplying and post-multiplying (6.60) by Π^{T} and Π, respectively, we obtain (6.15) immediately. This completes the proof.

Remark 6.5. It should be pointed out that although Theorem 6.2 and Theorem 6.3 are equivalent to each other according to Theorem 6.4, from the computational point of view, Theorem 6.2 is considered to be more desirable and elegant due to the fact that matrix variables in Theorem 6.2 are much fewer than those in Theorem 6.3. Hence, we will present a solution to the $\mathscr{L}_2 - \mathscr{L}_\infty$ filtering problem based on Theorem 6.2 in the following subsection.

6.3.2 $\mathscr{L}_2 - \mathscr{L}_\infty$ Filter Design

Theorem 6.6. *For a given scalar $\gamma > 0$, system (6.3) is exponentially admissible with $\mathscr{L}_2 - \mathscr{L}_\infty$ performance γ, if there exist matrices $S_1 > 0$, $S_2 > 0$, $R_1 > 0$, $R_2 > 0$, X, U, $\bar{A}_f$, $\bar{B}_f$ and $\bar{C}_f$ such that*

$$E^{\mathrm{T}}X = X^{\mathrm{T}}E \geqslant 0, \tag{6.62}$$

$$E^{\mathrm{T}}U = U^{\mathrm{T}}E \geqslant 0, \tag{6.63}$$

$$E^{\mathrm{T}}(X-U) = (X-U)^{\mathrm{T}}E \geqslant 0, \tag{6.64}$$

$$\begin{bmatrix} \Delta_1 & \Delta_2 & \Delta_4 & E^{\mathrm{T}}R_1E & (X-U)^{\mathrm{T}}B_\omega & d^2A^{\mathrm{T}}(R_1+R_2) \\ * & \Delta_3 & \Delta_5 & E^{\mathrm{T}}R_1E & X^{\mathrm{T}}B_\omega + \bar{B}_fD_\omega & d^2A^{\mathrm{T}}(R_1+R_2) \\ * & * & \Delta_6 & E^{\mathrm{T}}R_2E & 0 & d^2A_d^{\mathrm{T}}(R_1+R_2) \\ * & * & * & \Delta_7 & 0 & 0 \\ * & * & * & * & -I & d^2B_\omega^{\mathrm{T}}(R_1+R_2) \\ * & * & * & * & * & -4d^2(R_1+R_2) \end{bmatrix} < 0, \tag{6.65}$$

$$\begin{bmatrix} E^{\mathrm{T}}(X-U) & E^{\mathrm{T}}(X-U) & L^{\mathrm{T}} + \bar{C}_f^{\mathrm{T}} \\ * & E^{\mathrm{T}}X & L^{\mathrm{T}} \\ * & * & \gamma^2 I \end{bmatrix} \geqslant 0, \tag{6.66}$$

where

$$\begin{aligned} \Delta_1 &= A^{\mathrm{T}}(X-U) + (X-U)^{\mathrm{T}}A - E^{\mathrm{T}}R_1E + S_1, \\ \Delta_2 &= \bar{A}_f^{\mathrm{T}} - E^{\mathrm{T}}R_1E + S_1, \\ \Delta_3 &= A^{\mathrm{T}}X + C^{\mathrm{T}}\bar{B}_f^{\mathrm{T}} + X^{\mathrm{T}}A + \bar{B}_fC - E^{\mathrm{T}}R_1E + S_1, \\ \Delta_4 &= (X-U)^{\mathrm{T}}A_d, \\ \Delta_5 &= X^{\mathrm{T}}A_d + \bar{B}_fC_d, \\ \Delta_6 &= -S_2 - E^{\mathrm{T}}R_2E, \\ \Delta_7 &= -S_1 + S_2 - E^{\mathrm{T}}(R_1+R_2)E. \end{aligned}$$

Furthermore, if (6.62)-(6.66) *are solvable, a desired* $\mathscr{L}_2 - \mathscr{L}_\infty$ *filter in the form of* (6.2) *can be chosen with parameters as*

$$\begin{cases} E_f = E, \\ A_f = -U^{-\mathrm{T}}(\bar{A}_f - A^{\mathrm{T}}(X-U) - X^{\mathrm{T}}A - \bar{B}_f C), \\ B_f = U^{-\mathrm{T}}\bar{B}_f, \\ C_f = \bar{C}_f. \end{cases} \tag{6.67}$$

Proof. It can be seen from (6.65) that

$$\begin{bmatrix} I \\ I \\ I \end{bmatrix}^{\mathrm{T}} \begin{bmatrix} \Delta_1 & \Delta_4 & E^{\mathrm{T}}R_1E \\ * & \Delta_6 & E^{\mathrm{T}}R_2E \\ * & * & \Delta_7 \end{bmatrix} \begin{bmatrix} I \\ I \\ I \end{bmatrix} < 0, \tag{6.68}$$

which implies

$$(A + A_d)^{\mathrm{T}}(X - U) + (X - U)^{\mathrm{T}}(A + A_d) < 0. \tag{6.69}$$

Thus, $X - U$ is nonsingular. Define

$$P = \begin{bmatrix} X & U \\ U & U \end{bmatrix}, \ J_1 = \begin{bmatrix} (X-U)^{-1} & I \\ -(X-U)^{-1} & 0 \end{bmatrix}, \ \hat{E} = \begin{bmatrix} E & 0 \\ 0 & E \end{bmatrix}, \ H = \begin{bmatrix} I \\ 0 \end{bmatrix}^{\mathrm{T}}. \tag{6.70}$$

Then noting (6.62) and (6.63), we have

$$\hat{E}^{\mathrm{T}}P = \begin{bmatrix} E^{\mathrm{T}}X & E^{\mathrm{T}}U \\ E^{\mathrm{T}}U & E^{\mathrm{T}}U \end{bmatrix} = P^{\mathrm{T}}\hat{E}. \tag{6.71}$$

It can be deduced from (6.64) that

$$E^{\mathrm{T}}X - (E^{\mathrm{T}}U)(E^{\mathrm{T}}U)^{+}(E^{\mathrm{T}}U) = E^{\mathrm{T}}(X - U) \geqslant 0, \tag{6.72}$$

and

$$\begin{aligned} E^{\mathrm{T}}U(I - (E^{\mathrm{T}}U)(E^{\mathrm{T}}U)^{+}) &= E^{\mathrm{T}}U - (E^{\mathrm{T}}U)[(E^{\mathrm{T}}U)]^{\mathrm{T}}[(E^{\mathrm{T}}U)^{+}]^{\mathrm{T}} \\ &= E^{\mathrm{T}}U - (E^{\mathrm{T}}U)[(E^{\mathrm{T}}U)^{+}(E^{\mathrm{T}}U)]^{\mathrm{T}} \\ &= E^{\mathrm{T}}U - E^{\mathrm{T}}U \\ &= 0. \end{aligned} \tag{6.73}$$

Considering (6.62), (6.72) and (6.73), and using Lemma 1.3, we have

$$\hat{E}^{\mathrm{T}}P = P^{\mathrm{T}}\hat{E} \geqslant 0. \tag{6.74}$$

Now, pre- and post-multiplying (6.65) by $\mathrm{diag}\{(X-U)^{-\mathrm{T}}, I, I, I, I, I, I\}$ and its transpose, respectively, we obtain

$$\begin{bmatrix} \Sigma_1 & J_1^{\mathrm{T}} P^{\mathrm{T}} \bar{A}_d H^{\mathrm{T}} & J_1^{\mathrm{T}} \hat{E}^{\mathrm{T}} H^{\mathrm{T}} R_1 E & J_1^{\mathrm{T}} P^{\mathrm{T}} \bar{B}_\omega & d^2 J_1^{\mathrm{T}} \bar{A}^{\mathrm{T}} H^{\mathrm{T}} (R_1 + R_2) \\ * & \Delta_6 & E^{\mathrm{T}} R_2 E & 0 & d^2 H \bar{A}_d^{\mathrm{T}} H^{\mathrm{T}} (R_1 + R_2) \\ * & * & \Delta_7 & 0 & 0 \\ * & * & * & -I & d^2 \bar{B}_\omega^{\mathrm{T}} H^{\mathrm{T}} (R_1 + R_2) \\ * & * & * & * & -4d^2 (R_1 + R_2) \end{bmatrix} < 0, \tag{6.75}$$

where

$$\Sigma_1 = J_1^{\mathrm{T}} (P^{\mathrm{T}} \bar{A} + \bar{A}^{\mathrm{T}} P + H^{\mathrm{T}} S_1 H - \hat{E}^{\mathrm{T}} H^{\mathrm{T}} R_1 H \hat{E}) J_1.$$

and the matrices $\bar{A}$, $\bar{A}_d$ and $\bar{B}_\omega$ are given in (6.3) with the parameters E_f, A_f and B_f as in (6.67). Then pre- and post-multiplying (6.65) by $\mathrm{diag}\{J_1^{-\mathrm{T}}, I, I, I, I, I, I\}$ and its transpose, respectively, we obtain

$$\begin{bmatrix} \Upsilon_1 & P^{\mathrm{T}} \bar{A}_d H^{\mathrm{T}} & \hat{E}^{\mathrm{T}} H^{\mathrm{T}} R_1 E & P^{\mathrm{T}} \bar{B}_\omega & d^2 \bar{A}^{\mathrm{T}} H^{\mathrm{T}} (R_1 + R_2) \\ * & \Delta_6 & E^{\mathrm{T}} R_2 E & 0 & d^2 H \bar{A}_d^{\mathrm{T}} H^{\mathrm{T}} (R_1 + R_2) \\ * & * & \Delta_7 & 0 & 0 \\ * & * & * & -I & d^2 \bar{B}_\omega^{\mathrm{T}} H^{\mathrm{T}} (R_1 + R_2) \\ * & * & * & * & -4d^2 (R_1 + R_2) \end{bmatrix} < 0, \tag{6.76}$$

where

$$\Upsilon_1 = P^{\mathrm{T}} \bar{A} + \bar{A}^{\mathrm{T}} P + H^{\mathrm{T}} S_1 H - \hat{E}^{\mathrm{T}} H^{\mathrm{T}} R_1 H \hat{E}.$$

Then, a small enough scalar $\sigma > 0$ can always be find such that

$$\begin{bmatrix} \Lambda_1 & P^{\mathrm{T}} \bar{A}_d & \hat{E}^{\mathrm{T}} \bar{R}_1 \hat{E} & P^{\mathrm{T}} \bar{B}_\omega & d^2 \bar{A}^{\mathrm{T}} (\bar{R}_1 + \bar{R}_2) \\ * & \Lambda_2 & \hat{E}^{\mathrm{T}} \bar{R}_2 \hat{E} & 0 & d^2 \bar{A}_d^{\mathrm{T}} (\bar{R}_1 + \bar{R}_2) \\ * & * & \Lambda_3 & 0 & 0 \\ * & * & * & -I & d^2 \bar{B}_\omega^{\mathrm{T}} (\bar{R}_1 + \bar{R}_2) \\ * & * & * & * & -4d^2 (\bar{R}_1 + \bar{R}_2) \\ * & * & * & * & * \end{bmatrix} < 0, \tag{6.77}$$

where

$$\begin{aligned} \Lambda_1 &= P^{\mathrm{T}} \bar{A} + \bar{A}^{\mathrm{T}} P + \bar{S}_1 - \hat{E}^{\mathrm{T}} \bar{R}_1 \hat{E}, \\ \Lambda_2 &= -\bar{S}_2 - \hat{E}^{\mathrm{T}} \bar{R}_2 \hat{E}, \\ \Lambda_3 &= -\bar{S}_1 + \bar{S}_2 - \hat{E}^{\mathrm{T}} (\bar{R}_1 + \bar{R}_2) \hat{E}, \\ \bar{S}_1 &= \begin{bmatrix} S_1 & 0 \\ 0 & \sigma I \end{bmatrix}, k = 1, 2, \\ \bar{R}_l &= \begin{bmatrix} R_l & 0 \\ 0 & \sigma I \end{bmatrix}, l = 1, 2. \end{aligned}$$

Applying the same approach, we can find from (6.66) that

$$\begin{bmatrix} \hat{E}^{\mathrm{T}}P & \bar{L}^{\mathrm{T}} \\ * & \gamma^2 I \end{bmatrix} \geqslant 0. \tag{6.78}$$

Thus, by Theorem 6.2, the filtering error system (6.3) is exponentially admissible with $\mathscr{L}_2 - \mathscr{L}_\infty$ performance γ. This completes the proof.

Remark 6.7. It is noted that Theorem 6.6 provides a delay-dependent sufficient condition for the existence of the $\mathscr{L}_2 - \mathscr{L}_\infty$ filtering for SSs with time-delay. It is worth mentioning that the condition is expressed in terms of system matrices of (6.1), which means the design procedure involves no system decomposition, and thus avoids some certain numerical problems and makes the design procedure simple and reliable.

6.4 Numerical Examples

In this section, we provide some numerical examples to illustrate the reduced conservatism and usefulness of the proposed methods.

Example 6.8. Consider the following state-space time-delay system:

$$\begin{cases} \dot{x}(t) = Ax(t) + A_d x(t-d) + B_\omega \omega(t), \\ y(t) = Cx(t) + \omega(t), \\ z(t) = Lx(t), \end{cases} \tag{6.79}$$

where

$$A = \begin{bmatrix} -2 & 0 \\ 0 & -0.9 \end{bmatrix}, \ A_d = \begin{bmatrix} -1 & 0 \\ -1 & -0.9 \end{bmatrix},$$
$$B_\omega = \begin{bmatrix} 0 \\ 1 \end{bmatrix}, \ C = \begin{bmatrix} 1 & 0 \end{bmatrix}, \ L = \begin{bmatrix} 1 & 1.2 \end{bmatrix}.$$

In this example, the time delay d is assumed to be 1.1. The minimum $\mathscr{L}_2 - \mathscr{L}_\infty$ performance γ of the filtering system (6.3) achieved by [168] is 0.5460. However, applying Theorem 6.6 in this chapter, the achieved $\mathscr{L}_2 - \mathscr{L}_\infty$ performances of the filtering system (6.3) can be calculated as 0.4872, which is 10.77% smaller than that in [168]. Thus, for the above system, Theorem 6.6 in this chapter is less conservative than [168].

Example 6.9. Consider system (6.1) with the following matrices

$$E = \begin{bmatrix} 1 & 0 \\ 0 & 0 \end{bmatrix}, \ A = \begin{bmatrix} 0.4 & 0.2 \\ 0.1 & -1 \end{bmatrix}, \ A_d = \begin{bmatrix} -0.9 & 1 \\ 0 & 0.5 \end{bmatrix},$$
$$B_\omega = \begin{bmatrix} -1 \\ 0.4 \end{bmatrix}, \ C = \begin{bmatrix} 1 & -0.9 \end{bmatrix}, \ C_d = \begin{bmatrix} 0.5 & 0.4 \end{bmatrix},$$
$$D_\omega = 1, \ L = \begin{bmatrix} -1 & 0 \end{bmatrix}.$$

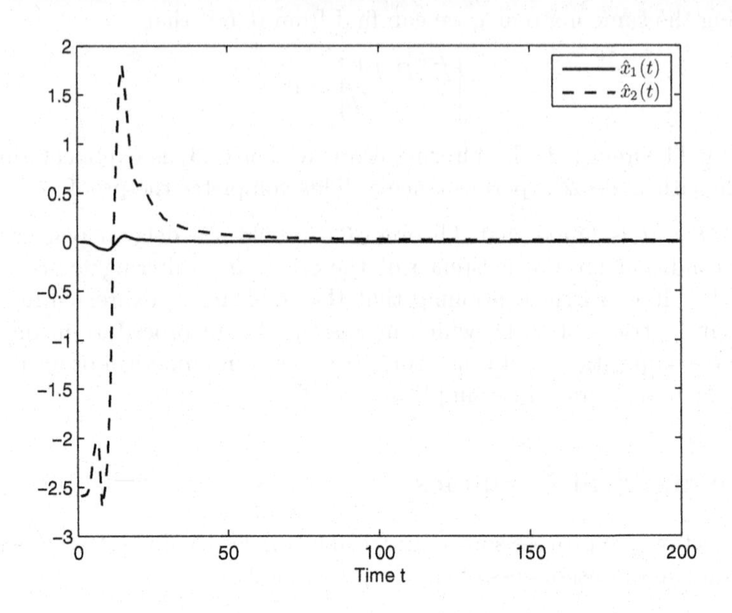

Fig. 6.1. Example 6.9: State responses of the designed filter

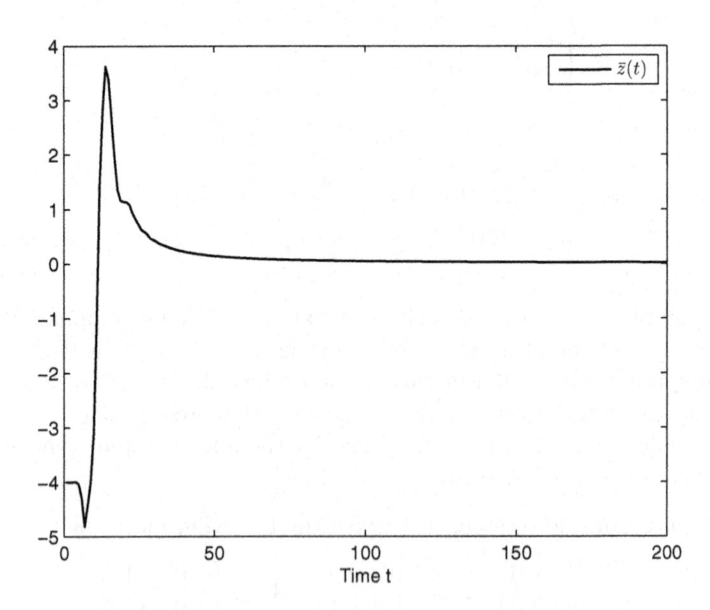

Fig. 6.2. Example 6.9: Error Response

For this example, the time delay $d = 0.6$ and $\mathscr{L}_2 - \mathscr{L}_\infty$ performance index $\gamma = 3.5$. Solving the matrix inequalities in (6.62)-(6.66), we get the desired $\mathscr{L}_2 - \mathscr{L}_\infty$ filter with the following parameter matrices,

$$E_f = \begin{bmatrix} 1 & 0 \\ 0 & 0 \end{bmatrix}, \; A_f = \begin{bmatrix} -1.2466 & 4.8477 \\ -0.3834 & -1.1635 \end{bmatrix},$$

$$B_f = \begin{bmatrix} 0.3736 \\ -0.4624 \end{bmatrix}, \; C_f = \begin{bmatrix} 0.3211 & 0 \end{bmatrix}.$$

The simulation result of the state response of the designed filter is given in Figure 6.1, where the exogenous disturbance input $\omega(t)$ is given as $\omega(t) = \frac{2.5}{2+5t}$, $t \geqslant 0$. Figure 6.2 is the simulation result of the error response of $z(t) - \hat{z}(t)$ of the designed filter. It can be seen that the designed $\mathscr{L}_2 - \mathscr{L}_\infty$ filter satisfies the specified requirements.

6.5 Conclusion

The problem of $\mathscr{L}_2 - \mathscr{L}_\infty$ filtering for SSs with time delays has been investigated in this chapter. A delay-dependent sufficient condition for the solvability of the problem has been obtained, which is formulated by a set of matrix inequalities. Several numerical examples have been provided to demonstrate the effectiveness of the proposed methods.

7

H_∞ Filtering for SSs with Time-Varying Delays

7.1 Introduction

In this chapter, we continue to devote our attention to filtering problem for singular time-delay systems. Addressing SSs with time-varying delays in a range, two delay-range-dependent BRLs are proposed by using LMI approach. A delay-dependent condition is proposed for the existence of H_∞ filter. The BRL and H_∞ filter results developed in this chapter are less conservative than the existing ones, which is demonstrated by providing some numerical examples.

7.2 Problem Formulation

Consider the following SS with time-varying delays

$$\begin{cases} E\dot{x}(t) = Ax(t) + A_d x(t-d(t)) + B_\omega \omega(t), \\ y(t) = Cx(t) + C_d x(t-d(t)) + D_\omega \omega(t), \\ z(t) = Lx(t), \\ x(t) = \phi(t),\ t \in [-d_2, 0], \end{cases} \tag{7.1}$$

where $x(t) \in \mathbb{R}^n$ is the state, $y(t) \in \mathbb{R}^s$ is the measurement, $z(t) \in \mathbb{R}^q$ is the signal to be estimated, $\omega(t) \in \mathbb{R}^p$ is the disturbance input that belongs to $\mathscr{L}_2[0,\infty)$, and $\phi(t) \in C_{n,d_2}$ is a compatible vector valued initial function. The matrix $E \in \mathbb{R}^{n\times n}$ may be singular and it is assumed that $\operatorname{rank} E = r \leqslant n$. A, A_d, B_ω, C, C_d, D_ω and L are known real constant matrices with appropriate dimensions. $d(t)$ is a time-varying continuous function that satisfies $0 < d_1 \leqslant d(t) \leqslant d_2$ and $\dot{d}(t) \leqslant \mu$, where d_1 and d_2 are the lower and upper bounds of time-varying delay $d(t)$, respectively, and $0 \leqslant \mu < 1$ is the variation rate of time-varying delay $d(t)$.

In this chapter, we aim at obtaining the estimate of $z(t)$. To be more specific, we are interested in constructing the following full-order filter:

Z.-G. Wu et al.: *Anal. & Synth. of Singular Syst. with Time-Delays*, LNCIS 443, pp. 89–102.
DOI: 10.1007/978-3-642-37497-5_7

$$\begin{cases} E_f\dot{\hat{x}}(t) = A_f\hat{x}(t) + B_f y(t), \\ \hat{z}(t) = C_f\hat{x}(t), \end{cases} \tag{7.2}$$

where $\hat{x} \in \mathbb{R}^n$, $\hat{z}(t) \in \mathbb{R}^q$, and the constant matrices E_f, A_f, B_f and C_f are the filter matrices with appropriate dimensions, which are to be designed.

Defining

$$\bar{z}(t) = z(t) - \hat{z}(t),$$
$$\bar{x}(t) = \left[x(t)^{\mathrm{T}}\ \hat{x}(t)^{\mathrm{T}}\right]^{\mathrm{T}},$$

and combining (7.1) and (7.2), we obtain the filtering error dynamics as follows:

$$\begin{cases} \bar{E}\dot{\bar{x}}(t) = \bar{A}\bar{x}(t) + \bar{A}_d\bar{x}(t-d(t)) + \bar{B}_\omega\omega(t), \\ \bar{z}(t) = \bar{L}\bar{x}(t), \end{cases} \tag{7.3}$$

where

$$\bar{E} = \begin{bmatrix} E & 0 \\ 0 & E_f \end{bmatrix}, \ \bar{A} = \begin{bmatrix} A & 0 \\ B_fC & A_f \end{bmatrix},$$
$$\bar{A}_d = \begin{bmatrix} A_d & 0 \\ B_fC_d & 0 \end{bmatrix}, \ \bar{B}_\omega = \begin{bmatrix} B_\omega \\ B_fD_\omega \end{bmatrix},$$
$$\bar{L} = \begin{bmatrix} L & -C_f \end{bmatrix}.$$

We are now in a position to formulate the H_∞ filter design problem to be addressed in this chapter as follows: for a given scalar $\gamma > 0$, design the parameters E_f, A_f, B_f and C_f for the filter (7.2) such that the filtering error system (7.3) is exponentially admissible under $\omega(t) = 0$ and satisfies

$$J_{\bar{z}\omega} = \int_0^\infty \left(\bar{z}(t)^{\mathrm{T}}\bar{z}(t) - \gamma^2\omega(t)^{\mathrm{T}}\omega(t)\right) \mathrm{d}t < 0, \tag{7.4}$$

under the zero initial condition for any nonzero $\omega(t) \in \mathscr{L}_2[0,\infty)$. In this case, the filtering error system (7.3) is said to be exponentially admissible with H_∞ performance γ.

7.3 Main Results

7.3.1 BRLs

In this subsection, we present two BRLs for system (7.1), one of which will be applied to design the filter (7.2).

Theorem 7.1. *For a given scalar $\gamma > 0$, system (7.1) is exponentially admissible with H_∞ performance γ, if there exist matrices $Q_1 > 0$, $Q_2 > 0$, $Q_3 > 0$, $R_1 > 0$, $R_2 > 0$ and P such that*

$$E^{\mathrm{T}}P = P^{\mathrm{T}}E \geqslant 0, \tag{7.5}$$

$$\Xi = \begin{bmatrix} \Xi_{11} & \Xi_{12} & 0 & 0 & P^{\mathrm{T}}B_{\omega} & A^{\mathrm{T}}W \\ * & \Xi_{22} & E^{\mathrm{T}}R_2E & \Xi_{24} & 0 & A_d^{\mathrm{T}}W \\ * & * & -Q_1 - E^{\mathrm{T}}R_2E & 0 & 0 & 0 \\ * & * & * & -Q_2 - \Xi_{24} & 0 & 0 \\ * & * & * & * & -\gamma^2 I & B_{\omega}^{\mathrm{T}}W \\ * & * & * & * & * & -W \end{bmatrix} < 0, \tag{7.6}$$

where $d_{12} = d_2 - d_1$, $W = d_{12}d_2^2R_1 + d_{12}^2R_2$, *and*

$$\begin{aligned} \Xi_{11} &= P^{\mathrm{T}}A + A^{\mathrm{T}}P + Q_1 + Q_2 + Q_3 - d_{12}E^{\mathrm{T}}R_1E + L^{\mathrm{T}}L, \\ \Xi_{12} &= P^{\mathrm{T}}A_d + d_{12}E^{\mathrm{T}}R_1E, \\ \Xi_{22} &= -(1-\mu)Q_3 - E^{\mathrm{T}}((d_{12}+d_2)R_1 + 2R_2)E, \\ \Xi_{24} &= E^{\mathrm{T}}(d_2R_1 + R_2)E. \end{aligned}$$

Proof. Firstly, we prove the exponential admissibility of system (7.1) with $\omega(t) = 0$. Since $\operatorname{rank} E = r \leqslant n$, there exist two nonsingular matrices G and H such that

$$GEH = \begin{bmatrix} I_r & 0 \\ 0 & 0 \end{bmatrix}. \tag{7.7}$$

Denote

$$GAH = \begin{bmatrix} A_1 & A_2 \\ A_3 & A_4 \end{bmatrix}, \; G^{-T}PH = \begin{bmatrix} \bar{P}_1 & \bar{P}_2 \\ \bar{P}_3 & \bar{P}_4 \end{bmatrix}. \tag{7.8}$$

From (7.5) and using the expressions in (7.7) and (7.8), it is easy to obtain that $P_2 = 0$. Then, pre-multiplying and post-multiplying $\Xi_{11} < 0$ by H^{T} and H, respectively, we have $A_4^{\mathrm{T}}\bar{P}_4 + \bar{P}_4^{\mathrm{T}}A_4 < 0$, which implies A_4 is nonsingular, and thus the pair (E, A) is regular and impulse free. Hence, by Definition 4.1, system (7.1) with $\omega(t) = 0$ is regular and impulse free.

Next, we will show the exponential stability of system (7.1) with $\omega(t) = 0$. To this end, define the following Lyapunov functional for system (7.1) with $\omega(t) = 0$,

$$\begin{aligned} V(x_t, t) = {} & x(t)^{\mathrm{T}}E^{\mathrm{T}}Px(t) + \sum_{k=1}^{2}\int_{t-d_k}^{t} x(\alpha)^{\mathrm{T}}Q_kx(\alpha)\,\mathrm{d}\alpha \\ & + \int_{t-d(t)}^{t} x(\alpha)^{\mathrm{T}}Q_3x(\alpha)\,\mathrm{d}\alpha \\ & + d_{12}d_2\int_{-d_2}^{0}\int_{t+\beta}^{t} \dot{x}(\alpha)^{\mathrm{T}}E^{\mathrm{T}}R_1E\dot{x}(\alpha)\,\mathrm{d}\alpha\mathrm{d}\beta \\ & + d_{12}\int_{-d_2}^{-d_1}\int_{t+\beta}^{t} \dot{x}(\alpha)^{\mathrm{T}}E^{\mathrm{T}}R_2E\dot{x}(\alpha)\,\mathrm{d}\alpha\mathrm{d}\beta, \end{aligned} \tag{7.9}$$

where $\{x_t = x(t+\theta), -2d_2 \leqslant \theta \leqslant 0\}$. Calculating the time derivative of $V(x_t,t)$ along the solution of system (7.1) with $\omega(t) \equiv 0$ yields

$$\begin{aligned}\dot{V}(x_t,t) \leqslant\ & 2x(t)^{\mathrm{T}}E^{\mathrm{T}}P\dot{x}(t) + \sum_{k=1}^{3} x(t)^{\mathrm{T}}Q_k x(t) - \sum_{k=1}^{2} x(t-d_k)^{\mathrm{T}}Q_k x(t-d_k) \\ & - (1-\mu)x(t-d(t))^{\mathrm{T}}Q_3 x(t-d(t)) + \dot{x}(t)^{\mathrm{T}}E^{\mathrm{T}}WE\dot{x}(t) \\ & - d_{12}d_2 \int_{t-d_2}^{t} \dot{x}(\alpha)^{\mathrm{T}}E^{\mathrm{T}}R_1E\dot{x}(\alpha)\,\mathrm{d}\alpha \\ & - d_{12} \int_{t-d_2}^{t-d_1} \dot{x}(\alpha)^{\mathrm{T}}E^{\mathrm{T}}R_2E\dot{x}(\alpha)\,\mathrm{d}\alpha. \end{aligned} \tag{7.10}$$

According to Jensen inequality, we have

$$\begin{aligned} & - d_{12}d_2 \int_{t-d_2}^{t} \dot{x}(\alpha)^{\mathrm{T}}E^{\mathrm{T}}R_1E\dot{x}(\alpha)\,\mathrm{d}\alpha - d_{12}\int_{t-d_2}^{t-d_1} \dot{x}(\alpha)^{\mathrm{T}}E^{\mathrm{T}}R_2E\dot{x}(\alpha)\,\mathrm{d}\alpha \\ = & - d_{12}d_2 \int_{t-d(t)}^{t} \dot{x}(\alpha)^{\mathrm{T}}E^{\mathrm{T}}R_1E\dot{x}(\alpha)\,\mathrm{d}\alpha - d_{12}\int_{t-d(t)}^{t-d_1} \dot{x}(\alpha)^{\mathrm{T}}E^{\mathrm{T}}R_2E\dot{x}(\alpha)\,\mathrm{d}\alpha \\ & - d_{12}\int_{t-d_2}^{t-d(t)} \dot{x}(\alpha)^{\mathrm{T}}E^{\mathrm{T}}(d_2R_1+R_2)E\dot{x}(\alpha)\,\mathrm{d}\alpha \\ \leqslant & - \left(\int_{t-d(t)}^{t} E\dot{x}(\alpha)\,\mathrm{d}\alpha\right)^{\mathrm{T}} d_{12}R_1 \int_{t-d(t)}^{t} E\dot{x}(\alpha)\,\mathrm{d}\alpha \\ & - \left(\int_{t-d(t)}^{t-d_1} E\dot{x}(\alpha)\,\mathrm{d}\alpha\right)^{\mathrm{T}} R_2 \int_{t-d(t)}^{t-d_1} E\dot{x}(\alpha)\,\mathrm{d}\alpha \\ & - \left(\int_{t-d_2}^{t-d(t)} E\dot{x}(\alpha)\,\mathrm{d}\alpha\right)^{\mathrm{T}} (d_2R_1+R_2) \int_{t-d_2}^{t-d(t)} E\dot{x}(\alpha)\,\mathrm{d}\alpha \\ = & \ \eta(t)^{\mathrm{T}} \begin{bmatrix} -d_{12}E^{\mathrm{T}}R_1E & d_{12}E^{\mathrm{T}}R_1E & 0 & 0 \\ * & -E^{\mathrm{T}}((d_{12}+d_2)R_1+2R_2)E & E^{\mathrm{T}}R_2E & \Xi_{24} \\ * & * & -E^{\mathrm{T}}R_2E & 0 \\ * & * & * & -\Xi_{24} \end{bmatrix} \eta(t), \end{aligned} \tag{7.11}$$

where

$$\eta(t) = \begin{bmatrix} x(t) \\ x(t-d(t)) \\ x(t-d_1) \\ x(t-d_2) \end{bmatrix}.$$

Thus, we have that

$$\dot{V}(x_t,t) \leqslant \eta(t)^{\mathrm{T}}\Psi\eta(t), \tag{7.12}$$

where

$$\Psi = \begin{bmatrix} \Xi_{11} - L^{\mathrm{T}}L & \Xi_{12} & 0 & 0 \\ * & \Xi_{22} & E^{\mathrm{T}}R_2E & \Xi_{24} \\ * & * & -Q_1 - E^{\mathrm{T}}R_2E & 0 \\ * & * & * & -Q_2 - \Xi_{24} \end{bmatrix} + \begin{bmatrix} A^{\mathrm{T}} \\ A_d^{\mathrm{T}} \\ 0 \\ 0 \end{bmatrix} W \begin{bmatrix} A^{\mathrm{T}} \\ A_d^{\mathrm{T}} \\ 0 \\ 0 \end{bmatrix}^{\mathrm{T}}.$$

Now, by (7.6) and Schur complement, it is easy to see that $\Psi < 0$, and thus there exits a scalar $\lambda > 0$ such that

$$\dot{V}(x_t, t) \leqslant -\lambda \|x(t)\|^2. \tag{7.13}$$

By using the similar analysis method applied in Theorem 4.2, we can find that system (7.1) with $\omega(t) = 0$ is exponentially stable.

Now, we focus on the H_∞ performance analysis of system (7.1). For this purpose, we consider the Lyapunov functional (7.9) and the index (3.3) for system (7.1). Under zero initial condition, it easy to see that

$$\begin{aligned} J_{z\omega} &\leqslant \int_0^\infty \left(z(s)^{\mathrm{T}} z(s) - \gamma^2 \omega(s)^{\mathrm{T}} \omega(s) + \dot{V}(x_s, s) \right) \mathrm{d}s \\ &\leqslant \int_0^\infty \begin{bmatrix} \eta(s) \\ \omega(s) \end{bmatrix}^{\mathrm{T}} \Theta \begin{bmatrix} \eta(s) \\ \omega(s) \end{bmatrix} \mathrm{d}s, \end{aligned} \tag{7.14}$$

where

$$\Theta = \begin{bmatrix} \Xi_{11} & \Xi_{12} & 0 & 0 & P^{\mathrm{T}}B_\omega \\ * & \Xi_{22} & E^{\mathrm{T}}R_2E & \Xi_{24} & 0 \\ * & * & -Q_1 - E^{\mathrm{T}}R_2E & 0 & 0 \\ * & * & * & -Q_2 - \Xi_{24} & 0 \\ * & * & * & * & -\gamma^2 I \end{bmatrix} + \begin{bmatrix} A^{\mathrm{T}} \\ A_d^{\mathrm{T}} \\ 0 \\ 0 \\ B_\omega^{\mathrm{T}} \end{bmatrix} W \begin{bmatrix} A^{\mathrm{T}} \\ A_d^{\mathrm{T}} \\ 0 \\ 0 \\ B_\omega^{\mathrm{T}} \end{bmatrix}^{\mathrm{T}}.$$

Hence, by Schur complement, (7.6) implies that $J_{z\omega} < 0$ for any nonzero $\omega(t) \in \mathscr{L}_2[0, \infty)$. This completes the proof.

Remark 7.2. Based on LMI approach, Theorem 7.1 provides a BRL for SSs with time-varying delays. It is noted that the proposed Lyapunov functional (7.9) not only makes use of the information on the upper bound of time-delay d_2, but also uses the information on the lower bound of time-delay $d_1 (> 0)$ and the interval of time-delay d_{12}. While, the involved time-varying delay of Lyapunov functional in [200] is from 0 to an upper bound, that is, the lower bound $d_1 = 0$. Meanwhile, the term $\int_{t-d_2}^{t} x(\alpha)^{\mathrm{T}} Q_2 x(\alpha)\, \mathrm{d}\alpha$ in the proposed Lyapunov functional (7.9) introduces the new state $x(t - d_2)$, which has been ignored in [200]. Thus, the proposed Lyapunov functional (7.9) makes more full use of the information on the considered time delay and is more general and elegant than that of [200]. On the other hand, in the proof of our result, the term $-\int_{t-d_2}^{t}\dot{x}(\alpha)^{\mathrm{T}}E^{\mathrm{T}}R_1E\dot{x}(\alpha)\,\mathrm{d}\alpha$ is written as $-\int_{t-d(t)}^{t}\dot{x}(\alpha)^{\mathrm{T}}E^{\mathrm{T}}R_1E\dot{x}(\alpha)\mathrm{d}\alpha-\int_{t-d_2}^{t-d(t)}\dot{x}(\alpha)^{\mathrm{T}}E^{\mathrm{T}}R_1E\dot{x}(\alpha)\,\mathrm{d}\alpha$, which is in contrast with the method of [200], where the term $-\int_{t-d_2}^{t-d(t)}\dot{x}(\alpha)^{\mathrm{T}}E^{\mathrm{T}}R_1E\dot{x}(\alpha)$ $\mathrm{d}\alpha$ has been ignored. The ignorance of the term $-\int_{t-d_2}^{t-d(t)}\dot{x}(\alpha)^{\mathrm{T}}E^{\mathrm{T}}R_1E\dot{x}(\alpha)$ $\mathrm{d}\alpha$ may lead to conservatism to some extent. Considering the above, our result derived here is expected to have less conservativeness than that of [200].

It is noted that in the proof of Theorem 7.1, we use the Jensen integral inequality to deal with the integral terms $\int_{t-d(t)}^{t}\dot{x}(\alpha)^{\mathrm{T}}E^{\mathrm{T}}R_1E\dot{x}(\alpha)\,\mathrm{d}\alpha$, $\int_{t-d(t)}^{t-d_1}\dot{x}(\alpha)^{\mathrm{T}}E^{\mathrm{T}}R_2E\dot{x}(\alpha)\,\mathrm{d}\alpha$ and $\int_{t-d_2}^{t-d(t)}\dot{x}(\alpha)^{\mathrm{T}}E^{\mathrm{T}}(d_2R_1+R_2)E\dot{x}(\alpha)\,\mathrm{d}\alpha$. Now adopting the free-weighting matrix method, we introduce the following null terms

$$\eta(t)^TN\left[Ex(t)-Ex(t-d(t))-\int_{t-d(t)}^{t}E\dot{x}(\alpha)\,\mathrm{d}\alpha\right]=0, \tag{7.15}$$

$$\eta(t)^TM\left[Ex(t-d_1)-Ex(t-d(t))-\int_{t-d(t)}^{t-d_1}E\dot{x}(\alpha)\,\mathrm{d}\alpha\right]=0, \tag{7.16}$$

$$\eta(t)^TS\left[Ex(t-d(t))-Ex(t-d_2)-\int_{t-d_2}^{t-d(t)}E\dot{x}(\alpha)\,\mathrm{d}\alpha\right]=0, \tag{7.17}$$

where

$$N=\begin{bmatrix}N_1\\N_2\\N_3\\N_4\end{bmatrix},\ M=\begin{bmatrix}M_1\\M_2\\M_3\\M_4\end{bmatrix},\ S=\begin{bmatrix}S_1\\S_2\\S_3\\S_4\end{bmatrix}$$

to deal with the above mentioned three integral terms and get another version of BRL for system (7.1) as follows.

Theorem 7.3. *For a given scalar $\gamma>0$, system (7.1) is exponentially admissible with H_∞ performance γ, if there exist matrices $Q_1>0$, $Q_2>0$, $Q_3>0$, $R_1>0$, $R_2>0$, P, N_i, M_i and S_i $(i=1,2,3,4)$ such that*

$$E^{\mathrm{T}}P=P^{\mathrm{T}}E\geqslant 0, \tag{7.18}$$

$$\hat{\Xi}+\Psi\Theta\Upsilon+\Upsilon^{\mathrm{T}}\Theta^{\mathrm{T}}\Psi^{\mathrm{T}}<0, \tag{7.19}$$

where d_{12} and W are given in Theorem 7.1, and

$$\hat{\Xi} = \begin{bmatrix} \hat{\Xi}_{11} & P^{\mathrm{T}}A_d & 0 & 0 & P^{\mathrm{T}}B_\omega & A^{\mathrm{T}}W & 0 & 0 & 0 \\ * & -(1-\mu)Q_3 & 0 & 0 & 0 & A_d^{\mathrm{T}}W & 0 & 0 & 0 \\ * & * & -Q_1 & 0 & 0 & 0 & 0 & 0 & 0 \\ * & * & * & -Q_2 & 0 & 0 & 0 & 0 & 0 \\ * & * & * & * & -\gamma^2 I & B_\omega^{\mathrm{T}}W & 0 & 0 & 0 \\ * & * & * & * & * & -W & 0 & 0 & 0 \\ * & * & * & * & * & * & -d_{12}R_1 & 0 & 0 \\ * & * & * & * & * & * & * & -R_2 & 0 \\ * & * & * & * & * & * & * & * & -d_2R_1 - R_2 \end{bmatrix},$$

$$\Xi_{11} = P^{\mathrm{T}}A + A^{\mathrm{T}}P + \sum_{k=1}^{3} Q_k + L^{\mathrm{T}}L,$$

$$\Psi = \begin{bmatrix} I & 0 & 0 & 0 \\ 0 & I & 0 & 0 \\ 0 & 0 & I & 0 \\ 0 & 0 & 0 & I \\ 0 & 0 & 0 & 0 \\ 0 & 0 & 0 & 0 \\ 0 & 0 & 0 & 0 \\ 0 & 0 & 0 & 0 \\ 0 & 0 & 0 & 0 \end{bmatrix}, \ \Theta = \begin{bmatrix} N_1 & M_1 & S_1 \\ N_2 & M_2 & S_2 \\ N_3 & M_3 & S_3 \\ N_4 & M_4 & S_4 \end{bmatrix}, \ \Upsilon = \begin{bmatrix} E & -E & 0 & 0 & 0 & 0 & I & 0 & 0 \\ 0 & -E & E & 0 & 0 & 0 & 0 & I & 0 \\ 0 & E & 0 & -E & 0 & 0 & 0 & 0 & I \end{bmatrix}.$$

Now, we are in the position to show the relationship between Theorem 7.1 and Theorem 7.3.

Theorem 7.4. *For a given scalar $\gamma > 0$, there exist matrices $Q_1 > 0$, $Q_2 > 0$, $Q_3 > 0$, $R_1 > 0$, $R_2 > 0$, P, N_i, M_i and S_i $(i = 1, 2, 3, 4)$ such that* (7.18) *and* (7.19) *hold if and only if there exist matrices $Q_1 > 0$, $Q_2 > 0$, $Q_3 > 0$, $R_1 > 0$, $R_2 > 0$ and P such that* (7.5) *and* (7.6) *hold.*

Proof. If there exist matrices $Q_1 > 0$, $Q_2 > 0$, $Q_3 > 0$, $R_1 > 0$, $R_2 > 0$ and P such that (7.5) and (7.6) hold, then setting $N_1 = -N_2 = -d_{12}E^TR_1$, $N_3 = N_4 = 0$, $M_2 = -M_3 = E^TR_2$, $M_1 = M_4 = 0$, $S_2 = -S_4 = -d_2E^TR_1 - E^TR_2$, $S_1 = S_3 = 0$, and applying Schur complement, we find (7.18) and (7.19) hold.

If there exist matrices $Q_1 > 0$, $Q_2 > 0$, $Q_3 > 0$, $R_1 > 0$, $R_2 > 0$, P, N_i, M_i and S_i $(i = 1, 2, 3, 4)$ such that (7.18) and (7.19) hold. By choosing

$$(\Upsilon^{\mathrm{T}})^{\perp} = \begin{bmatrix} I & 0 & 0 & 0 & 0 & 0 & -E^T & 0 & 0 \\ 0 & I & 0 & 0 & 0 & 0 & E^T & E^T & -E^T \\ 0 & 0 & I & 0 & 0 & 0 & 0 & -E^T & 0 \\ 0 & 0 & 0 & I & 0 & 0 & 0 & 0 & E^T \\ 0 & 0 & 0 & 0 & I & 0 & 0 & 0 & 0 \\ 0 & 0 & 0 & 0 & 0 & I & 0 & 0 & 0 \end{bmatrix}, \tag{7.20}$$

and using the well-known elimination procedure, we get

$$\Xi = (\Upsilon^{\mathrm{T}})^{\perp}\hat{\Xi}((\Upsilon^{\mathrm{T}})^{\perp})^{\mathrm{T}} < 0, \tag{7.21}$$

where Ξ is defined in (7.6). This completes the proof.

Remark 7.5. Notice that Theorem 7.4 implies that Theorem 7.1 and Theorem 7.3 are equivalent to each other. Therefore, the introduced free-weighting matrices N_i, M_i and S_i in Theorem 7.3 cannot help reduce the conservatism of this condition, that is, these free-weighting matrices in Theorem 7.3 are redundant, on the contrary, they mathematically complicate system analysis and synthesis, and consequently lead to a significant increase in the computational demand. It can be calculated that the condition in Theorem 7.1 involves $3.5n^2 + 2.5n$ variables, while the condition in Theorem 7.3 involves $15.5n^2 + 2.5n$ variables. Thus, we devote ourselves to the design of the H_∞ filter for system (7.1) based on Theorem 7.1 in the next subsection.

7.3.2 H_∞ Filter Design

Now we are in a position to present our main result on the design problem for H_∞ filter based on Theorem 7.1.

Theorem 7.6. *For a given scalar $\gamma > 0$, system (7.3) is exponentially admissible with H_∞ performance γ, if there exist matrices $S_1 > 0$, $S_2 > 0$, $S_3 > 0$, $R_1 > 0$, $R_2 > 0$, X, U, $\bar{A}_f$, $\bar{B}_f$ and $\bar{C}_f$ such that*

$$E^{\mathrm{T}}X = X^{\mathrm{T}}E \geqslant 0, \tag{7.22}$$

$$E^{\mathrm{T}}U = U^{\mathrm{T}}E \geqslant 0, \tag{7.23}$$

$$E^{\mathrm{T}}(X-U) = (X-U)^{\mathrm{T}}E \geqslant 0, \tag{7.24}$$

$$\Delta = \begin{bmatrix} \Delta_1 & \Delta_2 & \Delta_4 & 0 & 0 & (X-U)^{\mathrm{T}}B_\omega & A^{\mathrm{T}}V & L^{\mathrm{T}} - \bar{C}_f^{\mathrm{T}} \\ * & \Delta_3 & \Delta_5 & 0 & 0 & X^{\mathrm{T}}B_\omega + \bar{B}_f D_\omega & A^{\mathrm{T}}V & L^{\mathrm{T}} \\ * & * & \Delta_6 & E^{\mathrm{T}}R_2E & \Delta_8 & 0 & A_d^{\mathrm{T}}V & 0 \\ * & * & * & \Delta_7 & 0 & 0 & 0 & 0 \\ * & * & * & * & \Delta_9 & 0 & 0 & 0 \\ * & * & * & * & * & -\gamma^2 I & B_\omega^{\mathrm{T}}V & 0 \\ * & * & * & * & * & * & -V & 0 \\ * & * & * & * & * & * & * & -I \end{bmatrix} < 0, \tag{7.25}$$

where d_{12} is given in Theorem 7.1, $V = d_{12}d_2^2R_1 + d_{12}^2R_2$, and

$$\Delta_1 = A^{\mathrm{T}}(X-U) + (X-U)^{\mathrm{T}}A - d_{12}E^{\mathrm{T}}R_1E + \sum_{k=1}^{3} S_k,$$

$$\Delta_2 = \bar{A}_f^{\mathrm{T}} - d_{12}E^{\mathrm{T}}R_1E + \sum_{k=1}^{3} S_k,$$

$$\Delta_3 = A^{\mathrm{T}}X + C^{\mathrm{T}}\bar{B}_f^{\mathrm{T}} + X^{\mathrm{T}}A + \bar{B}_fC - d_{12}E^{\mathrm{T}}R_1E + \sum_{k=1}^{3} S_k,$$

$$\Delta_4 = (X-U)^{\mathrm{T}}A_d + d_{12}E^{\mathrm{T}}R_1E,$$

$$\Delta_5 = X^{\mathrm{T}}A_d + \bar{B}_fC_d + d_{12}E^{\mathrm{T}}R_1E,$$

$$\Delta_6 = -(1-\mu)S_3 - E^{\mathrm{T}}((d_{12}+d_2)R_1 + 2R_2)E,$$

$$\Delta_7 = -S_1 - E^{\mathrm{T}}R_2E,$$

$$\Delta_8 = E^{\mathrm{T}}(d_2R_1 + R_2)E,$$

$$\Delta_9 = -S_2 - E^{\mathrm{T}}(d_2R_1 + R_2)E.$$

Furthermore, if (7.22)-(7.25) *are solvable, the desired* H_∞ *filter in the form of* (7.2) *can be chosen with parameters as*

$$\begin{cases} E_f = E, \\ A_f = -U^{\mathrm{T}}(\bar{A}_f - A^{\mathrm{T}}(X-U) - X^{\mathrm{T}}A - \bar{B}_fC), \\ B_f = -U^{\mathrm{T}}\bar{B}_f, \\ C_f = \bar{C}_f. \end{cases} \tag{7.26}$$

Proof. It can be seen from (7.25) that

$$\begin{bmatrix} I \\ I \\ I \\ I \end{bmatrix}^{\mathrm{T}} \begin{bmatrix} \Delta_1 & \Delta_4 & 0 & 0 \\ * & \Delta_6 & E^{\mathrm{T}}Z_2E & \Delta_8 \\ * & * & \Delta_7 & 0 \\ * & * & * & \Delta_9 \end{bmatrix} \begin{bmatrix} I \\ I \\ I \\ I \end{bmatrix} < 0, \tag{7.27}$$

which implies

$$(A + A_d)^{\mathrm{T}}(X-U) + (X-U)^{\mathrm{T}}(A + A_d) < 0. \tag{7.28}$$

Thus, $X - U$ is nonsingular. Then by using the same line of arguments as in [185], it can be shown that matrix inequalities in (7.22)-(7.25) guarantee that there always exist nonsingular matrices S, $\tilde{S}$, W and $\tilde{W}$ such that

$$\begin{cases} E^{\mathrm{T}}\tilde{S} = S^{\mathrm{T}}E, \\ EW = \tilde{W}^{\mathrm{T}}E^{\mathrm{T}}, \\ X(X-U)^{-1} = I - \tilde{S}W, \\ (X-U)^{-1}X = I - \tilde{W}S. \end{cases} \tag{7.29}$$

Define

$$J_1 = \begin{bmatrix} (X-U)^{-1} & I \\ W & 0 \end{bmatrix}, \; J_2 = \begin{bmatrix} I & X \\ 0 & S \end{bmatrix}, \; \hat{E} = \begin{bmatrix} E & 0 \\ 0 & E \end{bmatrix}, \; H = \begin{bmatrix} I \\ 0 \end{bmatrix}^{\mathrm{T}}. \tag{7.30}$$

It is clear that J_1 and J_2 are nonsingular. Set

$$P = J_2 J_1^{-1}. \tag{7.31}$$

Then, it is also shown by the method in [185] that P is nonsingular, and

$$\hat{E}^{\mathrm{T}} P = P^{\mathrm{T}} \hat{E} \geqslant 0. \tag{7.32}$$

Now, pre- and post-multiplying (7.25) by $\mathrm{diag}\{(X-U)^{\mathrm{T}}, I, I, I, I, I, I, I\}$ and its transpose, respectively, we obtain

$$\begin{bmatrix} \Sigma_1 & \Sigma_2 & 0 & 0 & J_1^{\mathrm{T}} P^{\mathrm{T}} \bar{B}_\omega & J_1^{\mathrm{T}} \bar{A}^{\mathrm{T}} H^{\mathrm{T}} V & J_1^{\mathrm{T}} \bar{L}^{\mathrm{T}} \\ * & \Delta_6 & E^{\mathrm{T}} R_2 E & \Delta_8 & 0 & H \bar{A}_d^{\mathrm{T}} H^{\mathrm{T}} V & 0 \\ * & * & \Delta_7 & 0 & 0 & 0 & 0 \\ * & * & * & \Delta_9 & 0 & 0 & 0 \\ * & * & * & * & -\gamma^2 I & \bar{B}_\omega^{\mathrm{T}} H^{\mathrm{T}} V & 0 \\ * & * & * & * & * & -V & 0 \\ * & * & * & * & * & * & -I \end{bmatrix} < 0, \tag{7.33}$$

where

$$\begin{aligned} \Sigma_1 &= J_1^{\mathrm{T}} (P^{\mathrm{T}} \bar{A} + \bar{A}^{\mathrm{T}} P + \sum_{k=1}^{3} H^{\mathrm{T}} S_k H - d_{12} \hat{E}^{\mathrm{T}} H^{\mathrm{T}} R_1 H \hat{E}) J_1, \\ \Sigma_2 &= J_1^{\mathrm{T}} (P^{\mathrm{T}} \bar{A}_d H^{\mathrm{T}} + d_{12} \hat{E}^{\mathrm{T}} H^{\mathrm{T}} R_1 E), \end{aligned}$$

and the matrices $\bar{A}$, $\bar{A}_d$, $\bar{B}_\omega$ and $\bar{L}$ are given in (7.3) with the parameters E_f, A_f, B_f and C_f replaced by E, $\tilde{A}_f$, $\tilde{B}_f$ and $\tilde{C}_f$, respectively, where

$$\begin{aligned} \tilde{A}_f &= S^{\mathrm{T}} (\bar{A}_f - A^{\mathrm{T}} (X - U) - X^{\mathrm{T}} A - \bar{B}_f C)(X - U)^{-1} W^{-1}, \\ \tilde{B}_f &= S^{\mathrm{T}} \bar{B}_f, \\ \tilde{C}_f &= \bar{C}_f (X - U)^{-1} W^{-1}. \end{aligned}$$

Then, pre- and post-multiplying (7.37) by $\mathrm{diag}\{J_1^{\mathrm{T}}, I, I, I, I, I, I\}$ and its transpose, respectively, we obtain

$$\begin{bmatrix} \Upsilon_1 & \Upsilon_2 & 0 & 0 & P^{\mathrm{T}} \bar{B}_\omega & \bar{A}^{\mathrm{T}} H^{\mathrm{T}} V & \bar{L}^{\mathrm{T}} \\ * & \Delta_6 & E^{\mathrm{T}} R_2 E & \Delta_8 & 0 & H \bar{A}_d^{\mathrm{T}} H^{\mathrm{T}} V & 0 \\ * & * & \Delta_7 & 0 & 0 & 0 & 0 \\ * & * & * & \Delta_9 & 0 & 0 & 0 \\ * & * & * & * & -\gamma^2 I & \bar{B}_\omega^{\mathrm{T}} H^{\mathrm{T}} V & 0 \\ * & * & * & * & * & -V & 0 \\ * & * & * & * & * & * & -I \end{bmatrix} < 0, \tag{7.34}$$

where

$$\begin{aligned} \Upsilon_1 &= P^{\mathrm{T}} \bar{A} + \bar{A}^{\mathrm{T}} P + \sum_{k=1}^{3} H^{\mathrm{T}} S_k H - d_{12} \hat{E}^{\mathrm{T}} H^{\mathrm{T}} R_1 H \hat{E}, \\ \Upsilon_2 &= P^{\mathrm{T}} \bar{A}_d H^{\mathrm{T}} + d_{12} \hat{E}^{\mathrm{T}} H^{\mathrm{T}} R_1 E. \end{aligned}$$

Then, a small enough scalar $\sigma > 0$ can always be find such that

$$\begin{bmatrix} \Lambda_1 & \Lambda_2 & 0 & 0 & P^{\mathrm{T}}\bar{B}_\omega & \bar{A}^{\mathrm{T}}\bar{V} & \bar{L}^{\mathrm{T}} \\ * & \Lambda_3 & \hat{E}^{\mathrm{T}}\bar{R}_2\hat{E} & \hat{E}^{\mathrm{T}}(d_2\bar{R}_1+\bar{R}_2)\hat{E} & 0 & \bar{A}_d^{\mathrm{T}}\bar{V} & 0 \\ * & * & -\bar{S}_1-\hat{E}^{\mathrm{T}}\bar{R}_2\hat{E} & 0 & 0 & 0 & 0 \\ * & * & * & -\bar{S}_2-\hat{E}^{\mathrm{T}}(d_2\bar{R}_1+\bar{R}_2)\hat{E} & 0 & 0 & 0 \\ * & * & * & * & -\gamma^2 I & \bar{B}_\omega^{\mathrm{T}}\bar{V} & 0 \\ * & * & * & * & * & -\bar{V} & 0 \\ * & * & * & * & * & * & -I \end{bmatrix} < 0, \tag{7.35}$$

where

$$\begin{aligned} \Lambda_1 &= P^{\mathrm{T}}\bar{A}+\bar{A}^{\mathrm{T}}P+\sum_{k=1}^{3}\bar{S}_k-d_{12}\hat{E}^{\mathrm{T}}\bar{R}_1\hat{E}, \\ \Lambda_2 &= P^{\mathrm{T}}\bar{A}_d+d_{12}\hat{E}^{\mathrm{T}}\bar{R}_1\hat{E}, \\ \Lambda_3 &= -(1-\mu)\bar{S}_3-\hat{E}^{\mathrm{T}}((d_{12}+d_2)\bar{R}_1+2\bar{R}_2)\hat{E}, \\ \bar{S}_k &= \begin{bmatrix} Q_k & 0 \\ 0 & \sigma I \end{bmatrix}, \ k=1,2,3, \\ \bar{R}_l &= \begin{bmatrix} R_l & 0 \\ 0 & \sigma I \end{bmatrix}, \ l=1,2, \\ \bar{V} &= d_{12}d_2^2\bar{R}_1+d_{12}^2\bar{R}_2. \end{aligned}$$

Therefore, by Schur complement and Theorem 7.1, the following filter

$$\begin{cases} E\dot{x}_f(t) = \tilde{A}_f x_f(t)+\tilde{B}_f y(t), \\ z_f(t) = \tilde{C}_f x_f(t) \end{cases} \tag{7.36}$$

guarantees that system (7.3) is exponentially admissible with H_∞ performance γ. Next, performing an irreducible linear transformation $\hat{x}(t) = (X-U)^{-1}W^{-1}x_f(t)$ on the state in (7.36) yields

$$\begin{cases} E\dot{\hat{x}}(t) = A_f\hat{x}(t)+B_f y(t), \\ \hat{z}(t) = C_f\hat{x}(t), \end{cases} \tag{7.37}$$

where A_f, B_f and C_f are defined in (7.26). This completes the proof.

Remark 7.7. It is noted that Theorem 7.6 provides a delay-dependent sufficient condition for designing the H_∞ filter such that the resulting filtering error system is exponentially admissible with H_∞ performance γ, and the desired H_∞ filter can be constructed by solving the corresponding matrix inequalities. It can be found that the results proposed in [178, Theorem 9.8] and [199, 201] are delay-independent, and thus the proposed results are generally conservative, especially when the size of the delay is small. It is noted that [38] has provided a delay-dependent condition on designing the desired H_∞ filter.

However, system decomposition and transformation are absolutely necessary in [38], which is in contrast with Theorem 7.6 in this chapter, where the conditions are expressed by using the coefficient matrices of the original singular system, and some computational problems arising from the decomposition and transformation of SSs can be avoided. Moreover, the considered time delays are time-invariant in the above mentioned results, which limits the scope of applications of the proposed results. Thus, the results in this chapter is much more general and elegant than [178, Theorem 9.8] and [38, 199, 201].

7.4 Numerical Examples

This section presents several numerical examples to demonstrate the effectiveness of the methods.

Example 7.8. Consider system (7.1) with the following parameters:

$$E = \begin{bmatrix} 9 & 3 \\ 6 & 2 \end{bmatrix}, A = \begin{bmatrix} -13.1 & -13.7 \\ -15.4 & -23.8 \end{bmatrix},$$
$$A_d = \begin{bmatrix} -18.6 & -10.4 \\ -25.2 & -16.8 \end{bmatrix}, B_\omega = \begin{bmatrix} 1.9 \\ 1.8 \end{bmatrix}, L = \begin{bmatrix} 0.4 \\ -0.8 \end{bmatrix}^{\mathrm{T}}.$$

In this example, we suppose the variation rate $\mu = 0.2$ and the lower bound $d_1 \to 0$. The results on the maximum allowable upper bound d_2 for different $\gamma > 0$ by Theorem 7.1 and [200] are listed in Table 7.1, which implies our obtained BRL has less conservativeness than the one in [200].

Table 7.1. Example 7.8: Comparison of delay-dependent BRLs

γ	1	2	3	4
[200]	1.8863	2.3388	2.5358	2.6459
Theorem 7.1	1.8899	2.3430	2.5403	2.6506

Example 7.9. Consider the following system [210]:

$$\begin{cases} \dot{x}(t) = Ax(t) + A_d x(t - d(t)) + B_\omega \omega(t), \\ y(t) = Cx(t) + \omega(t), \\ z(t) = Lx(t), \end{cases} \tag{7.38}$$

where

$$A = \begin{bmatrix} -2 & 0 \\ 0 & -0.9 \end{bmatrix}, A_d = \begin{bmatrix} -1 & 0 \\ -1 & -1 \end{bmatrix},$$
$$B_\omega = \begin{bmatrix} 0 \\ 1 \end{bmatrix}, C = \begin{bmatrix} 1 & 0 \end{bmatrix}, L = \begin{bmatrix} 1 & 2 \end{bmatrix}.$$

From [210], we can get that the result of [210] has less conservatism than that of [39]. Thus, we only compare the result proposed in this chapter with [210] and assume that the lower bound $d_1 \to 0$. First, the upper bound d_2 is assumed to be one. In order to compare Theorem 7.6 in this chapter with the one in [210], we calculated the minimum H_∞ performance γ achieved by the filtering error system (7.3). For different variation rate μ, the obtained results by [210] and Theorem 7.6 in this chapter are listed in Table 7.2. Next, the variation rate μ is assumed to be 0.3, and for different d_2, the obtained results by [210] and Theorem 7.6 in this chapter can be found in Table 7.3. It is clear that Theorem 7.6 gives better results than those in [210].

Table 7.2. Example 7.9: The achieved H_∞ performance γ for different μ

μ	0	0.2	0.4	0.6	0.8
[210]	1.3522	1.4866	1.6837	1.9957	2.6813
Theorem 7.6	1.3522	1.4826	1.6559	1.8616	1.9914

Table 7.3. Example 7.9: The achieved H_∞ performance γ for different d_2

d_2	1.1	1.3	1.5	1.7	1.9
[210]	1.8480	2.5223	3.4457	4.7820	6.8784
Theorem 7.6	1.8335	2.4992	3.4083	4.7189	6.7632

Example 7.10. Consider system (7.1) with parameters as follows:

$$
\begin{aligned}
E &= \begin{bmatrix} 1 & 2 \\ 2 & 4 \end{bmatrix}, \; A = \begin{bmatrix} -1.1 & 1.4 \\ -2.4 & -3 \end{bmatrix}, \; A_d = \begin{bmatrix} 0.5 & -0.5 \\ -0.3 & -0.3 \end{bmatrix}, \\
B_\omega &= \begin{bmatrix} -1.2 \\ -2 \end{bmatrix}, \; C = \begin{bmatrix} -0.5 & 1.4 \end{bmatrix}, \; C_d = \begin{bmatrix} -0.91 & 0.1 \end{bmatrix}, \\
D_\omega &= 2, \; L = \begin{bmatrix} -1 & -1.1 \end{bmatrix}.
\end{aligned}
$$

In this example, we suppose the variation rate $\mu = 0.6$, the lower bound $d_1 = 0.1$, and the upper bound $d_2 = 0.95$, respectively. Using the Matlab LMI Control Toolbox to solve (7.22)-(7.25), the minimum H_∞ performance level γ obtained by Theorem 7.6 in this chapter is 0.5724. However, the conditions in [38, 178, 199, 201] all fail to conclude whether or not there exist filters for the above mentioned system. Especially, when $\gamma = 0.9$, the desired filter by Theorem 7.6 can be computed as

$$
\begin{cases}
\begin{bmatrix} 1 & 2 \\ 2 & 4 \end{bmatrix} \dot{\hat{x}}(t) = \begin{bmatrix} -9.1956 & -13.7952 \\ -14.0527 & -27.7061 \end{bmatrix} \hat{x}(t) + \begin{bmatrix} -0.7483 \\ 0.0060 \end{bmatrix} y(t), \\
\hat{z}(t) = \begin{bmatrix} -1.0000 & -1.1006 \end{bmatrix} \hat{x}(t).
\end{cases}
$$

7.5 Conclusion

In this chapter, the problem of H_∞ filtering for SSs with time-varying delays has been studied. The delay-range-dependent method for the design of full-order H_∞ filter has been developed by the LMI approach. The designed filter guarantees the exponential admissibility with a prescribed H_∞ performance level of the filtering error systems. The derived results are less conservative than the existing ones, which has been demonstrated by some illustrative examples.

8

Filter Design for Networked Discrete-Time SSs with Time-Varying Delays

8.1 Introduction

In recent years, increasing research interest has been paid to the study of networked control systems (NCSs) due to the fact that compared with the traditional point-to-point communication, NCSs have several advantages such as low cost, reduced weight and power requirements, simple installation and maintenance, and high reliability[209]. It is well known that in NCSs, the network-induced time delays (communication delays) and missing measurements (packet dropouts) are always inevitable that would degrade the system performance or even cause instability. It should be pointed out that in most results of the relevant literature, the network-induced time delays are assumed to be constant. Although the ideal assumption can simplify the analysis and synthesis of NCSs, it often cannot be satisfied because delays resulting from network transmissions are typically time-varying and random. Thus, more and more attention has been paid to model the network-induced time delays in various time-varying and probabilistic ways [24, 25, 43, 55, 56, 78, 83, 89, 99, 124, 192]. It also should be pointed out that up to now, most proposed results concerning missing measurements are based on an assumption that the measurement signal is either completely missing or completely available, and all the sensors have the same data missing probability [46, 123, 134, 212]. Such an assumption, however, inevitably limits the scope of applications of the given results, since it cannot be applied to some practical cases where multiple missing measurements occur, for example, the case when only partial information is missing and the case when the individual sensor has different missing probability. Very recently, a new way to describe the missing measurements has been proposed in [25, 26, 138], where the missing probability for each sensor is governed by an individual random variable satisfying a certain probabilistic distribution over the interval [0 1]. It is clear that the description of missing measurements in [25, 26, 138] is much more general and desirable than that in [46, 123, 134, 212].

Z.-G. Wu et al.: *Anal. & Synth. of Singular Syst. with Time-Delays*, LNCIS 443, pp. 103–114.
DOI: 10.1007/978-3-642-37497-5_8

In this chapter, the problem of H_∞ filtering is investigated for networked discrete-time SSs by using LMI method. The effects of multiple probabilistic time-varying communication delays and multiple missing measurements on the performance of the filtering error system are considered. The derived criteria for performance analysis of the filtering error systems and filter design are formulated by LMI. A numerical example shows the effectiveness of the design method.

8.2 Problem Formulation

Consider the following networked discrete-time SS with multiple stochastic communication delays:

$$\begin{cases} Ex(k+1) = Ax(k) + A_d \sum_{i=1}^{m} \alpha_i(k)x(k-d_i(k)) \\ \qquad + A_\tau \sum_{j=1}^{s} \beta_j(k) \sum_{v=1}^{\tau_j(k)} x(k-v) + B_\omega \omega(k), \\ z(k) = Lx(k), \end{cases} \tag{8.1}$$

where $x(k) \in \mathbb{R}^n$ is the state, $z(k) \in \mathbb{R}^q$ is the signal to be estimated, and $\omega(k) \in \mathbb{R}^p$ is the disturbance input that belongs to $l_2[0,\infty)$. E, A, A_d, A_τ, B_ω and L are known real constant matrices with appropriate dimensions. The matrix E may be singular and it is assumed that $\operatorname{rank} E = r \leqslant n$. $d_i(k)$ $(i = 1, 2, \cdots, m)$ denote the discrete time-varying delays satisfying

$$d_1^i \leqslant d_i(k) \leqslant d_2^i, \tag{8.2}$$

where d_1^i and d_2^i are constant positive integers representing the lower and upper bounds on the discrete delays, respectively. $\tau_j(k)$ $(j = 1, 2, \cdots, s)$ denote the distributed time-varying delays satisfying

$$\tau_1^j \leqslant \tau_j(k) \leqslant \tau_2^j, \tag{8.3}$$

where τ_1^j and τ_2^j are constant positive integers representing the lower and upper bounds on the distributed delays, respectively.

Remark 8.1. The communication delays considered in (8.1) are first introduced in [24, 25]. This type of delay widely exists in NCSs. However, SSs with such type of communication delays have not been fully considered.

The random variables $\alpha_i(k)$ $(i = 1, 2, \cdots, m)$ and $\beta_j(k)$ $(j = 1, 2, \cdots, s)$ in (8.1) are mutually uncorrelated Bernoulli distributed white sequences obeying the following probability distribution law[24]:

$$\begin{cases} \Pr\{\alpha_i(k) = 1\} = \mathscr{E}\{\alpha_i(k)\} = \bar{\alpha}_i, \Pr\{\alpha_i(k) = 0\} = 1 - \bar{\alpha}_i, \\ \Pr\{\beta_j(k) = 1\} = \mathscr{E}\{\beta_j(k)\} = \bar{\beta}_j, \Pr\{\beta_j(k) = 0\} = 1 - \bar{\beta}_j. \end{cases} \tag{8.4}$$

Definition 8.2

1. System (8.1) is said to be asymptotically stable in the mean square if, in the case of $\omega(k) = 0$, for each solution $x(k)$ of system (8.1), the following holds

$$\lim_{k\to\infty} \mathscr{E}\left\{||x(k)||^2\right\} = 0, \tag{8.5}$$

2. System (8.1) is said to be asymptotically admissible in the mean square, if it is regular, causal and asymptotically stable in the mean square.

In this chapter, the packet dropouts (missing measurement) are described by [25, 26, 138]

$$y(k) = \Omega C x(k) + D_\omega \omega(k) = \sum_{\iota=1}^{\rho} \chi_\iota C_\iota x(k) + D_\omega \omega(k), \tag{8.6}$$

where $y(k) \in \mathbb{R}^\rho$ is the measurement signal of (8.1),

$$\Omega = \mathrm{diag}\{\chi_1, \chi_2, \cdots, \chi_\rho\},$$

with χ_ι being ρ unrelated random variables which are also unrelated with $\alpha_i(k)$ and $\beta_j(k)$. It is assumed that χ_ι has the probabilistic density function $q_\iota(s)$ ($\iota = 1, 2, \cdots, \rho$) on the interval [0 1] with mathematical expectation μ_ι and variance σ_ι^2. C_ι is defined by [138]

$$C_\iota = \mathrm{diag}\{\underbrace{0, \cdots, 0}_{\iota-1}, 1, \underbrace{0, \cdots, 0}_{\rho-\iota}\}C. \tag{8.7}$$

Note that χ_ι could satisfy any discrete probabilistic distributions on the interval [0 1], which includes the widely used Bernoulli distribution as a special case. In the sequel, we denote $\bar{\Omega} = \mathscr{E}\{\Omega\}$.

In this chapter, we aim at obtaining the estimate of $z(k)$ from the measured output $y(k)$. To be more specific, we are interested in constructing the following full-order filter:

$$\begin{cases} \hat{x}(k+1) = A_f \hat{x}(k) + B_f y(k), \\ \hat{z}(k) = C_f \hat{x}(k), \end{cases} \tag{8.8}$$

where $\hat{x} \in \mathbb{R}^n$, $\hat{z}(t) \in \mathbb{R}^q$, and the matrices A_f, B_f and C_f are the filter matrices with appropriate dimensions, which are to be designed.

Defining

$$\begin{aligned} \bar{z}(k) &= z(k) - \hat{z}(k), \\ \bar{x}(k) &= \left[x(k)^{\mathrm{T}}\ \hat{x}(k)^{\mathrm{T}}\right]^{\mathrm{T}}, \end{aligned}$$

and combining (8.1), (8.6) and (8.8), we obtain the filtering error dynamics as follows:

$$\begin{cases} \bar{E}\bar{x}(k+1) = (\bar{A}+\hat{A})\bar{x}(k) + \sum_{i=1}^{m}(\bar{\alpha}_i\bar{A}_d + \hat{\alpha}_i(k)\bar{A}_d)x(k-d_i(k)) \\ \qquad + \sum_{j=1}^{s}(\bar{\beta}_j\bar{A}_\tau + \hat{\beta}_j(k)\bar{A}_\tau)\sum_{v=1}^{\tau_j(k)} x(k-v) + \bar{B}_\omega\omega(k), \\ \bar{z}(k) = \bar{L}\bar{x}(k), \end{cases} \tag{8.9}$$

where

$$\bar{E} = \begin{bmatrix} E & 0 \\ 0 & I \end{bmatrix}, \ \bar{A} = \begin{bmatrix} A & 0 \\ B_f\bar{\Omega}C & A_f \end{bmatrix},$$
$$\hat{A} = \begin{bmatrix} 0 & 0 \\ B_f(\Omega-\bar{\Omega})C & 0 \end{bmatrix}, \ \bar{A}_d = \begin{bmatrix} A_d \\ 0 \end{bmatrix},$$
$$\bar{A}_\tau = \begin{bmatrix} A_\tau \\ 0 \end{bmatrix}, \ \bar{B}_\omega = \begin{bmatrix} B_\omega \\ B_fD_\omega \end{bmatrix}, \ \bar{L} = \begin{bmatrix} L & -C_f \end{bmatrix},$$

where $\hat{\alpha}_i(k) = \alpha_i(k) - \bar{\alpha}_i$ and $\hat{\beta}_j(k) = \beta_j(k) - \bar{\beta}_j$. It is clear that $\mathscr{E}\{\hat{\alpha}_i(k)\} = 0$, $\mathscr{E}\{\hat{\beta}_j(k)\} = 0$, $\mathscr{E}\{\hat{\alpha}_i(k)\hat{\beta}_j(k)\} = 0$, and

$$\mathscr{E}\{\hat{\alpha}_i(k)\hat{\alpha}_{\hat{i}}(k)\} = \begin{cases} 0, & \hat{i} \neq i, \\ \bar{\alpha}_i(1-\bar{\alpha}_i) \triangleq \underline{\alpha}_i, & \hat{i} = i, \end{cases}$$

$$\mathscr{E}\{\hat{\beta}_j(k)\hat{\beta}_{\hat{j}}(k)\} = \begin{cases} 0, & \hat{j} \neq j, \\ \bar{\beta}_j(1-\bar{\beta}_j) \triangleq \underline{\beta}_j, & \hat{j} = j. \end{cases}$$

We are now in a position to formulate H_∞ filtering problem for networked discrete-time SSs with multiple stochastic communication delays and packet dropouts. More specifically, given a disturbance attenuation level γ, we will design the filter (8.8) for system (8.1) such that the filtering error system (8.9) with $\omega(k) = 0$ is asymptotically admissible in the mean square, and under zero initial condition,

$$\sum_{k=0}^{\infty}\mathscr{E}\{\|z(k)\|^2\} \leqslant \gamma^2\sum_{k=0}^{\infty}\|\omega(k)\|^2, \tag{8.10}$$

for all nonzero $\omega(k) \in l_2[0,\infty)$. In this case, the filtering error system (8.9) is said to be asymptotically admissible in the mean square with H_∞ performance γ.

8.3 Main Results

In this section, an LMI approach is developed to solve the problem of H_∞ filter design for networked discrete-time SSs. We first give the condition under which the asymptotical stability in the mean square and H_∞ performance of the filtering error system (8.9) are guaranteed. For presentation convenience, in the following, we denote

$$\begin{aligned}
\theta_i &= d_2^i - d_1^i + 1 \\
\vartheta_j &= \frac{\tau_2^j(\tau_2^j + \tau_1^j)(\tau_2^j - \tau_1^j + 1)}{2}, \\
\varphi_1(k) &= \left[x(k-d_1(k))^{\mathrm{T}}\ x(k-d_2(k))^{\mathrm{T}} \cdots x(k-d_m(k))^{\mathrm{T}}\right]^{\mathrm{T}}, \\
\varphi_2(k) &= \left[\sum_{v=1}^{\tau_1(k)} x(k-v)^{\mathrm{T}} \sum_{v=1}^{\tau_2(k)} x(k-v)^{\mathrm{T}} \cdots \sum_{v=1}^{\tau_s(k)} x(k-v)^{\mathrm{T}}\right]^{\mathrm{T}},
\end{aligned}$$

and $R \in \mathbb{R}^{2n\times(n-r)}$ is any matrix with full column and satisfies $\bar{E}^{\mathrm{T}}R = 0$.

Theorem 8.3. *For a given scalar $\gamma > 0$, system (8.9) is asymptotically admissible in the mean square with H_∞ performance γ, if there exist matrices $P > 0$, $Q_i > 0$ $(i = 1, 2, \cdots, m)$, $Z_j > 0$ $(j = 1, 2, \cdots, s)$ and S such that*

$$\Xi = \begin{bmatrix} \Xi_{11} & SR^{\mathrm{T}}\Xi_{26}^{\mathrm{T}} & SR^{\mathrm{T}}\Xi_{36}^{\mathrm{T}} & SR^{\mathrm{T}}\bar{B}_\omega & \bar{L}^{\mathrm{T}} & \bar{A}^{\mathrm{T}}P \\ * & \Xi_{22} & 0 & 0 & 0 & \Xi_{26}P \\ * & * & \Xi_{33} & 0 & 0 & \Xi_{36}P \\ * & * & * & -\gamma^2 I & 0 & \bar{B}_\omega^{\mathrm{T}}P \\ * & * & * & * & -I & 0 \\ * & * & * & * & * & -P \end{bmatrix} < 0, \tag{8.11}$$

where

$$\begin{aligned}
\Xi_{11} = &- \bar{E}^{\mathrm{T}}P\bar{E} + SR^{\mathrm{T}}\bar{A} + \bar{A}^{\mathrm{T}}RS^{\mathrm{T}} + \sum_{i=1}^{m}\theta_i H^{\mathrm{T}}Q_i H \\
&+ \sum_{j=1}^{s}\vartheta_j H^{\mathrm{T}}Z_j H + \sum_{\iota=1}^{\rho}\sigma_\iota^2\bar{C}_\iota^{\mathrm{T}}P\bar{C}_\iota, \\
\Xi_{22} = &\ \mathrm{diag}\{-Q_1 + \underline{\alpha}_1\bar{A}_d^{\mathrm{T}}P\bar{A}_d, -Q_2 + \underline{\alpha}_2\bar{A}_d^{\mathrm{T}}P\bar{A}_d, \cdots, -Q_m + \underline{\alpha}_m\bar{A}_d^{\mathrm{T}}P\bar{A}_d\}, \\
\Xi_{33} = &\ \mathrm{diag}\{-Z_1 + \underline{\beta}_1\bar{A}_\tau^{\mathrm{T}}P\bar{A}_\tau, -Z_2 + \underline{\beta}_2\bar{A}_\tau^{\mathrm{T}}P\bar{A}_\tau, \cdots, -Z_s + \underline{\beta}_s\bar{A}_\tau^{\mathrm{T}}P\bar{A}_\tau\}, \\
\Xi_{26} = &\ \left[\bar{\alpha}_1\bar{A}_d\ \bar{\alpha}_2\bar{A}_d \cdots \bar{\alpha}_m\bar{A}_d\right]^{\mathrm{T}}, \\
\Xi_{36} = &\ \left[\bar{\beta}_1\bar{A}_\tau\ \bar{\beta}_2\bar{A}_\tau \cdots \bar{\beta}_s\bar{A}_\tau\right]^{\mathrm{T}}, \\
\bar{C}_\iota = &\ \begin{bmatrix} 0 & 0 \\ B_f C_\iota & 0 \end{bmatrix},\ H = \begin{bmatrix} I & 0 \end{bmatrix}.
\end{aligned}$$

Proof. Under the given condition, we first show that system (8.9) with $\omega(k) = 0$ is regular and causal. Since $\mathrm{rank}\,\bar{E} = n + r$, we choose two nonsingular matrices M and G such that

$$M\bar{E}G = \begin{bmatrix} I_{n+r} & 0 \\ 0 & 0 \end{bmatrix}. \tag{8.12}$$

Set

$$M\bar{A}G = \begin{bmatrix} A_1 & A_2 \\ A_3 & A_4 \end{bmatrix}, \; G^{\mathrm{T}}S = \begin{bmatrix} S_{11} \\ S_{12} \end{bmatrix}, \; M^{-\mathrm{T}}R = \begin{bmatrix} 0 \\ I \end{bmatrix} F, \tag{8.13}$$

where $F \in \mathbb{R}^{(n-r)\times(n-r)}$ is any nonsingular matrix. It can be seen that $\Xi_{11} < 0$ implies

$$-\bar{E}^{\mathrm{T}}P\bar{E} + SR^{\mathrm{T}}\bar{A} + \bar{A}^{\mathrm{T}}RS^{\mathrm{T}} < 0. \tag{8.14}$$

Pre-multiplying and post-multiplying (8.14) by G^{T} and G, respectively, we have $S_{12}F^{\mathrm{T}}A_4 + A_4^{\mathrm{T}}FS_{12}^{\mathrm{T}} < 0$, which implies A_4 is nonsingular. Thus, the pair $(\bar{E}, \bar{A})$ is regular and causal. It can be found that

$$\begin{aligned} \det(z\bar{E} - (\bar{A} + \hat{A})) &= \det\left(\begin{bmatrix} zE - A & 0 \\ -B_f\Omega C & zI - A_f \end{bmatrix}\right) \\ &= \det(zE - A)\det(zI - A_f) \\ &= \det(z\bar{E} - \bar{A}), \end{aligned} \tag{8.15}$$

thus, $\det(z\bar{E} - (\bar{A} + \hat{A}))$ is not identically zero and $\deg(\det(z\bar{E} - (\bar{A} + \hat{A}))) = n + r = \operatorname{rank}\bar{E}$. According to Definition 2.2, system (8.9) with $\omega(k) = 0$ is regular and causal.

We now show that system (8.9) with $\omega(k) = 0$ is asymptotically stable in the mean square. To the end, we choose the following Lyapunov functional:

$$V(k) = V_1(k) + V_2(k) + V_3(k), \tag{8.16}$$

where

$$\begin{aligned} V_1(k) &= \bar{x}(k)^{\mathrm{T}}\bar{E}^{\mathrm{T}}P\bar{E}\bar{x}(k), \\ V_2(k) &= \sum_{i=1}^{m}\sum_{\beta=-d_2^i+1}^{-d_1^i+1}\sum_{\alpha=k-1+\beta}^{k-1} x(\alpha)^{\mathrm{T}}Q_i x(\alpha), \\ V_3(k) &= \sum_{j=1}^{s}\tau_2^j\sum_{\beta=\tau_1^j}^{\tau_2^j}\sum_{v=1}^{\beta}\sum_{\alpha=k-v}^{k-1} x(\alpha)^{\mathrm{T}}Z_j x(\alpha). \end{aligned}$$

Define $\mathscr{E}\{\Delta V(k)\} = \mathscr{E}\{V(k+1) - V(k)\}$. Then, along the solution of system (8.9) with $\omega(k) = 0$, we have

$$\begin{aligned} &\mathscr{E}\{\Delta V_1(k)\} \\ &= \mathscr{E}\left\{\bar{x}(k+1)^{\mathrm{T}}\bar{E}^{\mathrm{T}}P\bar{E}\bar{x}(k+1) - \bar{x}(k)^{\mathrm{T}}\bar{E}^{\mathrm{T}}P\bar{E}\bar{x}(k)\right\} \\ &= \mathscr{E}\left\{\bar{x}(k)^{\mathrm{T}}\bar{A}^{\mathrm{T}}P\bar{A}\bar{x}(k) + 2\bar{x}(k)^{\mathrm{T}}\bar{A}^{\mathrm{T}}P\Xi_{26}^{\mathrm{T}}\varphi_1(k) + 2\bar{x}(k)^{\mathrm{T}}\bar{A}^{\mathrm{T}}P\Xi_{36}^{\mathrm{T}}\varphi_2(k)\right. \\ &\quad + \varphi_1(k)^{\mathrm{T}}\Xi_{26}P\Xi_{26}^{\mathrm{T}}\varphi_1(k) + 2\varphi_1(k)^{\mathrm{T}}\Xi_{26}P\Xi_{36}^{\mathrm{T}}\varphi_2(k) + \varphi_2(k)^{\mathrm{T}}\Xi_{36}P\Xi_{36}^{\mathrm{T}}\varphi_2(k) \\ &\quad + \bar{x}(k)^{\mathrm{T}}\hat{A}^{\mathrm{T}}P\hat{A}\bar{x}(k) + \sum_{i=1}^{m}\hat{\alpha}_i(k)^2 x(k-d_i(k))^{\mathrm{T}}\bar{A}_d^{\mathrm{T}}P\bar{A}_d x(k-d_i(k)) \end{aligned}$$

$$+\sum_{j=1}^{s}\hat{\beta}_j(k)^2\sum_{v=1}^{\tau_j(k)}x(k-v)^{\mathrm{T}}\bar{A}_\tau^{\mathrm{T}}P\bar{A}_\tau\sum_{v=1}^{\tau_j(k)}x(k-v)-\bar{x}(k)^{\mathrm{T}}\bar{E}^{\mathrm{T}}P\bar{E}\bar{x}(k)\Bigg\}, \tag{8.17}$$

$$\begin{aligned}\mathscr{E}\{\Delta V_2(k)\} &= \sum_{i=1}^{m}\left(\theta_i x(k)^{\mathrm{T}}Q_i x(k)-\sum_{\alpha=k-d_2^i}^{k-d_1^i}x(\alpha)^{\mathrm{T}}Q_i x(\alpha)\right)\\ &\leqslant \sum_{i=1}^{m}\theta_i\bar{x}(k)^{\mathrm{T}}H^{\mathrm{T}}Q_iH\bar{x}(k)-\sum_{i=1}^{m}x(k-d_i(k))^{\mathrm{T}}Q_ix(k-d_i(k)),\end{aligned} \tag{8.18}$$

$$\begin{aligned}\mathscr{E}\{\Delta V_3(k)\} &= \sum_{j=1}^{s}\left(\vartheta_j x(k)^{\mathrm{T}}Z_j x(k)-\tau_2^j\sum_{\alpha=\tau_1^j}^{\tau_2^j}\sum_{v=1}^{\alpha}x(k-v)^{\mathrm{T}}Z_jx(k-v)\right)\\ &\leqslant \sum_{j=1}^{s}\vartheta_j\bar{x}(k)^{\mathrm{T}}H^{\mathrm{T}}Z_jH\bar{x}(k)-\sum_{j=1}^{s}\tau_2^j\sum_{v=1}^{\tau_j(k)}x(k-v)^{\mathrm{T}}Z_jx(k-v),\end{aligned} \tag{8.19}$$

On the other hand, it can be calculated that

$$\mathscr{E}\left\{\bar{x}(k)^{\mathrm{T}}\hat{A}^{\mathrm{T}}P\hat{A}\bar{x}(k)\right\}=\sum_{\iota=1}^{\rho}\sigma_\iota^2\bar{C}_\iota^{\mathrm{T}}P\bar{C}_\iota, \tag{8.20}$$

$$\begin{aligned}&\mathscr{E}\left\{\sum_{i=1}^{m}\hat{\alpha}_i(k)^2x(k-d_i(k))^{\mathrm{T}}\bar{A}_d^{\mathrm{T}}P\bar{A}_dx(k-d_i(k))\right\}\\ &=\sum_{i=1}^{m}\underline{\alpha}_ix(k-d_i(k))^{\mathrm{T}}\bar{A}_d^{\mathrm{T}}P\bar{A}_dx(k-d_i(k)),\end{aligned} \tag{8.21}$$

$$\begin{aligned}&\mathscr{E}\left\{\sum_{j=1}^{s}\hat{\beta}_j(k)^2\sum_{v=1}^{\tau_j(k)}x(k-v)^{\mathrm{T}}\bar{A}_\tau^{\mathrm{T}}P\bar{A}_\tau\sum_{v=1}^{\tau_j(k)}x(k-v)\right\}\\ &=\sum_{j=1}^{s}\underline{\beta}_j\sum_{v=1}^{\tau_j(k)}x(k-v)^{\mathrm{T}}\bar{A}_\tau^{\mathrm{T}}P\bar{A}_\tau\sum_{v=1}^{\tau_j(k)}x(k-v),\end{aligned} \tag{8.22}$$

and based on discretized Jensen inequality, we have that

$$-\tau_2^j\sum_{v=1}^{\tau_j(k)}x(k-v)^{\mathrm{T}}Z_jx(k-v)\leqslant-\sum_{v=1}^{\tau_j(k)}x(k-v)^{\mathrm{T}}Z_j\sum_{v=1}^{\tau_j(k)}x(k-v). \tag{8.23}$$

Furthermore, let $\kappa(k) = 2SR^{\mathrm{T}}\bar{E}\bar{x}(k+1)$. It is clear that $\kappa(k) = 0$. Thus, we can get that

$$\mathscr{E}\{\kappa(k)\} = 2SR^{\mathrm{T}}\bar{A}\bar{x}(k) + 2SR^{\mathrm{T}}\Xi_{26}^{\mathrm{T}}\varphi_1(k) + 2SR^{\mathrm{T}}\Xi_{36}^{\mathrm{T}}\varphi_2(k) = 0. \tag{8.24}$$

Using (8.17)-(8.24), we have that

$$\begin{aligned}\mathscr{E}\{\Delta V(k)\} &= \mathscr{E}\left\{\sum_{i=1}^{4}\Delta V_i(k) + \kappa(k)\right\} \\ &\leqslant \begin{bmatrix}\bar{x}(k)\\ \varphi_1(k)\\ \varphi_2(k)\end{bmatrix}^{\mathrm{T}} \Gamma_1 \begin{bmatrix}\bar{x}(k)\\ \varphi_1(k)\\ \varphi_2(k)\end{bmatrix},\end{aligned} \tag{8.25}$$

where

$$\Gamma_1 = \begin{bmatrix}\Xi_{11} & SR^{\mathrm{T}}\Xi_{26}^{\mathrm{T}} & SR^{\mathrm{T}}\Xi_{36}^{\mathrm{T}}\\ * & \Xi_{22} & 0\\ * & * & \Xi_{33}\end{bmatrix} + \begin{bmatrix}\bar{A}^{\mathrm{T}}\\ \Xi_{26}\\ \Xi_{36}\end{bmatrix} P \begin{bmatrix}\bar{A}^{\mathrm{T}}\\ \Xi_{26}\\ \Xi_{36}\end{bmatrix}^{\mathrm{T}}.$$

Applying Schur complement to (8.11), we can always find a scalar $\varsigma > 0$ such that $\Gamma_1 < -\mathrm{diag}\{\varsigma I, 0, 0\}$, and subsequently

$$\mathscr{E}\{\Delta V(k)\} \leqslant -\varsigma||\bar{x}(k)||^2. \tag{8.26}$$

Then, by using the similar analysis method employed in [90], we can confirm that system (8.9) is asymptotically stable in the mean square.

Next, we prove that system (8.9) has H_∞ performance γ. To this end, define

$$J_{\bar{z}\omega} = \mathscr{E}\left\{\sum_{k=0}^{\infty}\left[\bar{z}(k)^{\mathrm{T}}\bar{z}(k) - \gamma^2\omega(k)^{\mathrm{T}}\omega(k)\right]\right\}. \tag{8.27}$$

It can be found that under zero initial condition

$$\begin{aligned}J_{\bar{z}\omega} &\leqslant \mathscr{E}\left\{\sum_{k=0}^{\infty}\left[\Delta V(k) + \bar{z}(k)^{\mathrm{T}}\bar{z}(k) - \gamma^2\omega(k)^{\mathrm{T}}\omega(k)\right]\right\} \\ &\leqslant \begin{bmatrix}\bar{x}(k)\\ \varphi_1(k)\\ \varphi_2(k)\\ \omega(k)\end{bmatrix}^{\mathrm{T}} \Gamma_2 \begin{bmatrix}\bar{x}(k)\\ \varphi_1(k)\\ \varphi_2(k)\\ \omega(k)\end{bmatrix},\end{aligned} \tag{8.28}$$

where

$$\Gamma_2 = \begin{bmatrix}\Xi_{11} + \bar{L}^{\mathrm{T}}\bar{L} & SR^{\mathrm{T}}\Xi_{26}^{\mathrm{T}} & SR^{\mathrm{T}}\Xi_{36}^{\mathrm{T}} & SR^{\mathrm{T}}\bar{B}_\omega\\ * & \Xi_{22} & 0 & 0\\ * & * & \Xi_{33} & 0\\ * & * & * & -\gamma^2 I\end{bmatrix} + \begin{bmatrix}\bar{A}^{\mathrm{T}}\\ \Xi_{26}\\ \Xi_{36}\\ \bar{B}_\omega^{\mathrm{T}}\end{bmatrix} P \begin{bmatrix}\bar{A}^{\mathrm{T}}\\ \Xi_{26}\\ \Xi_{36}\\ \bar{B}_\omega^{\mathrm{T}}\end{bmatrix}^{\mathrm{T}}.$$

Applying Schur complement to (8.11) leads to that for all nonzero $\omega(k) \in l_2[0,\infty)$, $J_{\bar{z}\omega} < 0$, which implies (8.10) holds. This completes the proof.

With the help of Theorem 8.3, we are now to deal with the H_∞ filter design problem for networked discrete-time SSs with multiple stochastic communication delays and missing measurements. Similar to the idea in [45], by introducing the slack variable G, the equivalent form of (8.11) can be given in the following,

$$\begin{bmatrix} \Xi_{11} & SR^{\mathrm{T}}\Xi_{26}^{\mathrm{T}} & SR^{\mathrm{T}}\Xi_{36}^{\mathrm{T}} & SR^{\mathrm{T}}\bar{B}_\omega & \bar{L}^{\mathrm{T}} & \bar{A}^{\mathrm{T}}G^{\mathrm{T}} \\ * & \Xi_{22} & 0 & 0 & 0 & \Xi_{26}G^{\mathrm{T}} \\ * & * & \Xi_{33} & 0 & 0 & \Xi_{36}G^{\mathrm{T}} \\ * & * & * & -\gamma^2 I & 0 & \bar{B}_\omega^{\mathrm{T}}G^{\mathrm{T}} \\ * & * & * & * & -I & 0 \\ * & * & * & * & * & P-G-G^{\mathrm{T}} \end{bmatrix} < 0. \tag{8.29}$$

Choose

$$P = \begin{bmatrix} P_1 & P_2 \\ * & P_3 \end{bmatrix}, G = \begin{bmatrix} F_1 & V \\ F_2 & V \end{bmatrix}, S = \begin{bmatrix} S_1 \\ S_2 \end{bmatrix}, R = \begin{bmatrix} \bar{R} \\ 0 \end{bmatrix}. \tag{8.30}$$

where $\bar{R} \in \mathbb{R}^{n\times(n-r)}$ is any matrix with full column and satisfies $E^{\mathrm{T}}\bar{R} = 0$. Furthermore, letting

$$\hat{A}_f = VA_f, \hat{B}_f = VB_f, \hat{C}_f = C_f, \tag{8.31}$$

and using Schur complement, we can get from (8.29) that

$$\begin{bmatrix} \psi_{11} & \psi_{12} & S_1\bar{R}^{\mathrm{T}}\varphi_{37}^{\mathrm{T}} & S_1\bar{R}^{\mathrm{T}}\varphi_{47}^{\mathrm{T}} & S_1\bar{R}^{\mathrm{T}}B_\omega & L^{\mathrm{T}} & \psi_{17} & \psi_{18} & \psi_{19} \\ * & -P_3 & S_2\bar{R}^{\mathrm{T}}\varphi_{37}^{\mathrm{T}} & S_2\bar{R}^{\mathrm{T}}\varphi_{47}^{\mathrm{T}} & S_2\bar{R}^{\mathrm{T}}B_\omega & -\hat{C}_f^{\mathrm{T}} & \hat{A}_f^{\mathrm{T}} & \hat{A}_f^{\mathrm{T}} & 0 \\ * & * & \psi_{33} & 0 & 0 & 0 & \varphi_{37}F_1^{\mathrm{T}} & \varphi_{37}F_2^{\mathrm{T}} & 0 \\ * & * & * & \psi_{44} & 0 & 0 & \varphi_{47}F_1^{\mathrm{T}} & \varphi_{47}F_2^{\mathrm{T}} & 0 \\ * & * & * & * & -\gamma^2 I & 0 & \psi_{57} & \psi_{58} & 0 \\ * & * & * & * & * & -I & 0 & 0 & 0 \\ * & * & * & * & * & * & \psi_{77} & \psi_{78} & 0 \\ * & * & * & * & * & * & * & \psi_{88} & 0 \\ * & * & * & * & * & * & * & * & \psi_{99} \end{bmatrix} < 0, \tag{8.32}$$

where

$$\begin{aligned} \psi_{11} &= -E^{\mathrm{T}}P_1E + S_1\bar{R}^{\mathrm{T}}A + A^{\mathrm{T}}\bar{R}S_1^{\mathrm{T}} + \sum_{i=1}^{m}\theta_i Q_i + \sum_{j=1}^{s}\vartheta_j Z_j, \\ \psi_{12} &= A^{\mathrm{T}}\bar{R}S_2^{\mathrm{T}} - E^{\mathrm{T}}P_2, \\ \psi_{17} &= A^{\mathrm{T}}F_1^{\mathrm{T}} + C^{\mathrm{T}}\bar{\Omega}\hat{B}_f^{\mathrm{T}}, \\ \psi_{18} &= A^{\mathrm{T}}F_2^{\mathrm{T}} + C^{\mathrm{T}}\bar{\Omega}\hat{B}_f^{\mathrm{T}}, \end{aligned}$$

$$\psi_{37} = \begin{bmatrix}\bar{\alpha}_1 A_d \ \bar{\alpha}_2 A_d \ \cdots \ \bar{\alpha}_m A_d\end{bmatrix}^{\mathrm{T}},$$
$$\psi_{47} = \begin{bmatrix}\bar{\beta}_1 A_\tau \ \bar{\beta}_2 A_\tau \ \cdots \ \bar{\beta}_s A_\tau\end{bmatrix}^{\mathrm{T}},$$
$$\psi_{57} = B_\omega^{\mathrm{T}} F_1^{\mathrm{T}} + D_\omega^{\mathrm{T}} \hat{B}_f^{\mathrm{T}},$$
$$\psi_{58} = B_\omega^{\mathrm{T}} F_2^{\mathrm{T}} + D_\omega^{\mathrm{T}} \hat{B}_f^{\mathrm{T}},$$
$$\psi_{33} = \mathrm{diag}\{-Q_1 + \underline{\alpha}_1 A_d^{\mathrm{T}} P_1 A_d, -Q_2 + \underline{\alpha}_2 A_d^{\mathrm{T}} P_1 A_d, \cdots, -Q_m + \underline{\alpha}_m A_d^{\mathrm{T}} P_1 A_d\},$$
$$\psi_{44} = \mathrm{diag}\{-Z_1 + \underline{\beta}_1 A_\tau^{\mathrm{T}} P_1 A_\tau, -Z_2 + \underline{\beta}_2 A_\tau^{\mathrm{T}} P_1 A_\tau, \cdots, -Z_s + \underline{\beta}_s A_\tau^{\mathrm{T}} P_1 A_\tau\},$$
$$\psi_{77} = P_1 - F_1 - F_1^{\mathrm{T}},$$
$$\psi_{78} = P_2 - V - F_2^{\mathrm{T}},$$
$$\psi_{88} = P_3 - V - V^{\mathrm{T}},$$
$$\psi_{19} = \begin{bmatrix}\sigma_1 C_1^{\mathrm{T}} \hat{B}_f^{\mathrm{T}} \ \sigma_2 C_2^{\mathrm{T}} \hat{B}_f^{\mathrm{T}} \ \cdots \ \sigma_\rho C_\rho^{\mathrm{T}} \hat{B}_f^{\mathrm{T}}\end{bmatrix},$$
$$\psi_{99} = -\,\mathrm{diag}\{\underbrace{VP_3^{-1}V^{\mathrm{T}}, VP_3^{-1}V^{\mathrm{T}}, \cdots, VP_3^{-1}V^{\mathrm{T}}}_{\rho}\}.$$

According to the fact that $-VP_3^{-1}V^{\mathrm{T}} \leqslant P_3 - V - V^{\mathrm{T}}$, (8.32) holds implies the following LMI holds,

$$\begin{bmatrix}
\psi_{11} & \psi_{12} & S_1\bar{R}^{\mathrm{T}}\varphi_{37}^{\mathrm{T}} & S_1\bar{R}^{\mathrm{T}}\varphi_{47}^{\mathrm{T}} & S_1\bar{R}^{\mathrm{T}}B_\omega & L^{\mathrm{T}} & \psi_{17} & \psi_{18} & \psi_{19} \\
* & -P_3 & S_2\bar{R}^{\mathrm{T}}\varphi_{37}^{\mathrm{T}} & S_2\bar{R}^{\mathrm{T}}\varphi_{47}^{\mathrm{T}} & S_2\bar{R}^{\mathrm{T}}B_\omega & -\hat{C}_f^{\mathrm{T}} & \hat{A}_f^{\mathrm{T}} & \hat{A}_f^{\mathrm{T}} & 0 \\
* & * & \psi_{33} & 0 & 0 & 0 & \varphi_{37}F_1^{\mathrm{T}} & \varphi_{37}F_2^{\mathrm{T}} & 0 \\
* & * & * & \psi_{44} & 0 & 0 & \varphi_{47}F_1^{\mathrm{T}} & \varphi_{47}F_2^{\mathrm{T}} & 0 \\
* & * & * & * & -\gamma^2 I & 0 & \psi_{57} & \psi_{58} & 0 \\
* & * & * & * & * & -I & 0 & 0 & 0 \\
* & * & * & * & * & * & \psi_{77} & \psi_{78} & 0 \\
* & * & * & * & * & * & * & \psi_{88} & 0 \\
* & * & * & * & * & * & * & * & \hat{\psi}_{99}
\end{bmatrix} < 0, \tag{8.33}$$

where

$$\hat{\psi}_{99} = \mathrm{diag}\{\underbrace{P_3 - V - V^{\mathrm{T}}, P_3 - V - V^{\mathrm{T}}, \cdots, P_3 - V - V^{\mathrm{T}}}_{\rho}\}.$$

Immediately, we obtain the following sufficient condition for the existence of the desired filter (8.8).

Theorem 8.4. *For a given scalar $\gamma > 0$, system (8.9) is asymptotically admissible in the mean square with H_∞ performance γ, if there exist matrices $P = \begin{bmatrix} P_1 & P_2 \\ * & P_3 \end{bmatrix} > 0$, $Q_i > 0$ $(i = 1, 2, \cdots, m)$, $Z_j > 0$ $(j = 1, 2, \cdots, s)$, S_1, S_2, $\bar{A}_f$, $\bar{B}_f$, $\bar{C}_f$, F_1, F_2 and V such that (8.33) hold. Furthermore, if (8.33) is solvable, a desired H_∞ filter in the form of (8.8) can be chosen with parameters as*

$$A_f = V^{-1}\hat{A}_f, B_f = V^{-1}\hat{B}_f, C_f = \hat{C}_f. \tag{8.34}$$

8.4 Numerical Example

In this section, a numerical example is introduced to illustrate the usefulness and flexibility of the filter design method developed in this chapter. Consider system (8.1) with the following parameters:

$$E = \begin{bmatrix} 1 & 4 & 4 \\ 0 & 1 & 2 \\ 1 & 5 & 6 \end{bmatrix}, A = \begin{bmatrix} 0.7 & 1.1 & 0 \\ 0 & -0.6 & -1.8 \\ 0.5 & -0.3 & -3.4 \end{bmatrix},$$

$$A_d = \begin{bmatrix} 0.5 & 0.53 & -0.72 \\ -0.14 & -0.99 & -2.26 \\ 0.12 & -1.36 & -4.66 \end{bmatrix}, A_\tau = \begin{bmatrix} 0.3 & 0.7 & 0.4 \\ 0.1 & 0.2 & 0 \\ 0.4 & 0.9 & 0.4 \end{bmatrix},$$

$$B_\omega = \begin{bmatrix} 0.9 \\ 0.4 \\ 1.5 \end{bmatrix}, C = \begin{bmatrix} 1.7 & 4.9 & 4.4 \\ 0 & 1.5 & 4.2 \\ 0.3 & 0.8 & 0.6 \end{bmatrix}, D_\omega = \begin{bmatrix} 0.1 \\ 0.2 \\ 0.1 \end{bmatrix}, L = \begin{bmatrix} 0 \\ 0.1 \\ 0.4 \end{bmatrix}^{\mathrm{T}}.$$

In addition, the communication time-varying delays satisfy $2 \leqslant d_1(k) \leqslant 5$, $3 \leqslant d_2(k) \leqslant 6$, $4 \leqslant \tau_1(k) \leqslant 9$ and $2 \leqslant \tau_2(k) \leqslant 7$. It is assumed that $\bar{\alpha}_1 = 0.2$, $\bar{\alpha}_2 = 0.12$, $\bar{\beta}_1 = 0.2$ and $\bar{\beta}_2 = 0.3$. It can easily be calculated that $\underline{\alpha}_1 = 0.16$, $\underline{\alpha}_2 = 0.1056$, $\underline{\beta}_1 = 0.16$ and $\underline{\beta}_2 = 0.21$. Suppose that the probabilistic density functions of χ_1, χ_2 and χ_3 on the interval $[0\ 1]$ are described by

$$q_1(s_1) = \begin{cases} 0, & s_1 = 0 \\ 0.1, & s_1 = 0.5 \\ 0.9, & s_1 = 1, \end{cases}$$

$$q_2(s_2) = \begin{cases} 0.1, & s_2 = 0 \\ 0.1, & s_2 = 0.5 \\ 0.8, & s_2 = 1, \end{cases}$$

$$q_3(s_3) = \begin{cases} 0, & s_3 = 0 \\ 0.2, & s_3 = 0.5 \\ 0.8, & s_3 = 1. \end{cases}$$

from which the expectations and variances can be easily calculated as $\mu_1 = 0.95$, $\mu_2 = 0.85$, $\mu_3 = 0.9$, $\sigma_1 = 0.15$, $\sigma_2 = 0.32$ and $\sigma_3 = 0.2$.

Using Matlab LMI control Toolbox to solve the LMI in (8.33) with

$$R = \begin{bmatrix} -0.5 \\ -0.5 \\ 0.5 \end{bmatrix},$$

we can calculate the minimum H_∞ performance γ achieved by the filtering error system (8.9) is 0.9416, and the desired filter parameters are given by

$$A_f = \begin{bmatrix} -2.3762 & -3.4850 & 2.4840 \\ -2.0539 & -3.3584 & 2.3610 \\ -4.4291 & -6.8423 & 4.8440 \end{bmatrix},$$

$$B_f = \begin{bmatrix} -0.6977 & 1.9955 & -4.2189 \\ -0.1922 & 0.7591 & -1.3364 \\ -0.8899 & 2.7546 & -5.5552 \end{bmatrix},$$

$$C_f = \begin{bmatrix} -0.1042 & -0.1927 & 0.1336 \end{bmatrix}.$$

8.5 Conclusion

In this chapter, the H_∞ filtering problem has been studied for networked discrete-time SSs with both multiple stochastic time-varying communication delays and probabilistic missing measurements. The time-varying communication delays are assumed to occur in a random way, and contain discrete delays and distributed delays. A filter has been designed such that, for the admissible stochastic communication delays and random measurement missing, the filtering error system is asymptotically admissible in the mean square with H_∞ performance index. An illustrative example has been exploited to show the effectiveness of the proposed design procedure.

Part II

Singular Markov Jump Systems (SMJSs) with Time-Delays

Part II

9

Stability Analysis for Discrete-Time SMJSs with Time-Varying Delay

9.1 Introduction

In recent years, there has been increasing research interest on stability of SMJSs. For example, based on LMI approach, the delay-independent and delay-dependent sufficient conditions have been developed ensuring continuous-time SMJSs with time delay to be regular, impulse free and stochastically stable in [7, 9]. Some delay-dependent stochastic stability criteria have also been proposed in [159, 161], which have less conservatism than those in [7, 9]. In the discrete-time setting, the free-weighting matrix method has been employed in [104, 105] to obtain some delay-dependent conditions guaranteeing discrete-time singular Markov jump time-delay systems to be regular, causal and stochastically stable. However, the equivalent model transformation and decomposition of original system matrices is indispensable in [104, 105] and makes the analysis procedure indirect and complicated. Without resorting to the equivalent model transformation and decomposition of original system matrices, some delay-dependent stability conditions have been proposed in [100–102, 216] for discrete-time singular Markov jump time delay systems. It should be pointed out that the involved time delays of [101, 104] are time-invariant, which inevitably limits the scope of applications of the established results. It also should be pointed out that some important and useful terms are ignored in the Lyapunov functionals reported in [100, 102, 105, 216]. The ignorance of these terms may lead to conservatism to some extent. Therefore, it is significant and necessary to study further the discrete-time SMJSs with time-varying delay. It also should be mentioned that the results reported in [7, 9, 100, 101, 104, 105, 159, 161, 216] require the complete knowledge of transition probabilities in the Markov process or Markov chain. In fact, it is very hard and expensive to precisely obtain all the transition probabilities in the Markov process or Markov chain even for a simple system [204, 205, 207]. Thus, the ideal requirement on the precise knowledge of transition probabilities inevitably limits the applications of the established results.

Z.-G. Wu et al.: *Anal. & Synth. of Singular Syst. with Time-Delays*, LNCIS 443, pp. 117–132.
DOI: 10.1007/978-3-642-37497-5_9

The objective of this chapter is to study the problem of delay-dependent stability for discrete-time SMJSs with time-varying delay by using LMI approach. By constructing a new Lyapunov functional, which employs the integrated lower and upper bounds of the involved time-varying delay, some new delay-dependent conditions are established for the considered systems to be regular, causal and stochastically stable. Unlike the conditions of [104, 105], the newly obtained results here are formulated by the coefficient matrices of the considered system, which avoid the decomposition of the considered system. Furthermore, different from [100–102, 104, 105, 216], the considered transition probabilities in Markov chain here are assumed to be partially unknown, and thus our results are more powerful and have more wide applications. Examples are provided to demonstrate that the results in this chapter improve the existing ones.

9.2 Preliminaries

Fix a probability space $(\Omega, \mathcal{F}, \mathcal{P})$ and consider the following discrete-time SMJS with time-varying delay:

$$\begin{cases} Ex(k+1) = A(r(k))x(k) + A_d(r(k))x(k-d(k)), \\ x(k) = \phi(k),\ k \in \mathbb{N}[-d_2,\ 0], \end{cases} \tag{9.1}$$

where $x(k) \in \mathbb{R}^n$ is the state, $d(k)$ is a time-varying delay satisfying $d_1 \leqslant d(k) \leqslant d_2$, where d_1 and d_2 are prescribed positive integers representing the lower and upper bounds of time-varying delay, respectively, and $\phi(k)$ is the compatible initial condition. The matrix $E \in \mathbb{R}^{n\times n}$ may be singular and it is assumed that $\operatorname{rank} E = r \leqslant n$. $A(r(k))$ and $A_d(r(k))$ are known real constant matrices with appropriate dimensions. The parameter $r(k)$ represents a discrete-time homogeneous Markov chain taking values in a finite set $\mathcal{I} = \{1, 2, \cdots, N\}$ with transition probability matrix $\Pi \triangleq \{\pi_{ij}\}$ given by

$$\Pr\{r(k+1) = j | r(k) = i\} = \pi_{ij}, \tag{9.2}$$

where $0 \leqslant \pi_{ij} \leqslant 1$, $\forall i, j \in \mathcal{I}$, and $\sum_{j=1}^{N} \pi_{ij} = 1$, $\forall i \in \mathcal{I}$.

Generally speaking, for some practical systems, the transition probabilities in Markov chain we get will never be precise, that is, some elements in matrix Π are unknown. For instance, system (9.1) with four modes, the transition probability matrix Π may be

$$\Pi = \begin{bmatrix} \pi_{11} & ? & \pi_{13} & ? \\ ? & ? & ? & \pi_{24} \\ \pi_{31} & ? & \pi_{33} & ? \\ ? & ? & \pi_{43} & \pi_{44} \end{bmatrix}, \tag{9.3}$$

where "?" represents the unaccessible elements. For notation clarity, we denote that for any $i \in \mathcal{I}$

$$\mathcal{I}_{\mathcal{K}}^{i} \triangleq \{j : \pi_{ij} \text{ is known}\}, \; \mathcal{I}_{\mathcal{UK}}^{i} \triangleq \{j : \pi_{ij} \text{ is unknown}\}. \tag{9.4}$$

Remark 9.1. The transition probability matrix with partially unknown transition probabilities considered here is first introduced by [204, 205], which is rather different from the one of [100, 101, 104, 105, 216], where the transition probabilities are assumed to be completely available, and is also different from the one of [102], where the transition probabilities are unknown but bounded.

Definition 9.2. *[178]*

1. *System* (9.1) *is said to be regular and causal, if the pair* (E, A_i) *is regular and causal for any* $i \in \mathcal{I}$,
2. *System* (9.1) *is said to be stochastically stable, if for any initial state* $(\phi(k), r_0)$, *the following condition holds*

$$\mathscr{E}\left\{\sum_{k=0}^{\infty} ||x(k)||^2 |\phi, r_0\right\} < \infty, \tag{9.5}$$

3. *System* (9.1) *is said to be stochastically admissible, if it is regular, causal and stochastically stable.*

9.3 Main Results

In this section, we will develop the delay-dependent stability condition for system (9.1) with partially unknown transition probabilities. For presentation convenience, in the following, we denote

$$\begin{aligned}
\delta &= \frac{d_1 + d_2}{2} + \frac{\min\{(-1)^{d_1+d_2}, 0\}}{2}, \\
\tau_1 &= d_2 - \delta, \\
\tau_2 &= \delta - d_1, \\
m &= d_2 - d_1 + 1,
\end{aligned}$$

and define $R \in \mathbb{R}^{n\times(n-r)}$ as a matrix with the properties of $E^{\mathrm{T}}R = 0$ and $\mathrm{rank}R = n - r$.

For notational simplicity, in the sequel, for each possible $r(k) = i$, $i \in \mathcal{I}$, a matrix $M(r(k))$ will be denoted by M_i; for example, $A(r(k))$ is denoted by A_i, $A_d(r(k))$ by A_{di}, and so on.

Firstly, we propose the following criterion for system (9.1) under completely known transition probabilities.

Theorem 9.3. *System* (9.1) *with completely known transition probabilities is stochastically admissible, if there exist matrices* $P_i > 0$, $Q_m > 0$ $(m = 1, 2, 3, 4)$, $Z_l > 0$ $(l = 1, 2, 3)$ *and* $S_i = \begin{bmatrix} S_{i1}^{\mathrm{T}} & S_{i2}^{\mathrm{T}} & S_{i3}^{\mathrm{T}} & S_{i4}^{\mathrm{T}} & S_{i5}^{\mathrm{T}} \end{bmatrix}^{\mathrm{T}}$ *such that for any* $i \in \mathcal{I}$,

$$\Xi_i + \Psi_f < 0,\ f = 1, 2, 3, 4, \tag{9.6}$$

where $X_i = \sum_{j=1}^{N} \pi_{ij} P_j$, $\breve{Z}_l = E^{\mathrm{T}} Z_l E$, $U = \delta^2 Z_1 + \tau_1^2 Z_2 + \tau_2^2 Z_3$, *and*

$$\Xi_i = \Phi_{i11} + \Phi_{i12}^{\mathrm{T}} X_i \Phi_{i12} + \Phi_{i13}^{\mathrm{T}} U \Phi_{i13} + S_i R^{\mathrm{T}} \Phi_{i12} + \Phi_{i12}^{\mathrm{T}} R S_i^{\mathrm{T}},$$

$$\Phi_{i11} = \begin{bmatrix} \Xi_{i11} & 0 & 0 & 0 & \breve{Z}_1 \\ * & -Q_2 & 0 & 0 & 0 \\ * & * & -Q_3 & 0 & 0 \\ * & * & * & -Q_4 & 0 \\ * & * & * & * & -Q_1 - \breve{Z}_1 \end{bmatrix},$$

$$\Psi_1 = \begin{bmatrix} 0 & 0 & 0 & 0 & 0 \\ * & -3\breve{Z}_2 & \breve{Z}_2 & 0 & 2\breve{Z}_2 \\ * & * & -\breve{Z}_2 & 0 & 0 \\ * & * & * & -\breve{Z}_3 & \breve{Z}_3 \\ * & * & * & * & -2\breve{Z}_2 - \breve{Z}_3 \end{bmatrix},$$

$$\Psi_2 = \begin{bmatrix} 0 & 0 & 0 & 0 & 0 \\ * & -3\breve{Z}_2 & 2\breve{Z}_2 & 0 & \breve{Z}_2 \\ * & * & -2\breve{Z}_2 & 0 & 0 \\ * & * & * & -\breve{Z}_3 & \breve{Z}_3 \\ * & * & * & * & -\breve{Z}_2 - \breve{Z}_3 \end{bmatrix},$$

$$\Psi_3 = \begin{bmatrix} 0 & 0 & 0 & 0 & 0 \\ * & -3\breve{Z}_3 & 0 & \breve{Z}_3 & 2\breve{Z}_3 \\ * & * & -\breve{Z}_2 & 0 & \breve{Z}_2 \\ * & * & * & -\breve{Z}_3 & 0 \\ * & * & * & * & -2\breve{Z}_3 - \breve{Z}_2 \end{bmatrix},$$

$$\Psi_4 = \begin{bmatrix} 0 & 0 & 0 & 0 & 0 \\ * & -3\breve{Z}_3 & 0 & 2\breve{Z}_3 & \breve{Z}_3 \\ * & * & -\breve{Z}_2 & 0 & \breve{Z}_2 \\ * & * & * & -2\breve{Z}_3 & 0 \\ * & * & * & * & -\breve{Z}_3 - \breve{Z}_2 \end{bmatrix},$$

$$\begin{aligned} \Xi_{i11} &= -E^{\mathrm{T}} P_i E + Q_1 + m Q_2 + Q_3 + Q_4 - \breve{Z}_1, \\ \Phi_{i12} &= \begin{bmatrix} A_i & A_{di} & 0 & 0 & 0 \end{bmatrix}, \\ \Phi_{i13} &= \begin{bmatrix} A_i - E & A_{di} & 0 & 0 & 0 \end{bmatrix}. \end{aligned}$$

Proof. Under the given condition, we first show that system (9.1) is regular and causal. To this end, we choose two nonsingular matrices G and H such that

$$GEH = \begin{bmatrix} I_r & 0 \\ 0 & 0 \end{bmatrix}. \tag{9.7}$$

Write

$$GA_iH = \begin{bmatrix} A_{i1} & A_{i2} \\ A_{i3} & A_{i4} \end{bmatrix},\ H^{\mathrm{T}} S_{i1} = \begin{bmatrix} S_{i11} \\ S_{i12} \end{bmatrix},\ G^{-\mathrm{T}} R = \begin{bmatrix} 0 \\ I \end{bmatrix} M, \tag{9.8}$$

where $M \in \mathbb{R}^{(n-r)\times(n-r)}$ is any nonsingular matrix. Then, pre- and post-multiplying $\Xi_{i11} + S_{i1}R^{\mathrm{T}}A_i + A_i^{\mathrm{T}}RS_{i1}^{\mathrm{T}} < 0$ by H^{T} and H, respectively, we get

$$-H^{\mathrm{T}}E^{\mathrm{T}}P_iEH - H^{\mathrm{T}}\breve{Z}_1H + H^{\mathrm{T}}S_{i1}R^{\mathrm{T}}A_iH + H^{\mathrm{T}}A_i^{\mathrm{T}}RS_{i1}^{\mathrm{T}}H < 0, \tag{9.9}$$

which implies

$$\begin{aligned} -H^{\mathrm{T}}E^{\mathrm{T}}G^{\mathrm{T}}G^{-\mathrm{T}}P_iG^{-1}GEH - H^{\mathrm{T}}E^{\mathrm{T}}G^{\mathrm{T}}G^{-\mathrm{T}}Z_1G^{-1}GEH \\ +H^{\mathrm{T}}S_{i1}R^{\mathrm{T}}G^{-1}GA_iH + H^{\mathrm{T}}A_i^{\mathrm{T}}G^{\mathrm{T}}G^{-\mathrm{T}}RS_{i1}^{\mathrm{T}}H < 0. \end{aligned} \tag{9.10}$$

Now, substituting (9.7) and (9.8) to the above inequality gives

$$\begin{bmatrix} \star & \star \\ * & \chi_i \end{bmatrix} < 0, \tag{9.11}$$

where "$\star$" represents matrices that are not relevant in the following discussion, and $\chi_i = S_{i12}M^{\mathrm{T}}A_{i4} + A_{i4}^{\mathrm{T}}MS_{i12}^{\mathrm{T}}$. From the above inequality, it is easy to see that $\chi_i < 0$ for any $i \in \mathcal{I}$, which implies A_{i4} is nonsingular for any $i \in \mathcal{I}$. Thus, the pair (E, A_i) is regular and causal for any $i \in \mathcal{I}$. According to Definition 9.2, system (9.1) is regular and causal.

Next we will show that system (9.1) is stochastically stable. To this end, we define $\eta(k) = x(k+1) - x(k)$ and consider the following Lyapunov functional for system (9.1):

$$V(x(k), k, r(k)) = V_1(x(k), k, r(k)) + V_2(x(k), k, r(k)), \tag{9.12}$$

where

$$\begin{aligned} V_1(x(k), k, r(k)) &= x(k)^{\mathrm{T}}E^{\mathrm{T}}P(r(k))Ex(k) + \sum_{s=k-\delta}^{k-1} x(s)^{\mathrm{T}}Q_1x(s) \\ &\quad + \sum_{j=-d_2+1}^{-d_1+1}\ \sum_{s=k-1+j}^{k-1} x(s)^{\mathrm{T}}Q_2x(s) \\ &\quad + \delta\sum_{j=-\delta}^{-1}\ \sum_{s=k+j}^{k-1} \eta(s)^{\mathrm{T}}\breve{Z}_1\eta(s), \\ V_2(x(k), k, r(k)) &= \sum_{s=k-d_2}^{k-1} x(s)^{\mathrm{T}}Q_3x(s) + (d_2-\delta)\sum_{j=-d_2}^{-\delta-1}\ \sum_{s=k+j}^{k-1} \eta(s)^{\mathrm{T}}\breve{Z}_2\eta(s) \\ &\quad + \sum_{s=k-d_1}^{k-1} x(s)^{\mathrm{T}}Q_4x(s) + (\delta-d_1)\sum_{j=-\delta}^{-d_1-1}\ \sum_{s=k+j}^{k-1} \eta(s)^{\mathrm{T}}\breve{Z}_3\eta(s). \end{aligned}$$

Define $\mathscr{E}\{\Delta V(k)\} \triangleq \mathscr{E}\{V(x(k+1),k+1,r(k+1))|x(k),r(k)=i)\} - V(x(k),k,i)$. Then, along the solution of system (9.1), we have for any $i \in \mathcal{I}$:

$$\begin{aligned}\mathscr{E}\{\Delta V_1(k)\} \leqslant\, & x(k+1)^{\mathrm{T}}E^{\mathrm{T}}X_iEx(k+1) - x(k)^{\mathrm{T}}E^{\mathrm{T}}P_iEx(k) \\ & + x(k)^{\mathrm{T}}Q_1x(k) - x(k-\delta)^{\mathrm{T}}Q_1x(k-\delta) \\ & + mx(k)^{\mathrm{T}}Q_2x(k) - x(k-d(k))^{\mathrm{T}}Q_2x(k-d(k)) \\ & + \delta^2\eta(k)^{\mathrm{T}}\breve{Z}_1\eta(k) - \delta\sum_{s=k-\delta}^{k-1}\eta(s)^{\mathrm{T}}\breve{Z}_1\eta(s). \end{aligned} \tag{9.13}$$

Based on discretized Jensen inequality, it is easy to get that

$$\begin{aligned}-\delta\sum_{s=k-\delta}^{k-1}\eta(s)^{\mathrm{T}}\breve{Z}_1\eta(s) &\leqslant -\left[\sum_{s=k-\delta}^{k-1}\eta(s)^{\mathrm{T}}E^{\mathrm{T}}\right]Z_1\left[E\sum_{s=k-\delta}^{k-1}\eta(s)\right] \\ &= \begin{bmatrix} x(k) \\ x(k-\delta)\end{bmatrix}^{\mathrm{T}}\begin{bmatrix} -\breve{Z}_1 & \breve{Z}_1 \\ * & -\breve{Z}_1\end{bmatrix}\begin{bmatrix} x(k) \\ x(k-\delta)\end{bmatrix}. \end{aligned} \tag{9.14}$$

On the other hand, we have

$$\begin{aligned}\mathscr{E}\{\Delta V_2(k)\} =\, & x(k)^{\mathrm{T}}Q_3x(k) - x(k-d_2)^{\mathrm{T}}Q_3x(k-d_2) \\ & + x(k)^{\mathrm{T}}Q_4x(k) - x(k-d_1)^{\mathrm{T}}Q_4x(k-d_1) \\ & + (d_2-\delta)^2\eta(k)^{\mathrm{T}}\breve{Z}_2\eta(k) - (d_2-\delta)\sum_{s=k-d_2}^{k-\delta-1}\eta(s)^{\mathrm{T}}\breve{Z}_2\eta(s) \\ & + (\delta-d_1)^2\eta(k)^{\mathrm{T}}\breve{Z}_3\eta(k) - (\delta-d_1)\sum_{s=k-\delta}^{k-d_1-1}\eta(s)^{\mathrm{T}}\breve{Z}_3\eta(s). \end{aligned} \tag{9.15}$$

When $\delta < d(k) \leqslant d_2$, let $g(k) = \frac{d_2-d(k)}{d_2-\delta}$ [122], then it is easy to get that $0 \leqslant g(k) < 1$, $d(k)-\delta = (1-g(k))(d_2-\delta)$ and

$$\begin{aligned}& -(d_2-\delta)\sum_{s=k-d_2}^{k-\delta-1}\eta(s)^{\mathrm{T}}\breve{Z}_2\eta(s) \\ = & -(d_2-\delta)\sum_{s=k-d(k)}^{k-\delta-1}\eta(s)^{\mathrm{T}}\breve{Z}_2\eta(s) - (d_2-\delta)\sum_{s=k-d_2}^{k-d(k)-1}\eta(s)^{\mathrm{T}}\breve{Z}_2\eta(s) \\ = & -(d_2-d(k))\sum_{s=k-d(k)}^{k-\delta-1}\eta(s)^{\mathrm{T}}\breve{Z}_2\eta(s) - (d(k)-\delta)\sum_{s=k-d(k)}^{k-\delta-1}\eta(s)^{\mathrm{T}}\breve{Z}_2\eta(s) \\ & -(d_2-d(k))\sum_{s=k-d_2}^{k-d(k)-1}\eta(s)^{\mathrm{T}}\breve{Z}_2\eta(s) - (d(k)-\delta)\sum_{s=k-d_2}^{k-d(k)-1}\eta(s)^{\mathrm{T}}\breve{Z}_2\eta(s) \end{aligned}$$

$$
\begin{aligned}
&\leqslant -g(k)(d(k)-\delta)\sum_{s=k-d(k)}^{k-\delta-1}\eta(s)^{\mathrm{T}}\breve{Z}_2\eta(s)-(d(k)-\delta)\sum_{s=k-d(k)}^{k-\delta-1}\eta(s)^{\mathrm{T}}\breve{Z}_2\eta(s)\\
&\quad-(d_2-d(k))\sum_{s=k-d_2}^{k-d(k)-1}\eta(s)^{\mathrm{T}}\breve{Z}_2\eta(s)\\
&\quad-(1-g(k))(d_2-d(k))\sum_{s=k-d_2}^{k-d(k)-1}\eta(s)^{\mathrm{T}}\breve{Z}_2\eta(s)\\
&\leqslant -g(k)\sum_{s=k-d(k)}^{k-\delta-1}\eta(s)^{\mathrm{T}}\breve{Z}_2\sum_{s=k-d(k)}^{k-\delta-1}\eta(s)-\sum_{s=k-d(k)}^{k-\delta-1}\eta(s)^{\mathrm{T}}\breve{Z}_2\sum_{s=k-d(k)}^{k-\delta-1}\eta(s)\\
&\quad-\sum_{s=k-d_2}^{k-d(k)-1}\eta(s)^{\mathrm{T}}\breve{Z}_2\sum_{s=k-d_2}^{k-d(k)-1}\eta(s)-(1-g(k))\sum_{s=k-d_2}^{k-d(k)-1}\eta(s)^{\mathrm{T}}\breve{Z}_2\sum_{s=k-d_2}^{k-d(k)-1}\eta(s)\\
&=\begin{bmatrix}x(k-d(k))\\ x(k-\delta)\\ x(k-d_2)\end{bmatrix}^{\mathrm{T}}\begin{bmatrix}-2\breve{Z}_2 & \breve{Z}_2 & \breve{Z}_2\\ * & -\breve{Z}_2 & 0\\ * & * & -\breve{Z}_2\end{bmatrix}\begin{bmatrix}x(k-d(k))\\ x(k-\delta)\\ x(k-d_2)\end{bmatrix}\\
&\quad+g(k)\begin{bmatrix}x(k-d(k))\\ x(k-\delta)\end{bmatrix}^{\mathrm{T}}\begin{bmatrix}-\breve{Z}_2 & \breve{Z}_2\\ * & -\breve{Z}_2\end{bmatrix}\begin{bmatrix}x(k-d(k))\\ x(k-\delta)\end{bmatrix}\\
&\quad+(1-g(k))\begin{bmatrix}x(k-d(k))\\ x(k-d_2)\end{bmatrix}^{\mathrm{T}}\begin{bmatrix}-\breve{Z}_2 & \breve{Z}_2\\ * & -\breve{Z}_2\end{bmatrix}\begin{bmatrix}x(k-d(k))\\ x(k-d_2)\end{bmatrix},
\end{aligned}
\tag{9.16}
$$

and based on discretized Jensen inequality, we can also get that

$$
\begin{aligned}
-(\delta-d_1)\sum_{s=k-\delta}^{k-d_1-1}\eta(s)^{\mathrm{T}}\breve{Z}_3\eta(s)&\leqslant-\sum_{s=k-\delta}^{k-d_1-1}\eta(s)^{\mathrm{T}}\breve{Z}_3\sum_{s=k-\delta}^{k-d_1-1}\eta(s)\\
&=\begin{bmatrix}x(k-d_1)\\ x(k-\delta)\end{bmatrix}^{\mathrm{T}}\begin{bmatrix}-\breve{Z}_3 & \breve{Z}_3\\ * & -\breve{Z}_3\end{bmatrix}\begin{bmatrix}x(k-d_1)\\ x(k-\delta)\end{bmatrix}.
\end{aligned}
\tag{9.17}
$$

On the other hand, it is clear that

$$
2\xi(k)^{\mathrm{T}}S_iR^{\mathrm{T}}Ex(k+1)\equiv 0, \tag{9.18}
$$

where $\xi(k)=\left[x(k)^{\mathrm{T}}\ x(k-d(k))^{\mathrm{T}}\ x(k-d_2)^{\mathrm{T}}\ x(k-d_1)^{\mathrm{T}}\ x(k-\delta)^{\mathrm{T}}\right]^{\mathrm{T}}$, that is,

$$
f(k)=2\xi(k)^{\mathrm{T}}S_iR^{\mathrm{T}}\Phi_{i12}\xi(k)\equiv 0, \tag{9.19}
$$

Thus, it can be calculated that

$$
\mathscr{E}\{\Delta V(k)\}=\mathscr{E}\{\Delta V_1(k)\}+\mathscr{E}\{\Delta V_2(k)\}+f(k)
$$

$$\leqslant \xi(k)^{\mathrm{T}}(g(k)(\Xi_i + \Psi_1) + (1 - g(k))(\Xi_i + \Psi_2))\xi(k). \tag{9.20}$$

When $d_1 \leqslant d(k) \leqslant \delta$, let $g(k) = \frac{d(k)-d_1}{\delta-d_1}$, then it is easy to get that $0 \leqslant g(k) \leqslant 1$, $\delta - d(k) = (1-g(k))(\delta - d_1)$ and

$$\begin{aligned}
& -(\delta - d_1)\sum_{s=k-\delta}^{k-d_1-1} \eta(s)^{\mathrm{T}}\breve{Z}_3\eta(s) \\
= & -(\delta - d_1)\sum_{s=k-\delta}^{k-d(k)-1} \eta(s)^{\mathrm{T}}\breve{Z}_3\eta(s) - (\delta - d_1)\sum_{s=k-d(k)}^{k-d_1-1} \eta(s)^{\mathrm{T}}\breve{Z}_3\eta(s) \\
= & -(\delta - d(k))\sum_{s=k-\delta}^{k-d(k)-1} \eta(s)^{\mathrm{T}}\breve{Z}_3\eta(s) - (d(k) - d_1)\sum_{s=k-\delta}^{k-d(k)-1} \eta(s)^{\mathrm{T}}\breve{Z}_3\eta(s) \\
& -(\delta - d(k))\sum_{s=k-d(k)}^{k-d_1-1} \eta(s)^{\mathrm{T}}\breve{Z}_3\eta(s) - (d(k) - d_1)\sum_{s=k-d(k)}^{k-d_1-1} \eta(s)^{\mathrm{T}}\breve{Z}_3\eta(s) \\
\leqslant & -(\delta - d(k))\sum_{s=k-\delta}^{k-d(k)-1} \eta(s)^{\mathrm{T}}\breve{Z}_3\eta(s) - g(k)(\delta - d(k))\sum_{s=k-\delta}^{k-d(k)-1} \eta(s)^{\mathrm{T}}\breve{Z}_3\eta(s) \\
& -(1-g(k))(d(k) - d_1)\sum_{s=k-d(k)}^{k-d_1-1} \eta(s)^{\mathrm{T}}\breve{Z}_3\eta(s) \\
& -(d(k) - d_1)\sum_{s=k-d(k)}^{k-d_1-1} \eta(s)^{\mathrm{T}}\breve{Z}_3\eta(s) \\
\leqslant & -\sum_{s=k-\delta}^{k-d(k)-1} \eta(s)^{\mathrm{T}}\breve{Z}_3\sum_{s=k-\delta}^{k-d(k)-1} \eta(s) - g(k)\sum_{s=k-\delta}^{k-d(k)-1} \eta(s)^{\mathrm{T}}\breve{Z}_3\sum_{s=k-\delta}^{k-d(k)-1} \eta(s) \\
& -(1-g(k))\sum_{s=k-d(k)}^{k-d_1-1} \eta(s)^{\mathrm{T}}\breve{Z}_3\sum_{s=k-d(k)}^{k-d_1-1} \eta(s) \\
& -\sum_{s=k-d(k)}^{k-d_1-1} \eta(s)^{\mathrm{T}}\breve{Z}_3\sum_{s=k-d(k)}^{k-d_1-1} \eta(s) \\
= & \begin{bmatrix} x(k-d(k)) \\ x(k-\delta) \\ x(k-d_1) \end{bmatrix}^{\mathrm{T}} \begin{bmatrix} -2\breve{Z}_3 & \breve{Z}_3 & \breve{Z}_3 \\ * & -\breve{Z}_3 & 0 \\ * & * & -\breve{Z}_3 \end{bmatrix} \begin{bmatrix} x(k-d(k)) \\ x(k-\delta) \\ x(k-d_1) \end{bmatrix} \\
& + g(k)\begin{bmatrix} x(k-d(k)) \\ x(k-\delta) \end{bmatrix}^{\mathrm{T}} \begin{bmatrix} -\breve{Z}_3 & \breve{Z}_3 \\ * & -\breve{Z}_3 \end{bmatrix} \begin{bmatrix} x(k-d(k)) \\ x(k-\delta) \end{bmatrix} \\
& + (1-g(k))\begin{bmatrix} x(k-d(k)) \\ x(k-d_1) \end{bmatrix}^{\mathrm{T}} \begin{bmatrix} -\breve{Z}_3 & \breve{Z}_3 \\ * & -\breve{Z}_3 \end{bmatrix} \begin{bmatrix} x(k-d(k)) \\ x(k-d_1) \end{bmatrix},
\end{aligned} \tag{9.21}$$

and based on discretized Jensen inequality, we can also obtain that

$$-(d_2-\delta)\sum_{s=k-d_2}^{k-\delta-1}\eta(s)^{\mathrm{T}}\breve{Z}_2\eta(s)\leqslant-\sum_{s=k-d_2}^{k-\delta-1}\eta(s)^{\mathrm{T}}\breve{Z}_2\sum_{s=k-d_2}^{k-\delta-1}\eta(s)$$
$$=\begin{bmatrix}x(k-d_2)\\x(k-\delta)\end{bmatrix}^{\mathrm{T}}\begin{bmatrix}-\breve{Z}_2 & \breve{Z}_2\\ * & -\breve{Z}_2\end{bmatrix}\begin{bmatrix}x(k-d_2)\\x(k-\delta)\end{bmatrix}. \tag{9.22}$$

Thus, we can obtain that

$$\mathscr{E}\{\Delta V(k)\}\leqslant\xi(k)^{\mathrm{T}}(g(k)(\Xi_i+\Psi_3)+(1-g(k))(\Xi_i+\Psi_4))\xi(k). \tag{9.23}$$

Hence, we can get from (9.6), (9.20) and (9.23) that there exits a scalar $\alpha>0$ such that

$$\mathscr{E}\{\Delta V(k)\}\leqslant-\alpha||x(k)||^2, \tag{9.24}$$

By using the similar method of [6], it is easy to get from (9.12) and (9.24) that (9.5) holds, which means that system (9.1) is stochastically stable. This completes the proof.

Remark 9.4. It is noted that Theorem 9.3 proposes a new version of delay-dependent stability criterion for discrete-time SMJSs with time-varying delay. The obtained LMIs in (9.6) are formulated by the coefficient matrices of the original system, which is in contrast with the criteria in [104, 105], where the results are expressed by the coefficient matrices of the decomposed systems. Thus, the computational problems arising from decomposition and equivalent transformation of the original system can be avoided when using Theorem 9.3 of this chapter.

Remark 9.5. Different from the methods applied in [100, 102, 105, 216], the variation interval of the considered time-varying delay is divided into two subintervals, and the variation of the Lyapunov functional is checked when the time-varying delay belongs to different subinterval. The main advantages of such method are (I) it makes full use of the information on the considered time-varying delay and (II) the new state $x(k-\delta)$ is introduced, which has been ignored in [100, 102, 105, 216]. On the other hand, to further reduce the conservatism, the terms $-(d_2-\delta)\sum_{s=k-d_2}^{k-\delta-1}\eta(s)^{\mathrm{T}}\breve{Z}_2\eta(s)$ and $-(\delta-d_1)\sum_{s=k-\delta}^{k-d_1-1}\eta(s)^{\mathrm{T}}\breve{Z}_3\eta(s)$ are not simply enlarged as $-(d(k)-\delta)\sum_{s=k-d(k)}^{k-\delta-1}\eta(s)^{\mathrm{T}}\breve{Z}_2\eta(s)-(d_2-d(k))\sum_{s=k-d_2}^{k-d(k)-1}\eta(s)^{\mathrm{T}}\breve{Z}_2\eta(s)$ and $-(\delta-d(k))\sum_{s=k-\delta}^{k-d(k)-1}\eta(s)^{\mathrm{T}}\breve{Z}_3\eta(s)-(d(k)-d_1)\sum_{s=k-d(k)}^{k-d_1-1}\eta(s)^{\mathrm{T}}\breve{Z}_3\eta(s)$, respectively, but the terms $-g(k)(d(k)-\delta)\sum_{s=k-d(k)}^{k-\delta-1}\eta(s)^{\mathrm{T}}\breve{Z}_2\eta(s)-(1-g(k))(d_2-d(k))\sum_{s=k-d_2}^{k-d(k)-1}\eta(s)^{\mathrm{T}}\breve{Z}_2\eta(s)$ and $-g(k)(\delta-d(k))\sum_{s=k-\delta}^{k-d(k)-1}\eta(s)^{\mathrm{T}}\breve{Z}_3\eta(s)-(1-g(k))(d(k)-d_1)\sum_{s=k-d(k)}^{k-d_1-1}\eta(s)^{\mathrm{T}}\breve{Z}_3\eta(s)$ are taken into account as well.

Thus, the stability criterion derived here is expected to be less conservative than those in [100, 102, 105, 216], which will be demonstrated by a numerical example in the following section.

Next, the following theorem presents a sufficient condition for the regularity, causality and stochastic stability of system (9.1) under partially unknown transition probabilities.

Theorem 9.6. *System* (9.1) *with partially unknown transition probabilities is stochastically admissible, if there exist matrices* $P_i > 0$, $Q_m > 0$ $(m = 1, 2, 3, 4)$, $Z_l > 0$ $(l = 1, 2, 3)$ *and* $S_i = \begin{bmatrix} S_{i1}^{\mathrm{T}} & S_{i2}^{\mathrm{T}} & S_{i3}^{\mathrm{T}} & S_{i4}^{\mathrm{T}} & S_{i5}^{\mathrm{T}} \end{bmatrix}^{\mathrm{T}}$ *such that for any* $i \in \mathcal{I}$

$$\hat{\Xi}_i + \Psi_f < 0, \; f = 1, 2, 3, 4, \tag{9.25}$$

where

$$\hat{\Xi}_i = \Phi_{i11} + \Phi_{i12}^{\mathrm{T}} \hat{\Omega}_i \Phi_{i12} + \Phi_{i13}^{\mathrm{T}} U \Phi_{i13} + S_i R^{\mathrm{T}} \Phi_{i12} + \Phi_{i12}^{\mathrm{T}} R S_i^{\mathrm{T}},$$

$$\hat{\Omega}_i = \sum_{j \in \mathcal{I}_{\mathcal{K}}^i} \pi_{ij} P_j + (1 - \sum_{j \in \mathcal{I}_{\mathcal{K}}^i} \pi_{ij}) \sum_{j \in \mathcal{I}_{\mathcal{UK}}^i} P_j,$$

and the other matrices are given in Theorem 9.3.

Proof. It can be seen that

$$\sum_{j=1}^{N} \pi_{ij} P_j = \sum_{j \in \mathcal{I}_{\mathcal{K}}^i} \pi_{ij} P_j + \sum_{j \in \mathcal{I}_{\mathcal{UK}}^i} \pi_{ij} P_j \leqslant \hat{\Omega}_i. \tag{9.26}$$

Using (9.26), we can get that LMIs in (9.25) hold imply LMIs in (9.6) hold. This completes the proof.

Remark 9.7. It is noted that Theorem 9.6 establishes a stability condition for discrete-time SMJSs with partially unknown transition probabilities. It is worth mentioning that if for any $i \in \mathcal{I}$, $\mathcal{I}_{\mathcal{UK}}^i = \emptyset$, the considered system is the one with completely known transition probabilities and the correspondent LMIs in (9.25) are reduced to LMIs in (9.6). It can also be found that when all the transition probabilities are unaccessible, that is, for any $i \in \mathcal{I}$, $\mathcal{I}_{\mathcal{K}}^i = \emptyset$, Theorem 9.6 is still valid, which implies our result can be used to discrete-time SSs with time-varying delay with arbitrary switching. Therefore, the result of Theorem 9.6 is much more powerful and desirable than the results of [100, 101, 104, 105, 216], where the conditions can only be applied to discrete-time SMJSs with completely known transition probabilities.

9.4 Numerical Examples

In this section, some numerical examples are introduced to demonstrate the effectiveness and less conservatism of the proposed methods.

Example 9.8. Consider system (9.1) with

$$A_1 = \begin{bmatrix} 6.1 & 10.4 \\ 7.15 & 11.6 \end{bmatrix}, \ A_{d1} = \begin{bmatrix} -1.1 & -2 \\ -1.4 & -2.5 \end{bmatrix},$$
$$A_2 = \begin{bmatrix} 6.37 & 10.74 \\ 7.48 & 12.01 \end{bmatrix}, \ A_{d2} = \begin{bmatrix} -0.92 & -1.62 \\ -1.13 & -1.93 \end{bmatrix}.$$

In this example, set

$$E = \begin{bmatrix} 3 & 6 \\ 2 & 4 \end{bmatrix},$$

the transition probability matrix

$$\Pi = \begin{bmatrix} 0.45 & 0.55 \\ 0.7 & 0.3 \end{bmatrix},$$

and choose

$$R = \begin{bmatrix} -2 \\ 3 \end{bmatrix}.$$

For various d_1 and by using the methods of [100, 102, 105, 216] and our chapter, the allowable upper bound d_2 that guarantees the regularity, causality and stochastic stability of the considered system are presented in Table 9.1, which shows Theorem 9.3 and Theorem 9.6 in this chapter give much better than the approaches proposed in [100, 102, 105, 216]. In addition, the stability criteria proposed by [101, 104] are invalid for this example.

Table 9.1. Example 9.8: Comparison of the allowable upper bound d_2 for various d_1

d_1	2	4	6	8	10	12
[102, 216]	12	13	14	15	16	17
[100]	14	14	15	16	18	19
[105]	14	15	16	17	18	19
Theorem 9.3	19	21	23	25	27	29
Theorem 9.6	19	21	23	25	27	29

We assume that the time-varying delay $d(k)$ varies [2, 19], which is shown in Fig. 9.1. The possible realization of Markov chain $r(k)$ is plotted in Fig. 9.2, where the initial mode is assumed to be $r_0 = 1$. The state responses of system (9.1) with given parameters are shown in Fig. 9.3, from which we can find that system states asymptotically converge to zero.

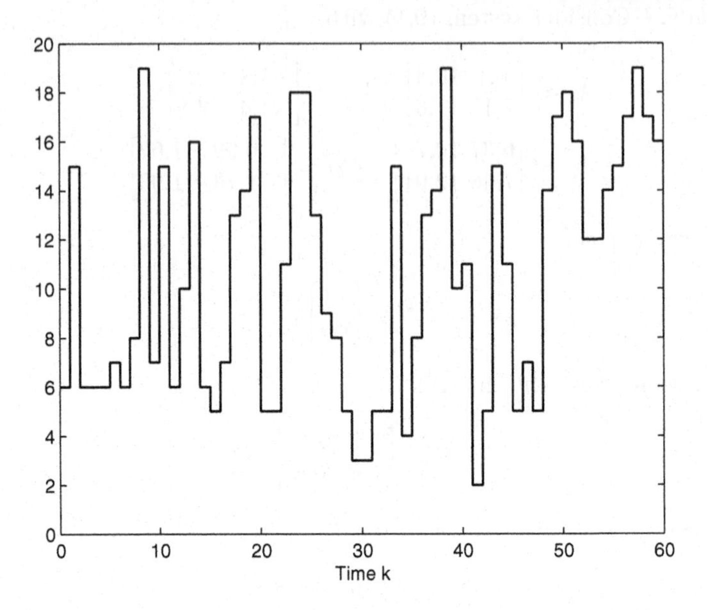

Fig. 9.1. Example 9.8: Time-varying delay $d(k)$

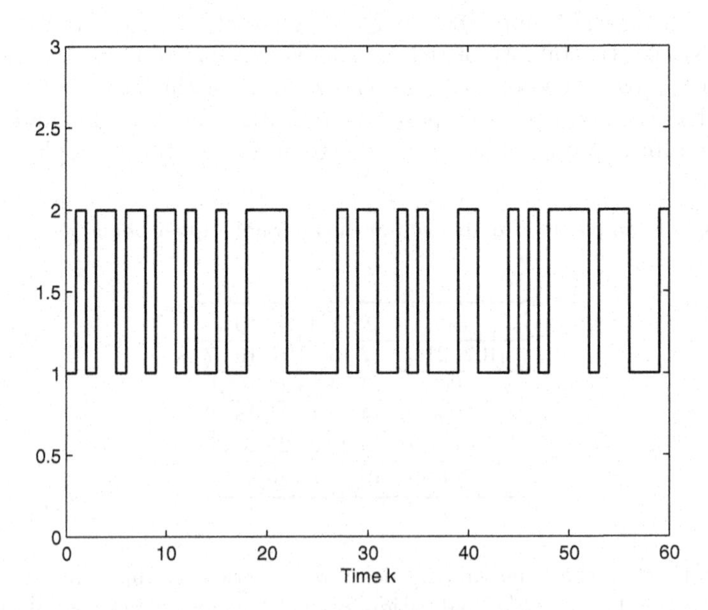

Fig. 9.2. Example 9.8: Markov chain $r(k)$

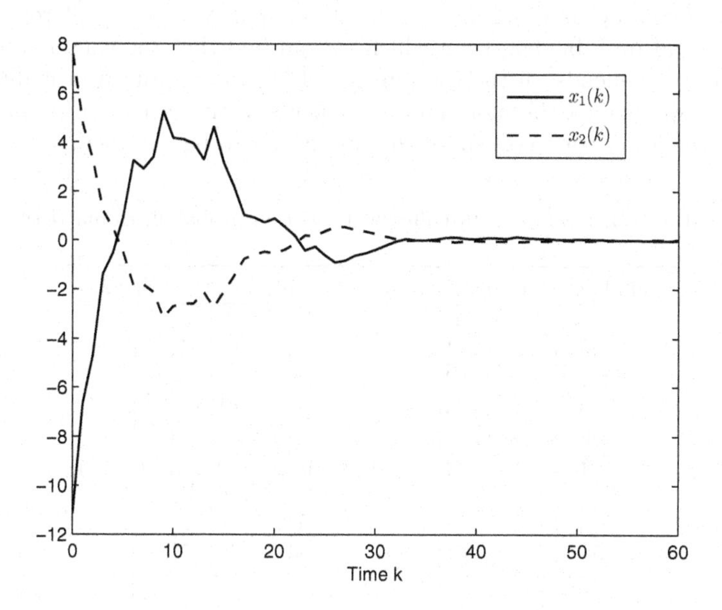

Fig. 9.3. Example 9.8: State trajectories of singular system

Example 9.9. Consider system (9.1) with

$$A_1 = \begin{bmatrix} -22.88 & -42.48 \\ 29.92 & 55.68 \end{bmatrix}, A_{d1} = \begin{bmatrix} 13.272 & 19.904 \\ -17.544 & -26.308 \end{bmatrix},$$

$$A_2 = \begin{bmatrix} -22 & -40.9 \\ 28.7 & 53.45 \end{bmatrix}, A_{d2} = \begin{bmatrix} 12.68 & 19.06 \\ -16.3 & -24.5 \end{bmatrix},$$

$$A_3 = \begin{bmatrix} -11.04 & -25.16 \\ 14.88 & 33.52 \end{bmatrix}, A_{d3} = \begin{bmatrix} 8.76 & 12.32 \\ -11.52 & -16.24 \end{bmatrix},$$

$$A_4 = \begin{bmatrix} -21.52 & -40.16 \\ 28.04 & 52.42 \end{bmatrix}, A_{d4} = \begin{bmatrix} 11.84 & 17.76 \\ -15.28 & -22.92 \end{bmatrix}.$$

In this example, we let

$$E = \begin{bmatrix} 4 & 6 \\ -8 & -12 \end{bmatrix},$$

and choose

$$R = \begin{bmatrix} 2 \\ 1 \end{bmatrix}.$$

The four cases for the transition probability matrix Π will be considered in this example as shown in Table 9.2. For various d_1 and by using Theorem 9.6 in our chapter, the allowable upper bound d_2 that guarantees the regularity, causality and stochastic stability of the considered system for different cases can be found in Table 9.3, from which we can find that the conservatism of Theorem 9.6 depends on the the number of known transition probabilities, that is, if we can get the more known elements in the transition probability matrix, the lower conservatism of the condition can be obtained.

Table 9.2. Example 9.9: Different transition probabilities matrices

Completely known (Case 1)

	1	2	3	4
1	0.3	0.2	0.1	0.4
2	0.3	0.2	0.3	0.2
3	0.1	0.1	0.5	0.3
4	0.2	0.2	0.1	0.5

Partially known (Case 2)

	1	2	3	4
1	0.3	0.2	0.1	0.4
2	?	?	0.3	0.2
3	0.1	0.1	0.5	0.3
4	0.2	?	?	?

Partially known (Case 3)

	1	2	3	4
1	0.3	?	?	?
2	?	?	0.3	0.2
3	0.1	0.1	?	?
4	0.2	?	?	?

Completely unknown(Case 4)

	1	2	3	4
1	?	?	?	?
2	?	?	?	?
3	?	?	?	?
4	?	?	?	?

Table 9.3. Example 9.9: The allowable upper bound d_2 for various d_1

d_1	1	3	5	7	9	11
Case 1	25	27	29	31	33	35
Case 2	20	22	24	26	28	30
Case 3	17	19	21	23	25	27
Case 4	7	9	11	13	15	17

Example 9.10. Consider the dynamic Leontief model of economic systems, which describes the time pattern of production sectors given by [21]

$$x(k) = Mx(k) + G(x(k+1) - x(k)) + \nu(k). \tag{9.27}$$

The elements of $x(k) \in \mathbb{R}^n$ are the levels of production in the sectors at time k. $M \in \mathbb{R}^{n\times n}$ is the input-output matrix, and $Mx(k)$ is the amount required as direct input for the current production. $G \in \mathbb{R}^{n\times n}$ is the capital coefficient matrix, and $G(x(k+1)-x(k))$ is the amount required for capacity expansion to be able to produce $x(k+1)$ in the next period. $\nu(k)$ is the amount of production going to demand. It is assumed that the amount of production $\nu(k)$ is, in turn, controlled by $u(k)$ such that $\nu(k) = Hu(k)$, where $u(k) \in \mathbb{R}^p$ $(1 \leqslant p < n)$. It is clear that (9.27) can be rewritten as

$$Gx(k+1) = (I - M + G)x(k) - Hu(k). \tag{9.28}$$

Typically the capital coefficient matrix G has nonzero elements in only a few rows, corresponding to the fact that capital is formed from only a few sectors. Thus, system (9.28) is a typical discrete-time SS, since G is often singular.

It is assumed that the control law is given by

$$u(k) = K_1 x(k) + K_2 x(k - d(k)), \tag{9.29}$$

where $K_1 \in \mathbb{R}^{p \times n}$ and $K_2 \in \mathbb{R}^{p \times n}$. In practical control system, actuators may fail during the course of system operation and the faults of the actuators may be random in nature [143]. We make use of the following fault model to represent the stochastic behavior of the actuator faults:

$$u^F(k) = F(r(k))u(k), \tag{9.30}$$

where $F(r(k)) = \text{diag}\{f_1(r(k)), f_2(r(k)), \ldots, f_p(r(k))\}$, $0 \leqslant f_q(r(k)) \leqslant 1$ $(q = 1, 2, \ldots, p)$, $\forall r(k) \in \mathcal{I}$. Obviously, when $f_q(r(k)) = 0$, the fault model (9.30) corresponds to the qth actuator outage case. When $0 < f_q(r(k)) < 1$, it corresponds to the case of partial failure of the qth actuator. When $f_q(r(k)) = 1$, it corresponds to the case of no fault in the qth actuator.

Here, we consider a Leontief model described by

$$G = \begin{bmatrix} 1 & 0 \\ 0 & 0 \end{bmatrix}, M = \begin{bmatrix} 2.04 & 1 \\ 0.8 & 1 \end{bmatrix}, H = \begin{bmatrix} -1 \\ 3.05 \end{bmatrix}.$$

Then system (9.28) can be rewritten as

$$\begin{bmatrix} 1 & 0 \\ 0 & 0 \end{bmatrix} x(k+1) = \begin{bmatrix} -0.04 & -1 \\ -0.8 & 0 \end{bmatrix} x(k) - \begin{bmatrix} -1 \\ 3.05 \end{bmatrix} u(k). \tag{9.31}$$

It can be seen that system (9.31) with $u(k) = 0$ is not regular and causal. Choose

$$K_1 = \begin{bmatrix} -0.01 & 0.6 \end{bmatrix}, K_2 = \begin{bmatrix} 0.1676 & 0.1170 \end{bmatrix}, \tag{9.32}$$

and $F_1 = 0.3$, $F_2 = 0.8$, $F_3 = 1$, and the transition probability matrix Π is given by

$$\Pi = \begin{bmatrix} 0.1 & 0.2 & 0.7 \\ ? & ? & 0.8 \\ 0.3 & ? & ? \end{bmatrix}. \tag{9.33}$$

Substituting $u^F(k)$ for $u(k)$ in (9.31), and considering (9.29), (9.30), (9.32) and (9.33), the resultant closed-loop system can be described by system (9.1) with

$$
\begin{aligned}
A_1 &= \begin{bmatrix} -0.0430 & -0.8200 \\ -0.7909 & -0.5490 \end{bmatrix}, A_{d1} = \begin{bmatrix} 0.0503 & 0.0351 \\ -0.1534 & -0.1071 \end{bmatrix}, \\
A_2 &= \begin{bmatrix} -0.0480 & -0.5200 \\ -0.7756 & -1.4640 \end{bmatrix}, A_{d2} = \begin{bmatrix} 0.1341 & 0.0936 \\ -0.4089 & -0.2855 \end{bmatrix}, \\
A_3 &= \begin{bmatrix} -0.0500 & -0.4000 \\ -0.7695 & -1.8300 \end{bmatrix}, A_{d3} = \begin{bmatrix} 0.1676 & 0.1170 \\ -0.5112 & -0.3569 \end{bmatrix}, \\
E &= \begin{bmatrix} 1 & 0 \\ 0 & 0 \end{bmatrix}.
\end{aligned}
$$

Now, choose $R = \begin{bmatrix} 0 & 1 \end{bmatrix}^{\mathrm{T}}$ and the lower bound $d_1 = 2$. We are in a position to find the upper bound d_2 such that the closed-loop system is regular, causal and stochastically stable. By using Theorem 9.6, it is found that the upper bound $d_2 = 8$, that is, the closed-loop system is regular, causal and stochastically stable for any time-varying delay $d(k)$ satisfying $2 \leqslant d(k) \leqslant 8$.

9.5 Conclusion

In terms of LMI approach, the delay-dependent stability problem for a class of discrete-time SMJSs with time-varying delay has been investigated in this chapter. The considered transition probabilities are partially unknown, which include the transition probabilities with completely known or completely unknown. By making use of a novel Lyapunov functional, the delay-dependent conditions have been established to ensure the considered system to be regular, causal and stochastic stable. The developed results are expressed by LMIs, and no decomposition and transformation of the original system is involved, which makes the analysis procedure relatively simple and reliable. Three examples have been given to illustrate the usefulness of the derived results.

10

H_∞ Control for SMJSs with Time Delays

10.1 Introduction

The delay-independent sufficient conditions on the stochastic stability and stochastic stabilizability have been developed for continuous-time SSs with time-delay and Markov jump parameters in [9] and a design algorithm for the desired state feedback controller, which guarantees the closed-loop dynamics systems to be regular, impulse free and stochastically stable, has been proposed by using LMI approach. The delay-dependent stochastic stabilization problem of continuous-time SMJSs has been discussed in [7] and two methods for designing the state feedback stabilizing controller have been derived. But it should be pointed out that in [7] the considered system is assumed to be necessarily regular and impulse free, moreover a matrix describing the relationship between fast and slow subsystems is needed and an improper choice of the matrix would make the results unreliable.

In this chapter, the delay-dependent H_∞ control problem is considered for singular time-delay systems with Markov jump parameters. The considered system is not assumed to be necessarily regular and impulse free. A delay-dependent BRL is provided for the considered system to be regular, impulse free and stochastically stable with H_∞ performance γ by LMIs. Based on this, a strict LMI-based approach is proposed to design a state feedback controller such that the resultant closed-loop system is delay-dependent stochastically admissible and satisfies a prescribed H_∞ performance. Numerical examples are provided to demonstrate the effectiveness and less conservatism of the proposed results.

10.2 Problem Formulation

Fix a probability space $(\Omega, \mathcal{F}, \mathcal{P})$ and consider the following SMJS with time delays:

Z.-G. Wu et al.: *Anal. & Synth. of Singular Syst. with Time-Delays*, LNCIS 443, pp. 133–147.
DOI: 10.1007/978-3-642-37497-5_10

$$\begin{cases} E\dot{x}(t) = A(r_t)x(t) + A_d(r_t)x(t-d) + B(r_t)u(t) + B_\omega(r_t)\omega(t), \\ z(t) = C(r_t)x(t) + C_d(r_t)x(t-d) + D(r_t)u(t) + D_\omega(r_t)\omega(t), \\ x(t) = \phi(t), \ t \in [-d, 0], \end{cases} \tag{10.1}$$

where $x(t) \in \mathbb{R}^n$ is the state, $u(t) \in \mathbb{R}^m$ is the control input, $\omega(t) \in \mathbb{R}^p$ is the disturbance input that belongs to $\mathscr{L}_2[0,\infty)$, $z(t) \in \mathbb{R}^s$ is the controlled output, d is the constant delay, and $\phi(t) \in C_{n,d}$ is a compatible vector valued initial function. $\{r_t, t \geqslant 0\}$ is a continuous-time Markov process with right continuous trajectories and takes values in a finite set $\mathcal{S} = \{1, 2, \cdots, s\}$ with transition rate matrix $\Pi \triangleq \{\pi_{ij}\}$ given by

$$\Pr\{r_{t+h} = j | r_t = i\} = \begin{cases} \pi_{ij}h + o(h), & j \neq i, \\ 1 + \pi_{ii}h + o(h), & j = i, \end{cases} \tag{10.2}$$

where $h > 0$,

$$\lim_{h \to 0} \frac{o(h)}{h} = 0, \tag{10.3}$$

$\pi_{ij} > 0$, for $j \neq i$, is the transition rate from mode i at time t to mode j at time $t + h$, and

$$\pi_{ii} = -\sum_{j=1, j \neq i}^{s} \pi_{ij}. \tag{10.4}$$

The matrix $E \in \mathbb{R}^{n \times n}$ may be singular and it is assumed that $\operatorname{rank} E = r \leqslant n$. $A(r_t)$, $A_d(r_t)$, $B(r_t)$, $B_\omega(r_t)$, $C(r_t)$, $C_d(r_t)$, $D(r_t)$ and $D_\omega(r_t)$ are known real constant matrices for each $r_t \in \mathcal{S}$.

Definition 10.1

1. System

$$\begin{cases} E\dot{x}(t) = A_i x(t) + A_{di} x(t-d), \\ x(t) = \phi(t), \ t \in [-d, 0] \end{cases} \tag{10.5}$$

is said to be regular and impulse free, if the pair (E, A_i) is regular and impulse free for every $i \in \mathcal{S}$.
2. System (10.5) is said to be stochastically stable, if there exists a scalar $M(r_0, \phi(\cdot))$ such that

$$\lim_{t \to \infty} \mathscr{E}\left\{\int_0^t \|x(s)\|^2 \mathrm{d}s | r_0, x(s) = \phi(s), s \in [-d, 0]\right\} \leqslant M(r_0, \phi(\cdot)), \tag{10.6}$$

3. System (10.5) is said to be stochastically admissible, if it is regular, impulse free and stochastically stable.

The problem to be addressed in this chapter can be formulated as follows: for a given scalar $\gamma > 0$, find a state feedback controller

$$u(t) = K(r_t)x(t) \tag{10.7}$$

such that the resultant closed-loop system with $\omega(t) = 0$ is stochastically admissible, and under zero initial condition,

$$\mathcal{E}\left\{\int_0^{\infty} z(t)^{\mathrm{T}} z(t)\,\mathrm{d}t\right\} \leqslant \gamma^2 \int_0^{\infty} \omega(t)^{\mathrm{T}} \omega(t)\,\mathrm{d}t, \tag{10.8}$$

holds for any non-zero $\omega(t) \in \mathscr{L}_2[0,\infty)$. In this case, the closed-loop system is said to be stochastically admissible with H_∞ performance γ.

10.3 Main Results

In this section, we will give a solution to the delay-dependent H_∞ control problem formulated previously by using LMI technique.

10.3.1 BRL

A BRL is proposed for the following system:

$$\begin{cases} E\dot{x}(t) = A_i x(t) + A_{di} x(t-d) + B_{\omega i}\omega(t), \\ z(t) = C_i x(t) + C_{di} x(t-d) + D_{\omega i}\omega(t), \\ x(t) = \phi(t),\ t \in [-d, 0], \end{cases} \tag{10.9}$$

which will play a key role in solving the delay-dependent H_∞ control problem.

Theorem 10.2. *For a given scalar $\gamma > 0$, system (10.9) is stochastically admissible with H_∞ performance γ, if there exist matrices $Q_i > 0$, $Q > 0$, $Z > 0$, P_i, M_i, N_i, S_i, R_i and T_i such that for every $i \in \mathcal{S}$,*

$$E^{\mathrm{T}} P_i = P_i^{\mathrm{T}} E \geqslant 0, \tag{10.10}$$

$$\Omega_i = \begin{bmatrix} \Xi_{i11} & \Xi_{i12} & -S_i E + M_i^{\mathrm{T}} A_{di} + E^{\mathrm{T}} T_i^{\mathrm{T}} & M_i^{\mathrm{T}} B_{\omega i} & dS_i & C_i^{\mathrm{T}} \\ * & \Xi_{i22} & -R_i E + N_i^{\mathrm{T}} A_{di} & N_i^{\mathrm{T}} B_{\omega i} & dR_i & 0 \\ * & * & -Q_i - T_i E - E^{\mathrm{T}} T_i^{\mathrm{T}} & 0 & dT_i & C_{di}^{\mathrm{T}} \\ * & * & * & -\gamma^2 I & 0 & D_{\omega i}^{\mathrm{T}} \\ * & * & * & * & -dZ & 0 \\ * & * & * & * & * & -I \end{bmatrix} < 0, \tag{10.11}$$

$$Q_i < Q, \tag{10.12}$$

where $\mu = \max\{|\pi_{ii}|, i \in \mathcal{S}\}$ and

$$\begin{aligned} \Xi_{i11} &= \sum_{j=1}^{s} \pi_{ij} E^{\mathrm{T}} P_j + M_i^{\mathrm{T}} A_i + A_i^{\mathrm{T}} M_i + S_i E + E^{\mathrm{T}} S_i^{\mathrm{T}} + Q_i + \mu d Q, \\ \Xi_{i12} &= P_i^{\mathrm{T}} - M_i^{\mathrm{T}} + A_i^{\mathrm{T}} N_i + E^{\mathrm{T}} R_i^{\mathrm{T}}, \\ \Xi_{i22} &= -N_i - N_i^{\mathrm{T}} + dZ. \end{aligned}$$

Proof. Firstly, we prove the stochastic admissibility of system (10.9). To this end, we consider system (10.5). From (10.10) and (10.11), it is easy to show that for every $i \in \mathcal{S}$,

$$\tilde{E}^{\mathrm{T}}\tilde{P}_i = \tilde{P}_i^{\mathrm{T}}\tilde{E} \geqslant 0, \tag{10.13}$$

$$\begin{bmatrix} \pi_{ii}\tilde{E}^{\mathrm{T}}\tilde{P}_i + \tilde{A}_i^{\mathrm{T}}\tilde{P}_i + \tilde{P}_i^{\mathrm{T}}\tilde{A}_i + \tilde{Q}_i + \tilde{S}_i\tilde{E} + \tilde{E}^{\mathrm{T}}\tilde{S}_i^{\mathrm{T}} & \tilde{P}_i^{\mathrm{T}}\tilde{A}_{di} - \tilde{S}_i\tilde{E} + \tilde{E}^{\mathrm{T}}\tilde{T}_i^{\mathrm{T}} \\ \tilde{A}_{di}^{\mathrm{T}}\tilde{P}_i - \tilde{E}^{\mathrm{T}}\tilde{S}_i^{\mathrm{T}} + \tilde{T}_i\tilde{E} & -\tilde{Q}_i - \tilde{E}^{\mathrm{T}}\tilde{T}_i^{\mathrm{T}} - \tilde{T}_i\tilde{E} \end{bmatrix} < 0, \tag{10.14}$$

where

$$\tilde{E} = \begin{bmatrix} E & 0 \\ 0 & 0 \end{bmatrix}, \quad \tilde{A}_i = \begin{bmatrix} 0 & I \\ A_i & -I \end{bmatrix}, \quad \tilde{A}_{di} = \begin{bmatrix} 0 & 0 \\ A_{di} & 0 \end{bmatrix},$$
$$\tilde{P}_i = \begin{bmatrix} P_i & 0 \\ M_i & N_i \end{bmatrix}, \quad \tilde{S}_i = \begin{bmatrix} S_i & 0 \\ R_i & 0 \end{bmatrix}, \quad \tilde{T}_i = \begin{bmatrix} T_i & 0 \\ 0 & 0 \end{bmatrix}, \quad \tilde{Q}_i = \begin{bmatrix} Q_i & 0 \\ 0 & dZ \end{bmatrix}.$$

Since $\operatorname{rank}\tilde{E} = \operatorname{rank}E = r \leqslant n$, there exist nonsingular matrices G and H such that

$$\hat{E} = G\tilde{E}H = \begin{bmatrix} I_r & 0 \\ 0 & 0 \end{bmatrix}. \tag{10.15}$$

Denote

$$G\tilde{A}_iH = \begin{bmatrix} A_{i11} & A_{i12} \\ A_{i21} & A_{i22} \end{bmatrix}, \quad G^{-T}\tilde{P}_iH = \begin{bmatrix} P_{i11} & P_{i12} \\ P_{i21} & P_{i22} \end{bmatrix}, \quad H^{\mathrm{T}}\tilde{S}_iG^{-1} = \begin{bmatrix} S_{i11} & S_{i12} \\ S_{i21} & S_{i22} \end{bmatrix}, \tag{10.16}$$

for every $i \in \mathcal{S}$. By (10.13), it can be shown that $P_{i12} = 0$ for every $i \in \mathcal{S}$. Pre- and post-multiplying

$$\pi_{ii}\tilde{E}^{\mathrm{T}}\tilde{P}_i + \tilde{A}_i^{\mathrm{T}}\tilde{P}_i + \tilde{P}_i^{\mathrm{T}}\tilde{A}_i + \tilde{S}_i\tilde{E} + \tilde{E}^{\mathrm{T}}\tilde{S}_i^{\mathrm{T}} < 0 \tag{10.17}$$

by H^{T} and H, respectively, we have

$$A_{i22}^{T}P_{i22} + P_{i22}^{T}A_{i22} < 0. \tag{10.18}$$

This implies A_{i22} is nonsingular for every $i \in \mathcal{S}$, and thus the pair $(\tilde{E}, \tilde{A}_i)$ is regular and impulse free for every $i \in \mathcal{S}$. Since $\det(sE - A_i) = \det(s\tilde{E} - \tilde{A}_i)$, we can easily see that the pair (E, A_i) is regular and impulse free for every $i \in \mathcal{S}$. According to Definition 10.1, system (10.5) is regular and impulse free.

Next, we will show the stochastic stability of system (10.5). Define a new process $\{(x_t, r_t), t \geqslant 0\}$ by $\{x_t = x(t+\theta), -2d \leqslant \theta \leqslant 0\}$, then $\{(x_t, r_t), t \geqslant d\}$ is a Markov process with initial state $(\phi(\cdot), r_0)$. Now, for $t \geqslant d$, define the following Lyapunov functional for system (10.5):

$$V(x_t, r_t, t) = V_1(x_t, r_t, t) + V_2(x_t, r_t, t) + V_3(x_t, r_t, t) + V_4(x_t, r_t, t), \tag{10.19}$$

where

$$
\begin{aligned}
V_1(x_t, r_t, t) =& x(t)^{\mathrm{T}} E^{\mathrm{T}} P(r_t) x(t), \\
V_2(x_t, r_t, t) =& \int_{t-d}^{t} x(\alpha)^{\mathrm{T}} Q(r_t) x(\alpha) \, \mathrm{d}\alpha, \\
V_3(x_t, r_t, t) =& \int_{-d}^{0} \int_{t+\beta}^{t} \dot{x}(\alpha)^{\mathrm{T}} E^{\mathrm{T}} Z E \dot{x}(\alpha) \, \mathrm{d}\alpha \mathrm{d}\beta, \\
V_4(x_t, r_t, t) =& \mu \int_{-d}^{0} \int_{t+\beta}^{t} x(\alpha)^{\mathrm{T}} Q x(\alpha) \, \mathrm{d}\alpha \mathrm{d}\beta.
\end{aligned}
$$

Let $\mathcal{A}$ be the weak infinitesimal generator of the random process $\{x_t, r_t\}$. Then, for each $i \in \mathcal{S}$, we have

$$
\begin{aligned}
\mathcal{A}V(x_t, i, t) =& x(t)^{\mathrm{T}} E^{\mathrm{T}} P_i \dot{x}(t) + x(t)^{\mathrm{T}} \left\{ \sum_{j=1}^{s} \pi_{ij} E^{\mathrm{T}} P_j \right\} x(t) + x(t)^{\mathrm{T}} Q_i x(t) \\
& - x(t-d)^{\mathrm{T}} Q_i x(t-d) + \int_{t-d}^{t} x(\alpha)^{\mathrm{T}} \left\{ \sum_{j=1}^{s} \pi_{ij} Q_j \right\} x(\alpha) \, \mathrm{d}\alpha \\
& + d\dot{x}(t)^{\mathrm{T}} E^{\mathrm{T}} Z E \dot{x}(t) - \int_{t-d}^{t} \dot{x}(\alpha)^{\mathrm{T}} E^{\mathrm{T}} Z E \dot{x}(\alpha) \, \mathrm{d}\alpha \\
& + \mu d x(t)^{\mathrm{T}} Q x(t) - \mu \int_{t-d}^{t} x(\alpha)^{\mathrm{T}} Q x(\alpha) \, \mathrm{d}\alpha \\
& + 2 \left[x(t)^{\mathrm{T}} S_i + (E\dot{x}(t))^{\mathrm{T}} R_i + x(t-d)^{\mathrm{T}} T_i \right] \\
& \times \left[Ex(t) - Ex(t-d) - \int_{t-d}^{t} E\dot{x}(\alpha) \, \mathrm{d}\alpha \right] \\
& + 2 \left[x(t)^{\mathrm{T}} M_i^{\mathrm{T}} + (E\dot{x}(t))^{\mathrm{T}} N_i^{\mathrm{T}} \right] \\
& \times \left[-E\dot{x}(t) + A_i x(t) + A_{di} x(t-d) \right].
\end{aligned} \tag{10.20}
$$

According to Jensen inequality, one can obtain

$$
- \int_{t-d}^{t} \dot{x}(\alpha)^{\mathrm{T}} E^{\mathrm{T}} Z E \dot{x}(\alpha) \, \mathrm{d}\alpha \leqslant \zeta(t)^{\mathrm{T}} (-dZ) \zeta(t), \tag{10.21}
$$

where $\zeta(t) = -\int_{t-d}^{t} \frac{1}{d} E\dot{x}(\alpha) \, \mathrm{d}\alpha$. Noting $Q_i < Q$, $\pi_{ij} > 0$ for $i \neq j$ and $-\mu \leqslant \pi_{ii} < 0$, we have

$$
\begin{aligned}
\int_{t-d}^{t} x(\alpha)^{\mathrm{T}} \left\{ \sum_{j=1}^{s} \pi_{ij} Q_j \right\} x(\alpha) \, \mathrm{d}\alpha \leqslant& \int_{t-d}^{t} x(\alpha)^{\mathrm{T}} \left\{ \sum_{j=1, j \neq i}^{s} \pi_{ij} Q_j \right\} x(\alpha) \, \mathrm{d}\alpha \\
\leqslant& \int_{t-d}^{t} x(\alpha)^{\mathrm{T}} \left\{ \sum_{j=1, j \neq i}^{s} \pi_{ij} Q \right\} x(\alpha) \, \mathrm{d}\alpha
\end{aligned}
$$

$$
\begin{aligned}
&= -\pi_{ii} \int_{t-d}^{t} x(\alpha)^{\mathrm{T}} Q x(\alpha)\,\mathrm{d}\alpha \\
&\leqslant \mu \int_{t-d}^{t} x(\alpha)^{\mathrm{T}} Q x(\alpha)\,\mathrm{d}\alpha. \qquad (10.22)
\end{aligned}
$$

Using this and (10.21), we have that, for every $i \in \mathcal{S}$,

$$
\mathcal{A}V(x_t,i,t) \leqslant \begin{bmatrix} x(t) \\ E\dot{x}(t) \\ x(t-d) \\ \zeta(t) \end{bmatrix}^{\mathrm{T}} \begin{bmatrix} \Xi_{i11} & \Xi_{i12} & -S_iE + M_i^{\mathrm{T}}A_{di} + E^{\mathrm{T}}T_i^{\mathrm{T}} & dS_i \\ * & \Xi_{i22} & -R_iE + N_i^{\mathrm{T}}A_{di} & dR_i \\ * & * & -Q_i - T_iE - E^{\mathrm{T}}T_i^{\mathrm{T}} & dT_i \\ * & * & * & -dZ \end{bmatrix} \times \begin{bmatrix} x(t) \\ E\dot{x}(t) \\ x(t-d) \\ \zeta(t) \end{bmatrix}. \qquad (10.23)
$$

From (10.11), it is easy to see that

$$
\begin{bmatrix} \Xi_{i11} & \Xi_{i12} & -S_iE + M_i^{\mathrm{T}}A_{di} + E^{\mathrm{T}}T_i^{\mathrm{T}} & dS_i \\ * & \Xi_{i22} & -R_iE + N_i^{\mathrm{T}}A_{di} & dR_i \\ * & * & -Q_i - T_iE - E^{\mathrm{T}}T_i^{\mathrm{T}} & dT_i \\ * & * & * & -dZ \end{bmatrix} < 0, \qquad (10.24)
$$

and thus there exits a scalar $\lambda > 0$ such that for every $i \in \mathcal{S}$,

$$
\mathcal{A}V(x_t,i,t) \leqslant -\lambda \|x(t)\|^2. \qquad (10.25)
$$

Therefore, for any $t \geqslant d$, by Dynkin's formula, we get

$$
\mathscr{E}\{V(x_t,i,t)\} - \mathscr{E}\{V(x_d,r_d,d)\} \leqslant -\lambda \mathscr{E}\left\{\int_d^t \|x(s)\|^2\,\mathrm{d}s\right\}, \qquad (10.26)
$$

which yields

$$
\mathscr{E}\left\{\int_d^t \|x(s)\|^2\,\mathrm{d}s\right\} \leqslant \lambda^{-1}\mathscr{E}\{V(x_d,r_d,d)\}. \qquad (10.27)
$$

Because of the regularity and the non-impulsiveness of the pair (E, A_i) for every $i \in \mathcal{S}$, we can choose two nonsingular matrices M and N such that

$$
MEN = \begin{bmatrix} I_r & 0 \\ 0 & 0 \end{bmatrix}, \quad MA_iN = \begin{bmatrix} A_{i1} & A_{i2} \\ A_{i3} & A_{i4} \end{bmatrix}, \qquad (10.28)
$$

where A_{i4} is nonsingular for every $i \in \mathcal{S}$. Set

$$
\hat{M} = \begin{bmatrix} I_r & -A_{i2}A_{i4}^{-1} \\ 0 & A_{i4}^{-1} \end{bmatrix} M. \qquad (10.29)
$$

It is easy to get

$$\hat{M}EN = \begin{bmatrix} I_r & 0 \\ 0 & 0 \end{bmatrix}, \ \hat{M}A_iN = \begin{bmatrix} \hat{A}_{i1} & 0 \\ \hat{A}_{i3} & I \end{bmatrix}, \tag{10.30}$$

where $\hat{A}_{i1} = A_{i1} - A_{i2}A_{i4}^{-1}A_{i3}$ and $\hat{A}_{i3} = A_{i4}^{-1}A_{i3}$. Denote

$$\hat{M}A_{di}N = \begin{bmatrix} A_{id1} & A_{id2} \\ A_{id3} & A_{id4} \end{bmatrix}. \tag{10.31}$$

Then, for every $i \in \mathcal{S}$, system (10.5) is equivalent to

$$\begin{cases} \dot{\zeta}_1(t) = \hat{A}_{i1}\zeta_1(t) + A_{id1}\zeta_1(t-d) + A_{id2}\zeta_2(t-d), \\ -\zeta_2(t) = \hat{A}_{i13}\zeta_1(t) + A_{id3}\zeta_1(t-d) + A_{id4}\zeta_2(t-d), \\ \zeta(t) = \psi(t) = N^{-1}\phi(t), \ t \in [-d, 0], \end{cases} \tag{10.32}$$

where

$$\zeta(t) = \begin{bmatrix} \zeta_1(t) \\ \zeta_2(t) \end{bmatrix} = N^{-1}x(t).$$

For any $t \geqslant 0$, it follows from (10.32) that

$$\begin{aligned} \|\zeta_1(t)\| &= \|\zeta_1(0) + \int_0^t \left[\hat{A}_{i1}\zeta_1(\alpha) + A_{id1}\zeta_1(\alpha-d) + A_{id2}\zeta_2(\alpha-d)\right] \mathrm{d}\alpha\| \\ &\leqslant \|\zeta_1(0)\| + k_1 \int_0^t [\|\zeta_1(\alpha)\| + \|\zeta_1(\alpha-d)\| + \|\zeta_2(\alpha-d)\|] \,\mathrm{d}\alpha, \end{aligned} \tag{10.33}$$

where $k_1 = \max_{i\in\mathcal{S}}\{\|\hat{A}_{i1}\|, \|A_{id1}\|, \|A_{id2}\|\} \geqslant 0$. Then, for any $0 \leqslant t \leqslant d$, we have

$$\|\zeta_1(t)\| \leqslant (2k_1d+1)\|\psi\|_d + k_1 \int_0^t \|\zeta_1(\alpha)\|\mathrm{d}\alpha. \tag{10.34}$$

Applying the Gronwall-Bellman Lemma, we obtain from (10.34) that for any $0 \leqslant t \leqslant d$,

$$\|\zeta_1(t)\| \leqslant (2k_1d+1)\|\psi\|_d e^{k_1d}. \tag{10.35}$$

Thus

$$\sup_{0\leqslant\alpha\leqslant d} \|\zeta_1(\alpha)\|^2 \leqslant (2k_1d+1)^2\|\psi\|_d^2 e^{2k_1d}. \tag{10.36}$$

We have from (10.32) that

$$\sup_{0\leqslant\alpha\leqslant d} \|\zeta_2(\alpha)\|^2 \leqslant k_2^2[(2k_1d+1)e^{k_1d}+2]^2\|\psi\|_d^2, \tag{10.37}$$

where $k_2 = \max_{i\in\mathcal{S}}\{\|\hat{A}_{i3}\|, \|A_{id3}\|, \|A_{id4}\|\}$. Hence,

$$\sup_{0\leqslant\alpha\leqslant d}\|\zeta(\alpha)\|^2 \leqslant \sup_{0\leqslant\alpha\leqslant d}\|\zeta_1(\alpha)\|^2 + \sup_{0\leqslant\alpha\leqslant d}\|\zeta_2(\alpha)\|^2 \leqslant k_3\|\psi\|_d^2, \tag{10.38}$$

where $k_3 = (2k_1d+1)^2e^{2k_1d} + k_2^2[(2k_1d+1)e^{k_1d}+2]^2$. Therefore,

$$\sup_{0\leqslant\alpha\leqslant d}\|x(\alpha)\|^2 \leqslant k_3\|N\|^2\|N^{-1}\|^2\|\phi\|_d^2. \tag{10.39}$$

Note that

$$\int_{-d}^{0}\int_{t+\beta}^{t}\dot{x}(\alpha)^{\mathrm{T}}E^{\mathrm{T}}ZE\dot{x}(\alpha)\,\mathrm{d}\alpha\mathrm{d}\beta \leqslant d\int_{t-d}^{t}\dot{x}(\alpha)^{\mathrm{T}}E^{\mathrm{T}}ZE\dot{x}(\alpha)\,\mathrm{d}\alpha, \tag{10.40}$$

and

$$\mu\int_{-d}^{0}\int_{t+\beta}^{t}x(\alpha)^{\mathrm{T}}Qx(\alpha)\,\mathrm{d}\alpha\mathrm{d}\beta \leqslant \mu d\int_{t-d}^{t}x(\alpha)^{\mathrm{T}}Qx(\alpha)\,\mathrm{d}\alpha. \tag{10.41}$$

Then we have from (10.19) and (10.39) that there exits a scalar ϱ such that

$$V(x_d, r_d, d) \leqslant \varrho\|\phi\|_d^2. \tag{10.42}$$

This together with (10.27) and (10.39) implies there exits a scalar ρ such that

$$\begin{aligned}\mathscr{E}\left\{\int_0^t\|x(s)\|^2\,\mathrm{d}s\right\} &= \mathscr{E}\left\{\int_0^d\|x(s)\|^2\,\mathrm{d}s\right\} + \mathscr{E}\left\{\int_d^t\|x(s)\|^2\,\mathrm{d}s\right\}\\ &\leqslant \rho\mathscr{E}\|\phi\|_d^2.\end{aligned} \tag{10.43}$$

Considering this and Definition 10.1, system (10.5) is stochastically stable.

In the following, we establish the H_∞ performance of system (10.9). For this purpose, we consider the following index:

$$J_{z\omega}(t) = \mathscr{E}\left\{\int_0^t\left[z(s)^{\mathrm{T}}z(s) - \gamma^2\omega(s)^{\mathrm{T}}\omega(s)\right]\,\mathrm{d}s\right\}. \tag{10.44}$$

Under zero initial condition, it easy to see that

$$\begin{aligned}J_{z\omega}(t) \leqslant &\mathscr{E}\left\{\int_0^t\left[z(s)^{\mathrm{T}}z(s) - \gamma^2\omega(s)^{\mathrm{T}}\omega(s) + \mathcal{A}V(x_s, i, s)\right]\,\mathrm{d}s\right\}\\ \leqslant &\mathscr{E}\left\{\int_0^t\mathcal{X}(s)^{\mathrm{T}}(\Theta_i + \Lambda_i^{\mathrm{T}}\Lambda_i)\mathcal{X}(s)\,\mathrm{d}s\right\},\end{aligned} \tag{10.45}$$

where

$$\mathcal{X}(t)^{\mathrm{T}} = \left[x^{\mathrm{T}}(t)\ (E\dot{x}(t))^{\mathrm{T}}\ x^{\mathrm{T}}(t-d)\ \omega^{\mathrm{T}}(t)\ \zeta^{\mathrm{T}}(t)\right],$$

$$\Theta_i = \begin{bmatrix}\Xi_{i11} & \Xi_{i12} & -S_iE + M_i^{\mathrm{T}}A_{di} + E^{\mathrm{T}}T_i^{\mathrm{T}} & M_i^{\mathrm{T}}B_{\omega i} & dS_i\\ * & \Xi_{i22} & -R_iE + N_i^{\mathrm{T}}A_{di} & N_i^{\mathrm{T}}B_{\omega i} & dR_i\\ * & * & -Q_i - T_iE - E^{\mathrm{T}}T_i^{\mathrm{T}} & 0 & dT_i\\ * & * & * & -\gamma^2I & 0\\ * & * & * & * & -dZ\end{bmatrix},$$

$$\Lambda_i = \left[C_i\ 0\ C_{di}\ D_{\omega i}\ 0\right].$$

Hence, by Schur complement, we can obtain from (10.11) that for all $t > 0$, $J_{z\omega}(t) < 0$. Therefore, we arrive at (10.8) for any nonzero $\omega(t) \in \mathscr{L}_2[0,\infty)$. This completes the proof.

Remark 10.3. A delay-dependent BRL is given in Theorem 10.2 for system (10.9) to be stochastically admissible with H_∞ performance γ. It should be pointed out that, during the proof process of Theorem 10.2, neither model transformation nor bounding technique for cross terms is required. Hence, the conservatism inherited from these ideas will no longer exist in Theorem 10.2.

Remark 10.4. It is clear that Theorem 10.2 can also be used to determine the regularity, absence of impulses and delay-dependent stochastic stability of system (10.5). A delay-dependent stability condition has also been proposed in [7] for the same system. But it should be pointed out that the considered system in [7] is assumed to be regular and impulse free, moreover, a matrix describing the relationship between fast and slow subsystems is needed and an improper choice of the matrix, which often can not be found, would make the results unreliable. Hence, Theorem 10.2 is much more desirable and powerful than the result of [7].

10.3.2 H_∞ Controller Design

We are now ready to deal with the design problem of the delay-dependent H_∞ controller for system (10.1). For this purpose, applying the state feedback control (10.7) to system (10.1), the resultant closed-loop system is described by

$$\begin{cases} E\dot{x}(t) = (A_i + B_i K_i)x(t) + A_{di}x(t-d) + B_{\omega i}\omega(t), \\ z(t) = (C_i + D_i K_i)x(t) + C_{di}x(t-d) + D_{\omega i}\omega(t), \\ x(t) = \phi(t), \; t \in [-d, 0]. \end{cases} \tag{10.46}$$

Using Theorem 10.2 to the above system, we get

$$\begin{bmatrix} \mathcal{Q}_i & \tilde{P}_i^{\mathrm{T}} \begin{bmatrix} 0 \\ A_{di} \end{bmatrix} - \begin{bmatrix} S_i \\ R_i \end{bmatrix} E + \begin{bmatrix} E^{\mathrm{T}} \\ 0 \end{bmatrix} T_i^{\mathrm{T}} & \tilde{P}_i^{\mathrm{T}} \begin{bmatrix} 0 \\ B_\omega \end{bmatrix} & d \begin{bmatrix} S_i \\ R_i \end{bmatrix} & (C_i + D_i K_i)^{\mathrm{T}} \\ * & -Q_i - T_i E - E^{\mathrm{T}} T_i^{\mathrm{T}} & 0 & dT_i & C_{di}^{\mathrm{T}} \\ * & * & -\gamma^2 I & 0 & D_{\omega i}^{\mathrm{T}} \\ * & * & * & -dZ & 0 \\ * & * & * & * & -I \end{bmatrix} < 0, \tag{10.47}$$

where

$$\begin{aligned}\mathcal{Q}_i =& \tilde{P}_i^{\mathrm{T}} \begin{bmatrix} 0 & I \\ A_i + B_iK_i & -I \end{bmatrix} + \begin{bmatrix} 0 & I \\ A_i + B_iK_i & -I \end{bmatrix}^{\mathrm{T}} \tilde{P}_i \\ &+ \begin{bmatrix} \sum_{j=1}^{s} \pi_{ij} E^{\mathrm{T}} P_j + Q_i + \mu d Q & 0 \\ 0 & dZ \end{bmatrix} + \begin{bmatrix} S_i \\ R_i \end{bmatrix} [E\ 0] + \begin{bmatrix} E^{\mathrm{T}} \\ 0 \end{bmatrix} [S_i^{\mathrm{T}}\ R_i^{\mathrm{T}}], \\ \tilde{P}_i =& \begin{bmatrix} P_i & 0 \\ M_i & N_i \end{bmatrix}.\end{aligned}$$

From the proof of Theorem 10.2, we know that $\tilde{P}_i$ is nonsingular for every $i \in \mathcal{S}$. Define

$$\tilde{P}_i^{-1} = \mathcal{L}_i = \begin{bmatrix} L_i & 0 \\ \bar{M}_i & \bar{N}_i \end{bmatrix}, \begin{bmatrix} S_i \\ R_i \end{bmatrix} = \delta_{1i} \tilde{P}_i^{\mathrm{T}} \begin{bmatrix} 0 \\ I \end{bmatrix}, T_i = \delta_{2i} Q_i. \tag{10.48}$$

Pre- and post-multiplying (10.10) by L_i^{T} and L_i, respectively, we get

$$EL_i = L_i^{\mathrm{T}} E^{\mathrm{T}} \geqslant 0. \tag{10.49}$$

Similarly, pre- and post-multiply (10.47) by $\mathrm{diag}\{\mathcal{L}_i^{\mathrm{T}}, Q_i^{-1}, I, I, I\}$ and its transposition, respectively, and introduce change of variables such that

$$\bar{Q}_i = Q_i^{-1},\ \bar{Z} = Z^{-1},\ V_i = K_i L_i. \tag{10.50}$$

After some manipulation including Schur complement, we can obtain from (10.47) and (10.12) that

$$\begin{bmatrix} \Xi_{i11} & \Xi_{i12} & \delta_{2i} L_i^{\mathrm{T}} E^{\mathrm{T}} & 0 & 0 & V_i^{\mathrm{T}} D_i^{\mathrm{T}} + L_i^{\mathrm{T}} C_i^{\mathrm{T}} & L_i^{\mathrm{T}} & \mu d L_i^{\mathrm{T}} & d M_i^{\mathrm{T}} \\ * & \Xi_{i22} & \Xi_{i23} & B_{\omega i} & d\delta_{1i} \bar{Z} & 0 & 0 & 0 & d N_i^{\mathrm{T}} \\ * & * & \Xi_{i33} & 0 & d\delta_{2i} \bar{Z} & C_{di}^{\mathrm{T}} & 0 & 0 & 0 \\ * & * & * & -\gamma^2 I & 0 & D_{\omega i}^{\mathrm{T}} & 0 & 0 & 0 \\ * & * & * & * & -d\bar{Z} & 0 & 0 & 0 & 0 \\ * & * & * & * & * & -I & 0 & 0 & 0 \\ * & * & * & * & * & * & -\bar{Q}_i & 0 & 0 \\ * & * & * & * & * & * & * & -\mu d \bar{Q} & 0 \\ * & * & * & * & * & * & * & * & d\bar{Z} \end{bmatrix} < 0, \tag{10.51}$$

$$\bar{Q} < \bar{Q}_i, \tag{10.52}$$

where

$$\begin{aligned}\Xi_{i11} =& \bar{M}_i + \bar{M}_i^{\mathrm{T}} + \pi_{ii} L_i^{\mathrm{T}} E^{\mathrm{T}} + \sum_{j=1, j\neq i}^{s} \pi_{ij} L_i^{\mathrm{T}} E^{\mathrm{T}} L_j^{-1} L_i, \\ \Xi_{i12} =& \bar{N}_i - \bar{M}_i^{\mathrm{T}} + \delta_{1i} L_i^{\mathrm{T}} E^{\mathrm{T}} + (A_i L_i + B_i V_i)^T, \\ \Xi_{i22} =& -\bar{N}_i - \bar{N}_i^{\mathrm{T}}, \\ \Xi_{i23} =& (-\delta_{1i} E + A_{di}) \bar{Q}_i, \\ \Xi_{i33} =& -\bar{Q}_i - \delta_{2i} E \bar{Q}_i - \delta_{2i} \bar{Q}_i E^{\mathrm{T}}.\end{aligned}$$

It is noted that the conditions in (10.51) are no longer LMI conditions because of the term $\sum_{j=1,j\neq i}^{s} \pi_{ij} L_i^{\mathrm{T}} E^{\mathrm{T}} L_j^{-1} L_i$. As a result, unfortunately in this case, it is rather difficult to solve them. In order to obtain an LMI-based design method of the desired state feedback controller, without loss of generality, in the following discussion we assume that

$$E = \begin{bmatrix} I_r & 0 \\ 0 & 0 \end{bmatrix}. \tag{10.53}$$

Then (10.49) implies

$$L_i = \begin{bmatrix} L_{i1} & 0 \\ L_{i2} & L_{i3} \end{bmatrix}, \tag{10.54}$$

where $L_{i1} > 0$. After some manipulation, we find

$$\sum_{j=1,j\neq i}^{s} \pi_{ij} L_i^{\mathrm{T}} E^{\mathrm{T}} L_j^{-1} L_i = \sum_{j=1,j\neq i}^{s} \pi_{ij} \begin{bmatrix} L_{i1} \\ 0 \end{bmatrix} L_{j1}^{-1} \begin{bmatrix} L_{i1} & 0 \end{bmatrix}. \tag{10.55}$$

Now, applying Schur complement to (10.51), we get

$$\begin{bmatrix} \bar{\Xi}_{i11} & \Xi_{i12} & \delta_{2i} L_i E^{\mathrm{T}} & 0 & 0 & V_i^{\mathrm{T}} D_i^{\mathrm{T}} + L_i^{\mathrm{T}} C_i^{\mathrm{T}} & L_i^{\mathrm{T}} & \mu d L_i^{\mathrm{T}} & d M_i^{\mathrm{T}} & \mathcal{Z}_i \\ * & \Xi_{i22} & \Xi_{i23} & B_{\omega i} & d\delta_{1i} \bar{Z} & 0 & 0 & 0 & d N_i^{\mathrm{T}} & 0 \\ * & * & \Xi_{i33} & 0 & d\delta_{2i} \bar{Z} & C_{di}^{\mathrm{T}} & 0 & 0 & 0 & 0 \\ * & * & * & -\gamma^2 I & 0 & D_{\omega i}^{\mathrm{T}} & 0 & 0 & 0 & 0 \\ * & * & * & * & -d\bar{Z} & 0 & 0 & 0 & 0 & 0 \\ * & * & * & * & * & -I & 0 & 0 & 0 & 0 \\ * & * & * & * & * & * & -\bar{Q}_i & 0 & 0 & 0 \\ * & * & * & * & * & * & * & -\mu d \bar{Q} & 0 & 0 \\ * & * & * & * & * & * & * & * & -d\bar{Z} & 0 \\ * & * & * & * & * & * & * & * & * & -\mathcal{F}_i \end{bmatrix} < 0, \tag{10.56}$$

where

$$\begin{aligned} \bar{\Xi}_{i11} &= \bar{M}_i + \bar{M}_i^{\mathrm{T}} + \pi_{ii} L_i^{\mathrm{T}} E^{\mathrm{T}}, \\ \mathcal{Z}_i &= \left[\sqrt{\pi_{i1}} \begin{bmatrix} L_{i1} \\ 0 \end{bmatrix} \cdots \sqrt{\pi_{i(i-1)}} \begin{bmatrix} L_{i1} \\ 0 \end{bmatrix} \sqrt{\pi_{i(i+1)}} \begin{bmatrix} L_{i1} \\ 0 \end{bmatrix} \cdots \sqrt{\pi_{is}} \begin{bmatrix} L_{i1} \\ 0 \end{bmatrix} \right], \\ \mathcal{F}_i &= \mathrm{diag}\left\{ L_{11} \cdots L_{(i-1)1} \; L_{(i+1)1} \cdots L_{s1} \right\}. \end{aligned}$$

Them, we have the following result on the delay-dependent H_∞ controller for system (10.1).

Theorem 10.5. *For given scalars $\gamma > 0$, δ_{1i} and δ_{2i}, system (10.1) is stochastically admissible with H_∞ performance γ, if there exist matrices $\bar{Q}_i > 0$, $\bar{Q} > 0$, $\bar{Z} > 0$, $L_{i1} > 0$, $L_i = \begin{bmatrix} L_{i1} & 0 \\ L_{i2} & L_{i3} \end{bmatrix}$, V_i, $\bar{M}_i$ and $\bar{N}_i$ such that for every $i \in \mathcal{S}$, (10.52) and (10.56) hold. Furthermore, if (10.52) and (10.56) are solvable, the desired controller gain is given as*

$$K_i = V_i L_i^{-1}. \tag{10.57}$$

Remark 10.6. Without any additional assumption on the regularity, absence of impulses of the considered system, Theorem 10.5 provides a sufficient condition for the solvability of the delay-dependent H_∞ control problem for SMJSs. Observe that the conditions in (10.52) and (10.56) are strict LMIs that can be readily solved. It is also worth noting that the desired controller can be constructed by solving the LMIs in (10.52) and (10.56).

Remark 10.7. Based on Theorem 10.5 in this chapter, we can readily obtain the delay-dependent result on the solvability of the delay-dependent stabilization problem for system (10.1) with $\omega(t) = 0$. The stabilization problem of SMJSs with time-delay and H_∞ problem of SMJSs with delay free have been studied in [7, 9] and [178, Theorem 10.4], respectively. However, the considered system of [7] is assumed to be regular and impulse free, hence the effect of the obtained controller upon the regularity and absence of impulses of the closed-loop system has not been considered. Furthermore, in order to get the design methods of the desired state feedback controller, [7] has introduced the contain equation constraints, and [9] and [178, Theorem 10.4] have introduced scalars ρ_i such that $0 \leqslant EL_i = L_i^{\mathrm{T}} E^{\mathrm{T}} \leqslant \rho_i I$. Clearly, such methods will result in some numerical problems and increase the conservatism and complexity of the design procedure. Therefore, our result is much more desirable and elegant than [178, Theorem 10.4] and [7, 9].

10.4 Numerical Examples

This section provides several numerical examples that demonstrate the effectiveness and less conservatism of the results presented in this chapter.

Example 10.8. Consider system (10.9) with $E = I$, two modes and the following parameters:

$$
\begin{aligned}
A_1 &= \begin{bmatrix} -4.5 & 0.9 \\ -0.7 & -3 \end{bmatrix}, \; A_2 = \begin{bmatrix} -3.5 & 0.5 \\ 1.4 & -0.2 \end{bmatrix}, \\
A_{d1} &= \begin{bmatrix} -1 & -1.4 \\ -0.9 & -2 \end{bmatrix}, \; A_{d2} = \begin{bmatrix} -3 & 0.4 \\ -1 & -1.1 \end{bmatrix}, \\
B_{\omega 1} &= \begin{bmatrix} 0.05 \\ 0.6 \end{bmatrix}, \; B_{\omega 2} = \begin{bmatrix} 0.5 \\ -0.3 \end{bmatrix}, \\
C_1 &= \begin{bmatrix} -0.4 & -0.3 \end{bmatrix}, \; C_2 = \begin{bmatrix} -1.5 & -0.2 \end{bmatrix}, \\
C_{d1} &= C_{d2} = 0, \; D_{\omega 1} = 0.12, \; D_{\omega 2} = 0.31.
\end{aligned}
$$

It is assumed that $\pi_{11} = -0.1$ and $\pi_{22} = -0.8$. By solving the LMIs in Theorem 10.2, we can calculate the maximum allowed time delay d for given $\gamma > 0$ and the minimum allowed γ for given $d > 0$. Table 10.1 and Table 10.2 provide the comparison results, respectively, via the methods in [61, 144] and Theorem 10.2 in our chapter. It can be seen that these comparison results

Table 10.1. Example 10.8: Comparison of delay-dependent BRLs

γ	1.6	1.4	1.2	1.0	0.8
[61, 144]	0.5302	0.5236	0.5145	0.5014	0.4809
Theorem 10.2	0.5708	0.5597	0.5444	0.5252	0.5037

Table 10.2. Example 10.8: Comparison of delay-dependent BRLs

d	0.40	0.45	0.50	0.55	0.60
[61, 144]	0.4553	0.6179	0.9829	2.9275	—
Theorem 10.2	0.4500	0.5145	0.7705	1.2658	2.9182

show that the BRL in Theorem 10.2 for regular time-delay systems in this chapter is less conservative than those in [61, 144].

Example 10.9. Consider system (10.9) with two modes. The mode switching is governed by the transition rate matrix

$$\Pi = \begin{bmatrix} -0.5 & 0.5 \\ 0.3 & -0.3 \end{bmatrix}.$$

The system parameters are described as follows.
For mode 1

$$A_1 = \begin{bmatrix} 0.4972 & 0 \\ 0 & -0.9541 \end{bmatrix}, \; A_{d1} = \begin{bmatrix} -1.010 & 1.5415 \\ 0 & 0.5449 \end{bmatrix},$$

$$B_{\omega 1} = \begin{bmatrix} 0.4212 \\ -0.3211 \end{bmatrix}, \; C_1 = \begin{bmatrix} -0.1252 & 0.4523 \end{bmatrix}, \; C_{d1} = 0, \; D_{\omega 1} = 0.2.$$

For mode 2

$$A_2 = \begin{bmatrix} 0.5121 & 0 \\ 0 & -0.7215 \end{bmatrix}, \; A_{d2} = \begin{bmatrix} -0.8521 & 1.9721 \\ 0 & 0.4321 \end{bmatrix},$$

$$B_{\omega 2} = \begin{bmatrix} 0.5100 \\ -0.3100 \end{bmatrix}, \; C_2 = \begin{bmatrix} -0.1987 & 0.4921 \end{bmatrix}, \; C_{d2} = 0, \; D_{\omega 2} = 0.1.$$

The singular matrix E is given by the following expression:

$$E = \begin{bmatrix} 1 & 0 \\ 0 & 0 \end{bmatrix}.$$

Using Theorem 10.2, it can be shown that system with $\omega(t) = 0$ is regular, impulse free and stochastically stable for any constant time delay d satisfying $0 \leqslant d \leqslant 1.0878$. Although the result of [7] fails to determine the stochastic admissibility of the above system. When taking disturbance $\omega(t)$ into account, Table 10.3 and Table 10.4 provide the maximum allowed time-delay d for different $\gamma > 0$ and the minimum allowed γ for various $d > 0$, respectively.

Table 10.3. Example 10.9: Maximum allowed time delay d for different γ

γ	0.85	1.00	1.15	1.30	1.45	1.60	1.75
d	0.3945	0.4987	0.5862	0.6597	0.7217	0.7739	0.8179

Table 10.4. Example 10.9: Minimum allowed γ for different d

d	0.4	0.5	0.6	0.7	0.8	0.9	1.0
γ	0.8573	1.0021	1.1762	1.3945	1.6857	2.1266	3.2477

Example 10.10. Consider system (10.1) with two modes. The system parameters are described as follows.
For mode 1

$$
\begin{aligned}
A_1 &= \begin{bmatrix} 2.7 & 1.5 & 2 \\ -1.2 & 1.6 & 1 \\ 1.2 & 1 & -0.5 \end{bmatrix}, \; A_{d1} = \begin{bmatrix} -2 & 2 & 1 \\ 2.2 & 1.3 & 1.9 \\ 1.3 & 1 & -0.2 \end{bmatrix}, \\
B_1 &= \begin{bmatrix} 4.5 & 2 \\ 1 & -1 \\ 1 & 0.7 \end{bmatrix}, \; B_{\omega 1} = \begin{bmatrix} 0.02 \\ 0.01 \\ -0.1 \end{bmatrix}, \\
C_1 &= \begin{bmatrix} 0.1 & 0.5 & 0 \end{bmatrix}, \; C_{d1} = 0, \\
D_1 &= \begin{bmatrix} 1.5 & -2 \end{bmatrix}, \; D_{\omega 1} = 0.
\end{aligned}
$$

For mode 2

$$
\begin{aligned}
A_2 &= \begin{bmatrix} 2.5 & 2 & 2 \\ -1.5 & -2 & 2 \\ 1 & 1.2 & -0.2 \end{bmatrix}, \; A_{d2} = \begin{bmatrix} 1 & 1.1 & 1.5 \\ 2.5 & -2.2 & 1 \\ -2 & 1.5 & -0.1 \end{bmatrix}, \\
B_2 &= \begin{bmatrix} 5.5 & 1 \\ 1 & 1 \\ 2 & 2 \end{bmatrix}, \; B_{\omega 2} = \begin{bmatrix} 0.02 \\ 0.01 \\ -0.1 \end{bmatrix}, \\
C_2 &= \begin{bmatrix} 0.5 & 0.2 & -1 \end{bmatrix}, \; C_{d2} = 0, \\
D_2 &= \begin{bmatrix} 0.5 & 1 \end{bmatrix}, \; D_{\omega 2} = 0.
\end{aligned}
$$

The singular matrix E is given by the following expression:

$$
E = \begin{bmatrix} 1 & 0 & 0 \\ 0 & 1 & 0 \\ 0 & 0 & 0 \end{bmatrix},
$$

and the transition rate matrix

$$
\Pi = \begin{bmatrix} -0.5 & 0.5 \\ 1 & -1 \end{bmatrix}.
$$

Solving the LMIs (10.52) and (10.56) with $d = 0.93$ and γ=1.1, we get the following gains:

$$K_1 = \begin{bmatrix} -8.5807 & -30.8279 & -1.8836 \\ -35.2273 & 53.8544 & -2.7768 \end{bmatrix},$$

$$K_2 = \begin{bmatrix} -15.0834 & 55.4537 & 3.5584 \\ -11.2341 & 234.3990 & 22.0586 \end{bmatrix}.$$

Thus, the controller with the above given gain matrices can stabilize the considered system and at the same time guarantee the disturbance rejection with a level $\gamma = 1.1$.

10.5 Conclusion

The problem of delay-dependent H_∞ control for SMJSs with time delays has been studied. By using LMI approach, a delay-dependent BRL has been established such that the considered system is stochastically admissible with H_∞ performance γ. An LMI approach has been developed to design state feedback controller such that both the delay-dependent stochastic admissibility and a prescribed H_∞ performance level of the resultant closed-loop system are guaranteed. Three numerical examples have been introduced to illustrate the reduced conservatism and effectiveness of the results proposed.

11

Passivity Analysis for SMJSs with Time-Varying Delays

11.1 Introduction

In Chapter 4, the problem of passive control has been investigated for SSs with time-varying delays and actuator failures. In this chapter, we will investigate the problem of delay-dependent passivity analysis for SMJSs with time delays. By LMI approach and delay partitioning method, some delay-dependent passivity conditions are established, which not only depend upon the time delay, but also depend upon the partitioning. The proposed Lyapunov functionals include some mode-dependent double integral terms, which make full use of the information of Markov process, and thus lead to some improved results. Four numerical examples are proposed to show the improvement of the obtained results.

11.2 Preliminaries

Let $\{r_t, t \geqslant 0\}$ be a continuous-time Markov process with right continuous trajectories and follow the same definition as that in Chapter 10. Fix a probability space $(\Omega, \mathcal{F}, \mathcal{P})$ and consider the following SMJS with time delays:

$$\begin{cases} E\dot{x}(t) = A(r_t)x(t) + A_d(r_t)x(t-d(t)) + B_\omega(r_t)\omega(t), \\ z(t) = C(r_t)x(t) + C_d(r_t)x(t-d(t)) + D_\omega(r_t)\omega(t), \\ x(t) = \phi(t), t \in [-d_2, 0], \end{cases} \tag{11.1}$$

where $x(t) \in \mathbb{R}^n$ is the state, $z(t) \in \mathbb{R}^s$ is the controlled output, $\omega(t) \in \mathbb{R}^p$ is the disturbance input that belongs to $\mathscr{L}_2[0,\infty)$, and $\phi(t) \in C_{n,d_2}$ is a compatible vector valued initial function. The matrix $E \in \mathbb{R}^{n\times n}$ may be singular and it is assumed that $\operatorname{rank} E = r \leqslant n$. $A(r_t)$, $A_d(r_t)$, $B_\omega(r_t)$, $C(r_t)$, $C_d(r_t)$ and $D_\omega(r_t)$ are known real constant matrices with appropriate dimensions. $d(t)$ is a time-varying continuous function that satisfies $0 < d_1 \leqslant d(t) \leqslant d_2$, $\dot{d}(t) \leqslant \mu$, where d_1 and d_2 are the lower and upper bounds of

Z.-G. Wu et al.: *Anal. & Synth. of Singular Syst. with Time-Delays*, LNCIS 443, pp. 149–164.
DOI: 10.1007/978-3-642-37497-5_11

time-varying delay $d(t)$, respectively, and $0 \leqslant \mu < 1$ is the variation rate of time-varying delay $d(t)$.

Definition 11.1. *System (11.1) is said to be passive (in the mean square sense) if there exists a scalar $\gamma > 0$ such that following inequality*

$$\mathcal{E}\left\{2\int_0^{t^*} z(t)^{\mathrm{T}}\omega(t)\mathrm{d}t\right\} \geqslant -\gamma\int_0^{t^*} \omega(t)^{\mathrm{T}}\omega(t)\mathrm{d}t,\ \forall t^* \geqslant 0 \tag{11.2}$$

holds under zero initial condition.

11.3 Main Results

In this section, some delay-dependent conditions will be proposed to ensure the passivity of the considered system (11.1) based on the LMI approach and the delay partitioning method. For presentation convenience, we denote

$$\begin{aligned}
\Upsilon(t) &= \left[x(t)^{\mathrm{T}}\ x(t-\tfrac{1}{m}d_1)^{\mathrm{T}}\ x(t-\tfrac{2}{m}d_1)^{\mathrm{T}}\ \ldots\ x(t-\tfrac{m-1}{m}d_1)^{\mathrm{T}}\right]^{\mathrm{T}},\\
\eta(t) &= \left[\Upsilon(t)^{\mathrm{T}}\ x(t-d_1)^{\mathrm{T}}\right]^{\mathrm{T}},\\
\zeta(t) &= \left[\eta(t)^{\mathrm{T}}\ x(t-d(t))^{\mathrm{T}}\ x(t-d_2)^{\mathrm{T}}\ \omega(t)^{\mathrm{T}}\right]^{\mathrm{T}},\\
\hat{W}_1 &= \left[I_{mn}\ 0_{mn\times n}\right],\\
\hat{W}_2 &= \left[0_{mn\times n}\ I_{mn}\right],\\
g_l &= \left[0_{n\times(l-1)n}\ I_n\ 0_{n\times(m-l+1)n}\right],\ l=1,2,\cdots,m+1.
\end{aligned}$$

Then system (11.1) can be rewritten as

$$\begin{cases}
E\dot{x}(t) = A(r_t)g_1\eta(t) + A_d(r_t)x(t-d(t)) + B_\omega(r_t)\omega(t),\\
z(t) = C(r_t)g_1\eta(t) + C_d(r_t)x(t-d(t)) + D_\omega(r_t)\omega(t),\\
x(t) = \phi(t),\ t\in[-d_2,0].
\end{cases} \tag{11.3}$$

Theorem 11.2. *For a given integer $m > 0$, system (11.1) is stochastically admissible and passive, if there exist matrices $P_i > 0$, $Q_i > 0$, $Z_{1i} > 0$, $Z_{2i} > 0$, $R_i > 0$, $G_1 > 0$, $G_2 > 0$, $G_3 > 0$, $S_{li} > 0$, $U_l > 0$ $(l = 1, 2, \cdots, m)$, W_i, and a scalar $\gamma > 0$ such that the following LMIs hold for any $i \in \hat{\mathcal{S}}$:*

$$\begin{bmatrix}
\Xi_{11i} & \Xi_{12i} & 0 & \Xi_{14i} & g_1^{\mathrm{T}}A_i^{\mathrm{T}}\mathcal{D}_i\\
* & \Xi_{22i} & \Xi_{23i} & -C_{di}^{\mathrm{T}} & A_{di}^{\mathrm{T}}\mathcal{D}_i\\
* & * & \Xi_{33i} & 0 & 0\\
* & * & * & \Xi_{44i} & B_{\omega i}^{\mathrm{T}}\mathcal{D}_i\\
* & * & * & * & -\mathcal{D}_i
\end{bmatrix} < 0, \tag{11.4a}$$

$$\sum_{j=1}^{s}\pi_{ij}Q_j \leqslant G_1, \tag{11.4b}$$

$$\sum_{j=1}^{s} \pi_{ij} Z_{1j} \leqslant G_2, \tag{11.4c}$$

$$\sum_{j=1}^{s} \pi_{ij} (Z_{1j} + Z_{2j}) \leqslant G_2, \tag{11.4d}$$

$$\sum_{j=1}^{s} \pi_{ij} S_{lj} \leqslant U_l, \tag{11.4e}$$

$$\sum_{j=1}^{s} \pi_{ij} R_j \leqslant G_3, \tag{11.4f}$$

where $d_{12} = d_2 - d_1$, $\mathcal{D}_i = \sum_{l=1}^{m} \left[\left(\frac{d_1}{m}\right)^2 S_{li} + \frac{(2l-1)d_1^3}{2m^3} U_l \right] + d_{12}^2 R_i + \frac{d_{12}(d_2^2 - d_1^2)}{2} G_3$, $R \in \mathbb{R}^{n \times (n-r)}$ *is any matrix with full column satisfying* $E^{\mathrm{T}} R = 0$, *and*

$$\begin{aligned}
\Xi_{11i} =& g_1^{\mathrm{T}} (E^{\mathrm{T}} P_i + W_i R^{\mathrm{T}}) A_i g_1 + g_1^{\mathrm{T}} A_i^{\mathrm{T}} (P_i E + R W_i^{\mathrm{T}}) g_1 \\
&+ g_1^{\mathrm{T}} \left[\sum_{j=1}^{s} \pi_{ij} E^{\mathrm{T}} P_j E \right] g_1 + \hat{W}_1^{\mathrm{T}} (Q_i + \frac{d_1}{m} G_1) \hat{W}_1 - \hat{W}_2^{\mathrm{T}} Q_i \hat{W}_2 \\
&+ g_{m+1}^{\mathrm{T}} Z_{1i} g_{m+1} + g_{m+1}^{\mathrm{T}} Z_{2i} g_{m+1} + d_{12} g_1^{\mathrm{T}} G_2 g_1 \\
&- \sum_{l=1}^{m} (g_l - g_{l+1})^{\mathrm{T}} E^{\mathrm{T}} S_{li} E (g_l - g_{l+1}) - g_{m+1}^{\mathrm{T}} E^{\mathrm{T}} R_i E g_{m+1}, \\
\Xi_{12i} =& g_1^{\mathrm{T}} (E^{\mathrm{T}} P_i + W_i R^{\mathrm{T}}) A_{di} + g_{m+1}^{\mathrm{T}} E^{\mathrm{T}} R_i E, \\
\Xi_{14i} =& g_1^{\mathrm{T}} (E^{\mathrm{T}} P_i + W_i R^{\mathrm{T}}) B_{\omega i} - g_1^{\mathrm{T}} C_i^{\mathrm{T}}, \\
\Xi_{22i} =& -(1-\mu) Z_{2i} - 2E^{\mathrm{T}} R_i E, \\
\Xi_{23i} =& E^{\mathrm{T}} R_i E, \\
\Xi_{33i} =& -Z_{1i} - E^{\mathrm{T}} R_i E, \\
\Xi_{44i} =& -\gamma I - D_{\omega i} - D_{\omega i}^{\mathrm{T}}.
\end{aligned}$$

Proof. We are now in a position to prove the regularity and absence of impulses of system (11.1) with $\omega(t) = 0$. Since $\operatorname{rank} E = r \leqslant n$, there exist two nonsingular matrices G and H such that

$$GEH = \begin{bmatrix} I_r & 0 \\ 0 & 0 \end{bmatrix}. \tag{11.5}$$

Then

$$R = G^{\mathrm{T}} \begin{bmatrix} 0 \\ I \end{bmatrix} M, \tag{11.6}$$

where $M \in \mathbb{R}^{(n-r) \times (n-r)}$ is any nonsingular matrix. Denote

$$GA_iH = \begin{bmatrix} A_{1i} & A_{2i} \\ A_{3i} & A_{4i} \end{bmatrix}, \; G^{-\mathrm{T}} P_i G^{-1} = \begin{bmatrix} \bar{P}_{1i} & \bar{P}_{2i} \\ * & \bar{P}_{3i} \end{bmatrix}, \; H^{\mathrm{T}} W_i = \begin{bmatrix} W_{1i} \\ W_{2i} \end{bmatrix}. \tag{11.7}$$

It can be found from $\Xi_{11i} < 0$ that

$$(E^{\mathrm{T}}P_i + W_iR^{\mathrm{T}})A_i + A_i^{\mathrm{T}}(P_iE + RW_i^{\mathrm{T}}) + \sum_{j=1}^{s}\pi_{ij}E^{\mathrm{T}}P_jE - E^{\mathrm{T}}S_{1i}E < 0. \tag{11.8}$$

Then, pre- and post-multiplying (11.8) by H^{T} and H, respectively, we have

$$A_{4i}^{\mathrm{T}}MW_{2i}^{\mathrm{T}} + W_{2i}M^{\mathrm{T}}A_{4i} < 0, \tag{11.9}$$

which implies A_{4i} is nonsingular and thus the pair (E, A_i) is regular and impulse free. Hence, by Definition 10.1, system (11.1) with $\omega(t) = 0$ is regular and impulse free.

Next we will prove that the stochastic stability and passivity of system (11.1). To the end, we define a new process $\{(x_t, r_t), t \geqslant 0\}$ by $\{x_t = x(t+\theta), -2d_2 \leqslant \theta \leqslant 0\}$, then $\{(x_t, r_t), t \geqslant 0\}$ is a Markov process with initial state $(\phi(\cdot), r_0)$, and consider the following Lyapunov functional for system (11.1):

$$V(x_t, r_t) = \sum_{f=1}^{5} V_f(x_t, r_t) \tag{11.10}$$

where

$$\begin{aligned}
V_1(x_t, r_t) &= x(t)^{\mathrm{T}}E^{\mathrm{T}}P(r_t)Ex(t),\\
V_2(x_t, r_t) &= \int_{t-\frac{d_1}{m}}^{t}\Upsilon(s)^{\mathrm{T}}Q(r_t)\Upsilon(s)\mathrm{d}s + \int_{-\frac{d_1}{m}}^{0}\int_{t+\beta}^{t}\Upsilon(s)^{\mathrm{T}}G_1\Upsilon(s)\mathrm{d}s\mathrm{d}\beta,\\
V_3(x_t, r_t) &= \int_{t-d_2}^{t-d_1}x(s)^{\mathrm{T}}Z_1(r_t)x(s)\mathrm{d}s + \int_{t-d(t)}^{t-d_1}x(s)^{\mathrm{T}}Z_2(r_t)x(s)\mathrm{d}s\\
&\quad + \int_{-d_2}^{-d_1}\int_{t+\beta}^{t}x(s)^{\mathrm{T}}G_2x(s)\mathrm{d}s\mathrm{d}\beta,\\
V_4(x_t, r_t) &= \frac{d_1}{m}\sum_{l=1}^{m}\int_{-\frac{l}{m}d_1}^{-\frac{l-1}{m}d_1}\int_{t+\beta}^{t}\dot{x}(s)^{\mathrm{T}}E^{\mathrm{T}}S_l(r_t)E\dot{x}(s)\mathrm{d}s\mathrm{d}\beta\\
&\quad + \frac{d_1}{m}\sum_{l=1}^{m}\int_{-\frac{l}{m}d_1}^{-\frac{l-1}{m}d_1}\int_{\theta}^{0}\int_{t+\beta}^{t}\dot{x}(s)^{\mathrm{T}}E^{\mathrm{T}}U_lE\dot{x}(s)\mathrm{d}s\mathrm{d}\beta\mathrm{d}\theta,\\
V_5(x_t, r_t) &= d_{12}\int_{-d_2}^{-d_1}\int_{t+\beta}^{t}\dot{x}(s)^{\mathrm{T}}E^{\mathrm{T}}R(r_t)E\dot{x}(s)\mathrm{d}s\mathrm{d}\beta\\
&\quad + d_{12}\int_{-d_2}^{-d_1}\int_{\theta}^{0}\int_{t+\beta}^{t}\dot{x}(s)^{\mathrm{T}}E^{\mathrm{T}}G_3E\dot{x}(s)\mathrm{d}s\mathrm{d}\beta\mathrm{d}\theta.
\end{aligned}$$

Let $\mathcal{A}$ be the weak infinitesimal generator of the random process $\{x_t, r_t\}$. Then we have

$$
\begin{aligned}
\mathcal{A}V_1(x_t,r_t) &= 2x(t)^{\mathrm{T}}E^{\mathrm{T}}P_iE\dot{x}(t) + x(t)^{\mathrm{T}}\left[\sum_{j=1}^{s}\pi_{ij}E^{\mathrm{T}}P_jE\right]x(t)\\
&= 2\eta(t)^{\mathrm{T}}g_1^{\mathrm{T}}E^{\mathrm{T}}P_i(A_ig_1\eta(t) + A_{di}x(t-d(t)) + B_{\omega i}\omega(t))\\
&\quad + \eta(t)^{\mathrm{T}}g_1^{\mathrm{T}}\left[\sum_{j=1}^{s}\pi_{ij}E^{\mathrm{T}}P_jE\right]g_1\eta(t), \qquad (11.11)
\end{aligned}
$$

$$
\begin{aligned}
\mathcal{A}V_2(x_t,r_t) &= \Upsilon(t)^{\mathrm{T}}Q_i\Upsilon(t) - \Upsilon\left(t-\frac{d_1}{m}\right)^{\mathrm{T}}Q_i\Upsilon\left(t-\frac{d_1}{m}\right)\\
&\quad + \sum_{j=1}^{s}\pi_{ij}\int_{t-\frac{d_1}{m}}^{t}\Upsilon(s)^{\mathrm{T}}Q_j\Upsilon(s)\mathrm{d}s\\
&\quad + \frac{d_1}{m}\Upsilon(t)^{\mathrm{T}}G_1\Upsilon(t) - \int_{t-\frac{d_1}{m}}^{t}\Upsilon(s)^{\mathrm{T}}G_1\Upsilon(s)\mathrm{d}s\\
&\leqslant \eta(t)^{\mathrm{T}}\hat{W}_1^{\mathrm{T}}\left(Q_i + \frac{d_1}{m}G_1\right)\hat{W}_1\eta(t) - \eta(t)^{\mathrm{T}}\hat{W}_2^{\mathrm{T}}Q_i\hat{W}_2\eta(t)\\
&\quad + \sum_{j=1}^{s}\pi_{ij}\int_{t-\frac{d_1}{m}}^{t}\Upsilon(s)^{\mathrm{T}}Q_j\Upsilon(s)\mathrm{d}s - \int_{t-\frac{d_1}{m}}^{t}\Upsilon(s)^{\mathrm{T}}G_1\Upsilon(s)\mathrm{d}s, \qquad (11.12)
\end{aligned}
$$

$$
\begin{aligned}
\mathcal{A}V_3(x_t,r_t) &\leqslant x(t-d_1)^{\mathrm{T}}Z_{1i}x(t-d_1) - x(t-d_2)^{\mathrm{T}}Z_{1i}x(t-d_2)\\
&\quad + \sum_{j=1}^{s}\pi_{ij}\int_{t-d_2}^{t-d(t)}x(s)^{\mathrm{T}}Z_{1j}x(s)\mathrm{d}s + d_{12}x(t)^{\mathrm{T}}G_2x(t)\\
&\quad + x(t-d_1)^{\mathrm{T}}Z_{2i}x(t-d_1) - (1-\mu)x(t-d(t))^{\mathrm{T}}Z_{2i}x(t-d(t))\\
&\quad + \sum_{j=1}^{s}\pi_{ij}\int_{t-d(t)}^{t-d_1}x(s)^{\mathrm{T}}(Z_{1j}+Z_{2j})x(s)\mathrm{d}s\\
&\quad - \int_{t-d(t)}^{t-d_1}x(s)^{\mathrm{T}}G_2x(s)\mathrm{d}s - \int_{t-d_2}^{t-d(t)}x(s)^{\mathrm{T}}G_2x(s)\mathrm{d}s\\
&= \eta(t)^{\mathrm{T}}g_{m+1}^{\mathrm{T}}Z_{1i}g_{m+1}\eta(t) + \eta(t)^{\mathrm{T}}g_{m+1}^{\mathrm{T}}Z_{2i}g_{m+1}\eta(t)\\
&\quad - x(t-d_2)^{\mathrm{T}}Z_{1i}x(t-d_2) - (1-\mu)x(t-d(t))^{\mathrm{T}}Z_{2i}x(t-d(t))\\
&\quad + \sum_{j=1}^{s}\pi_{ij}\int_{t-d_2}^{t-d(t)}x(s)^{\mathrm{T}}Z_{1j}x(s)\mathrm{d}s - \int_{t-d_2}^{t-d(t)}x(s)^{\mathrm{T}}G_2x(s)\mathrm{d}s\\
&\quad + \sum_{j=1}^{s}\pi_{ij}\int_{t-d(t)}^{t-d_1}x(s)^{\mathrm{T}}(Z_{1j}+Z_{2j})x(s)\mathrm{d}s\\
&\quad + d_{12}\eta(t)^{\mathrm{T}}g_1^{\mathrm{T}}G_2g_1\eta(t) - \int_{t-d(t)}^{t-d_1}x(s)^{\mathrm{T}}G_2x(s)\mathrm{d}s, \qquad (11.13)
\end{aligned}
$$

$$
\begin{aligned}
\mathcal{A}V_4(x_t, r_t) = & \left(\frac{d_1}{m}\right)^2 \sum_{l=1}^{m} \dot{x}(t)^{\mathrm{T}} E^{\mathrm{T}} S_{li} E\dot{x}(t) \\
& - \frac{d_1}{m} \sum_{l=1}^{m} \int_{t-\frac{l}{m}d_1}^{t-\frac{l-1}{m}d_1} \dot{x}(s)^{\mathrm{T}} E^{\mathrm{T}} S_{li} E\dot{x}(s)\mathrm{d}s \\
& + \frac{d_1}{m} \sum_{l=1}^{m} \sum_{j=1}^{s} \pi_{ij} \int_{-\frac{l}{m}d_1}^{-\frac{l-1}{m}d_1} \int_{t+\beta}^{t} \dot{x}(s)^{\mathrm{T}} E^{\mathrm{T}} S_{lj} E\dot{x}(s)\mathrm{d}s\mathrm{d}\beta \\
& + \sum_{l=1}^{m} \frac{(2l-1)d_1^3}{2m^3} \dot{x}(s)^{\mathrm{T}} E^{\mathrm{T}} U_l E\dot{x}(s) \\
& - \frac{d_1}{m} \sum_{l=1}^{m} \int_{-\frac{l}{m}d_1}^{-\frac{l-1}{m}d_1} \int_{t+\beta}^{t} \dot{x}(s)^{\mathrm{T}} E^{\mathrm{T}} U_l E\dot{x}(s)\mathrm{d}s\mathrm{d}\beta \\
\leqslant & \; \dot{x}(t)^{\mathrm{T}} E^{\mathrm{T}} \sum_{l=1}^{m} \left[\left(\frac{d_1}{m}\right)^2 S_{li} + \frac{(2l-1)d_1^3}{2m^3} U_l\right] E\dot{x}(t) \\
& - \sum_{l=1}^{m} \eta(t)^{\mathrm{T}} (g_l - g_{l+1})^{\mathrm{T}} E^{\mathrm{T}} S_{li} E (g_l - g_{l+1}) \eta(t) \\
& + \frac{d_1}{m} \sum_{l=1}^{m} \sum_{j=1}^{s} \pi_{ij} \int_{-\frac{l}{m}d_1}^{-\frac{l-1}{m}d_1} \int_{t+\beta}^{t} \dot{x}(s)^{\mathrm{T}} E^{\mathrm{T}} S_{lj} E\dot{x}(s)\mathrm{d}s\mathrm{d}\beta \\
& - \frac{d_1}{m} \sum_{l=1}^{m} \int_{-\frac{l}{m}d_1}^{-\frac{l-1}{m}d_1} \int_{t+\beta}^{t} \dot{x}(s)^{\mathrm{T}} E^{\mathrm{T}} U_l E\dot{x}(s)\mathrm{d}s\mathrm{d}\beta \\
= & \; (A_i g_1 \eta(t) + A_{di} x(t-d(t)) + B_{\omega i}\omega(t))^{\mathrm{T}} \\
& \times \sum_{l=1}^{m} \left[\left(\frac{d_1}{m}\right)^2 S_{li} + \frac{(2l-1)d_1^3}{2m^3} U_l\right] \\
& \times (A_i g_1 \eta(t) + A_{di} x(t-d(t)) + B_{\omega i}\omega(t)) \\
& - \sum_{l=1}^{m} \eta(t)^{\mathrm{T}} (g_l - g_{l+1})^{\mathrm{T}} E^{\mathrm{T}} S_{li} E (g_l - g_{l+1}) \eta(t) \\
& + \frac{d_1}{m} \sum_{l=1}^{m} \sum_{j=1}^{s} \pi_{ij} \int_{-\frac{l}{m}d_1}^{-\frac{l-1}{m}d_1} \int_{t+\beta}^{t} \dot{x}(s)^{\mathrm{T}} E^{\mathrm{T}} S_{lj} E\dot{x}(s)\mathrm{d}s\mathrm{d}\beta \\
& - \frac{d_1}{m} \sum_{l=1}^{m} \int_{-\frac{l}{m}d_1}^{-\frac{l-1}{m}d_1} \int_{t+\beta}^{t} \dot{x}(s)^{\mathrm{T}} E^{\mathrm{T}} U_l E\dot{x}(s)\mathrm{d}s\mathrm{d}\beta, \qquad (11.14)
\end{aligned}
$$

$$
\begin{aligned}
\mathcal{A}V_5(x_t,r_t) &= d_{12}^2\dot{x}(t)^{\mathrm{T}}E^{\mathrm{T}}R_iE\dot{x}(t) - d_{12}\int_{t-d_2}^{t-d_1}\dot{x}(s)^{\mathrm{T}}E^{\mathrm{T}}R_iE\dot{x}(s)\mathrm{d}s \\
&\quad + d_{12}\sum_{j=1}^{s}\pi_{ij}\int_{-d_2}^{-d_1}\int_{t+\beta}^{t}\dot{x}(s)^{\mathrm{T}}E^{\mathrm{T}}R_jE\dot{x}(s)\mathrm{d}s\mathrm{d}\beta \\
&\quad + \frac{d_{12}(d_2^2-d_1^2)}{2}\dot{x}(t)^{\mathrm{T}}E^{\mathrm{T}}G_3E\dot{x}(t) \\
&\quad - d_{12}\int_{-d_2}^{-d_1}\int_{t+\beta}^{t}\dot{x}(s)^{\mathrm{T}}E^{\mathrm{T}}G_3E\dot{x}(s)\mathrm{d}s\mathrm{d}\beta \\
&\leqslant \dot{x}(t)^{\mathrm{T}}E^{\mathrm{T}}\left[d_{12}^2R_i + \frac{d_{12}(d_2^2-d_1^2)}{2}G_3\right]E\dot{x}(t) \\
&\quad - d_{12}\int_{t-d(t)}^{t-d_1}\dot{x}(s)^{\mathrm{T}}E^{\mathrm{T}}R_iE\dot{x}(s)\mathrm{d}s \\
&\quad - d_{12}\int_{t-d_2}^{t-d(t)}\dot{x}(s)^{\mathrm{T}}E^{\mathrm{T}}R_iE\dot{x}(s)\mathrm{d}s \\
&\quad + d_{12}\sum_{j=1}^{s}\pi_{ij}\int_{-d_2}^{-d_1}\int_{t+\beta}^{t}\dot{x}(s)^{\mathrm{T}}E^{\mathrm{T}}R_jE\dot{x}(s)\mathrm{d}s\mathrm{d}\beta \\
&\quad - d_{12}\int_{-d_2}^{-d_1}\int_{t+\beta}^{t}\dot{x}(s)^{\mathrm{T}}E^{\mathrm{T}}G_3E\dot{x}(s)\mathrm{d}s\mathrm{d}\beta \\
&\leqslant (A_ig_1\eta(t) + A_{di}x(t-d(t)) + B_{\omega i}\omega(t))^{\mathrm{T}} \\
&\quad \times \left[d_{12}^2R_i + \frac{d_{12}(d_2^2-d_1^2)}{2}G_3\right] \\
&\quad \times (A_ig_1\eta(t) + A_{di}x(t-d(t)) + B_{\omega i}\omega(t)) \\
&\quad - (g_{m+1}\eta(t) - x(t-d(t)))^{\mathrm{T}}E^{\mathrm{T}}R_iE(g_{m+1}\eta(t) - x(t-d(t))) \\
&\quad - (x(t-d(t)) - x(t-d_2))^{\mathrm{T}}E^{\mathrm{T}}R_iE(x(t-d(t)) - x(t-d_2)) \\
&\quad + d_{12}\sum_{j=1}^{s}\pi_{ij}\int_{-d_2}^{-d_1}\int_{t+\beta}^{t}\dot{x}(s)^{\mathrm{T}}E^{\mathrm{T}}R_jE\dot{x}(s)\mathrm{d}s\mathrm{d}\beta \\
&\quad - d_{12}\int_{-d_2}^{-d_1}\int_{t+\beta}^{t}\dot{x}(s)^{\mathrm{T}}E^{\mathrm{T}}G_3E\dot{x}(s)\mathrm{d}s\mathrm{d}\beta,
\end{aligned}
\tag{11.15}
$$

where Jensen inequality is applied. Using (11.4b)-(11.4f), we can get

$$
\sum_{j=1}^{s}\pi_{ij}\int_{t-\frac{d_1}{m}}^{t}\Upsilon(s)^{\mathrm{T}}Q_j\Upsilon(s)\mathrm{d}s \leqslant \int_{t-\frac{d_1}{m}}^{t}\Upsilon(s)^{\mathrm{T}}G_1\Upsilon(s)\mathrm{d}s, \tag{11.16}
$$

$$
\sum_{j=1}^{s}\pi_{ij}\int_{t-d_2}^{t-d(t)}x(s)^{\mathrm{T}}Z_{1j}x(s)\mathrm{d}s \leqslant \int_{t-d_2}^{t-d(t)}x(s)^{\mathrm{T}}G_2x(s)\mathrm{d}s, \tag{11.17}
$$

$$\sum_{j=1}^{s}\pi_{ij}\int_{t-d(t)}^{t-d_1}x(s)^{\mathrm{T}}(Z_{1j}+Z_{2j})x(s)\mathrm{d}s\leqslant\int_{t-d(t)}^{t-d_1}x(s)^{\mathrm{T}}G_2x(s)\mathrm{d}s, \quad (11.18)$$

$$\begin{aligned}&\frac{d_1}{m}\sum_{l=1}^{m}\sum_{j=1}^{s}\pi_{ij}\int_{-\frac{l}{m}d_1}^{-\frac{l-1}{m}d_1}\int_{t+\beta}^{t}\dot{x}(s)^{\mathrm{T}}E^{\mathrm{T}}S_{lj}E\dot{x}(s)\mathrm{d}s\mathrm{d}\beta\\&\quad\leqslant\frac{d_1}{m}\sum_{l=1}^{m}\int_{-\frac{l}{m}d_1}^{-\frac{l-1}{m}d_1}\int_{t+\beta}^{t}\dot{x}(s)^{\mathrm{T}}E^{\mathrm{T}}U_lE\dot{x}(s)\mathrm{d}s\mathrm{d}\beta,\end{aligned} \quad (11.19)$$

and

$$\begin{aligned}&d_{12}\sum_{j=1}^{s}\pi_{ij}\int_{-d_2}^{-d_1}\int_{t+\beta}^{t}\dot{x}(s)^{\mathrm{T}}E^{\mathrm{T}}R_jE\dot{x}(s)\mathrm{d}s\mathrm{d}\beta\\&\quad\leqslant d_{12}\int_{-d_2}^{-d_1}\int_{t+\beta}^{t}\dot{x}(s)^{\mathrm{T}}E^{\mathrm{T}}G_3E\dot{x}(s)\mathrm{d}s\mathrm{d}\beta.\end{aligned} \quad (11.20)$$

On the other hand, we have that

$$0=2\eta(t)^{\mathrm{T}}g_1^{\mathrm{T}}W_iR^{\mathrm{T}}(A_ig_1\eta(t)+A_{di}x(t-d(t))+B_{\omega i}\omega(t)). \quad (11.21)$$

Then, adding the right hand side of (11.21) to $\mathcal{A}V(x_t,r_t)$ and applying (11.16)-(11.20), we get from (11.11)-(11.15) that

$$\begin{aligned}&\mathcal{A}V(x_t,r_t)-2z(t)^{\mathrm{T}}\omega(t)-\gamma\omega(t)^{\mathrm{T}}\omega(t)\\&\quad\leqslant 2\eta(t)^{\mathrm{T}}g_1^{\mathrm{T}}E^{\mathrm{T}}P_i(A_ig_1\eta(t)+A_{di}x(t-d(t))+B_{\omega i}\omega(t))\\&\qquad+\eta(t)^{\mathrm{T}}g_1^{\mathrm{T}}\left[\sum_{j=1}^{s}\pi_{ij}E^{\mathrm{T}}P_jE\right]g_1\eta(t)+d_{12}\eta(t)^{\mathrm{T}}g_1^{\mathrm{T}}G_2g_1\eta(t)\\&\qquad+\eta(t)^{\mathrm{T}}\hat{W}_1^{\mathrm{T}}(Q_i+\frac{d_1}{m}G_1)\hat{W}_1\eta(t)-\eta(t)^{\mathrm{T}}\hat{W}_2^{\mathrm{T}}Q_i\hat{W}_2\eta(t)\\&\qquad+\eta(t)^{\mathrm{T}}g_{m+1}^{\mathrm{T}}Z_{1i}g_{m+1}\eta(t)+\eta(t)^{\mathrm{T}}g_{m+1}^{\mathrm{T}}Z_{2i}g_{m+1}\eta(t)\\&\qquad-x(t-d_2)^{\mathrm{T}}Z_{1i}x(t-d_2)-(1-\mu)x(t-d(t))^{\mathrm{T}}Z_{2i}x(t-d(t))\\&\qquad+(A_ig_1\eta(t)+A_{di}x(t-d(t))+B_{\omega i}\omega(t))^{\mathrm{T}}\\&\qquad\times\sum_{l=1}^{m}\left[\left(\frac{d_1}{m}\right)^2S_{li}+\frac{(2l-1)d_1^3}{2m^3}U_l\right]\\&\qquad\times(A_ig_1\eta(t)+A_{di}x(t-d(t))+B_{\omega i}\omega(t))\\&\qquad-\sum_{l=1}^{m}\eta(t)^{\mathrm{T}}(g_l-g_{l+1})^{\mathrm{T}}E^{\mathrm{T}}S_{li}E(g_l-g_{l+1})\eta(t)\end{aligned}$$

$$
\begin{aligned}
&+ (A_i g_1 \eta(t) + A_{di} x(t-d(t)) + B_{\omega i}\omega(t))^{\mathrm{T}} \\
&\times \left[d_{12}^2 R_i + \frac{d_{12}(d_2^2 - d_1^2)}{2} G_3 \right] \\
&\times (A_i g_1 \eta(t) + A_{di} x(t-d(t)) + B_{\omega i}\omega(t)) \\
&- (g_{m+1}\eta(t) - x(t-d(t)))^{\mathrm{T}} E^{\mathrm{T}} R_i E (g_{m+1}\eta(t) - x(t-d(t))) \\
&- (x(t-d(t)) - x(t-d_2))^{\mathrm{T}} E^{\mathrm{T}} R_i E (x(t-d(t)) - x(t-d_2)) \\
&+ 2\eta(t)^{\mathrm{T}} g_1^{\mathrm{T}} W_i R^{\mathrm{T}} (A_i g_1 \eta(t) + A_{di} x(t-d(t)) + B_{\omega i}\omega(t)) \\
\leqslant\ & \zeta(t)^{\mathrm{T}} \Sigma_i \zeta(t),
\end{aligned}
\tag{11.22}
$$

where

$$
\Sigma_i = \begin{bmatrix} \Xi_{11i} & \Xi_{12i} & \Xi_{13i} & \Xi_{14i} \\ * & \Xi_{22i} & \Xi_{23i} & -C_{di}^{\mathrm{T}} \\ * & * & \Xi_{33i} & 0 \\ * & * & * & \Xi_{44i} \end{bmatrix} + \begin{bmatrix} g_1^{\mathrm{T}} A_i^{\mathrm{T}} \\ A_{di}^{\mathrm{T}} \\ 0 \\ B_{\omega i}^{\mathrm{T}} \end{bmatrix} \mathcal{D}_i \begin{bmatrix} g_1^{\mathrm{T}} A_i^{\mathrm{T}} \\ A_{di}^{\mathrm{T}} \\ 0 \\ B_{\omega i}^{\mathrm{T}} \end{bmatrix}^{\mathrm{T}}.
$$

Applying Schur complement to (11.4a), we can get that $\Sigma_i < 0$, thus

$$
\mathcal{A}V(x_t, r_t) - 2z(t)^{\mathrm{T}}\omega(t) - \gamma\omega(t)^{\mathrm{T}}\omega(t) \leqslant 0, \tag{11.23}
$$

which implies that for any $t^* \geqslant 0$

$$
\mathscr{E}\left\{ V(x_{t^*}, r_{t^*}) - V(x_0, r_0) - 2\int_0^{t^*} z(t)^{\mathrm{T}}\omega(t)\mathrm{d}t - \gamma\int_0^{t^*} \omega(t)^{\mathrm{T}}\omega(t)\mathrm{d}t \right\} \leqslant 0. \tag{11.24}
$$

It is noted that $\mathscr{E}\{V(x_{t^*}, r_{t^*})\} \geqslant 0$ and $V(x_0, r_0) = 0$ under the zero initial condition. Thus, we can find from (11.24) that (11.2) holds. According to Definition 11.1, system (11.1) is passive. On the other hand, applying Schur complement to (11.4a) again, we can obtain that

$$
\Gamma_i = \begin{bmatrix} \Xi_{11i} & \Xi_{12i} & 0 \\ * & \Xi_{22i} & \Xi_{23i} \\ * & * & \Xi_{33i} \end{bmatrix} + \begin{bmatrix} g_1^{\mathrm{T}} A_i^{\mathrm{T}} \\ A_{di}^{\mathrm{T}} \\ 0 \end{bmatrix} \mathcal{D}_i \begin{bmatrix} g_1^{\mathrm{T}} A_i^{\mathrm{T}} \\ A_{di}^{\mathrm{T}} \\ 0 \end{bmatrix}^{\mathrm{T}} < 0, \tag{11.25}
$$

By using the similar proof method, we can get

$$
\mathcal{A}V(x_t, r_t) \leqslant \begin{bmatrix} \eta(t) \\ x(t-d(t)) \\ x(t-d_2) \end{bmatrix}^{\mathrm{T}} \Gamma_i \begin{bmatrix} \eta(t) \\ x(t-d(t)) \\ x(t-d_2) \end{bmatrix} \tag{11.26}
$$

in case of $\omega(t) \equiv 0$. Thus, there exists a scalar $\lambda > 0$ such that

$$
\mathcal{A}V(x_t, r_t) \leqslant -\lambda\|x(t)\|^2. \tag{11.27}
$$

Therefore, by Dynkin's formula, we get for any $t \geqslant 0$,

$$\mathscr{E}\{V(x_t, r_t)\} - \mathscr{E}\{V(x_0, r_0)\} \leqslant -\lambda \mathscr{E}\left\{\int_0^t \|x(s)\|^2 \mathrm{d}s\right\}, \tag{11.28}$$

which yields

$$\mathscr{E}\left\{\int_0^t \|x(s)\|^2 \mathrm{d}s\right\} \leqslant \lambda^{-1}\mathscr{E}\{V(x_0, r_0)\}. \tag{11.29}$$

According to Definition 10.1, system (11.1) with $\omega(t) = 0$ is stochastically stable. This completes the proof.

Remark 11.3. It is noted that Theorem 11.2 proposes a delay-dependent passivity condition of system (11.1) based on the idea of delay partitioning and LMI approach. It should be pointed out that in order to obtain less conservative results, a novel Lyapunov functional is introduced in (11.10) including the following mode-dependent double integral terms

$$\frac{d_1}{m}\sum_{l=1}^{m}\int_{-\frac{l}{m}d_1}^{-\frac{l-1}{m}d_1}\int_{t+\beta}^{t} \dot{x}(s)^{\mathrm{T}}E^{\mathrm{T}}S_l(r_t)E\dot{x}(s)\mathrm{d}s\mathrm{d}\beta$$

and

$$d_{12}\int_{-d_2}^{-d_1}\int_{t+\beta}^{t} \dot{x}(s)^{\mathrm{T}}E^{\mathrm{T}}R(r_t)E\dot{x}(s)\mathrm{d}s\mathrm{d}\beta.$$

The obvious advantage of the above terms is that the information of the underlying Markov process r_t is fully applied. It is also noted that several triple integral terms are introduced to deal with the derivative terms of the mode-dependent double integral terms[91]. In [31, 162, 182, 215], the involved double integral terms are all mode-independent. Thus, our result has less conservatism than those in [31, 162, 182, 215].

Based on Theorem 11.2, the stochastic stability criterion for system

$$\begin{cases} E\dot{x}(t) = A(r_t)x(t) + A_d(r_t)x(t - d(t)), \\ x(t) = \phi(t),\ t \in [-d_2, 0] \end{cases} \tag{11.30}$$

can be easily obtained as follows.

Corollary 11.4. *For a given integer $m > 0$, system (11.30) is stochastically admissible, if there exist matrices $P_i > 0$, $Q_i > 0$, $Z_{1i} > 0$, $Z_{2i} > 0$, $R_i > 0$, $G_1 > 0$, $G_2 > 0$, $G_3 > 0$, $S_{li} > 0$, $U_l > 0$ $(l = 1, 2, \cdots, m)$ and W_i such that for any $i \in \hat{\mathcal{S}}$, (11.4b)-(11.4f) and (11.31) hold:*

$$\begin{bmatrix} \Xi_{11i} & \Xi_{12i} & 0 & g_1^{\mathrm{T}}A_i^{\mathrm{T}}\mathcal{D}_i \\ * & \Xi_{22i} & \Xi_{23i} & A_{di}^{\mathrm{T}}\mathcal{D}_i \\ * & * & \Xi_{33i} & 0 \\ * & * & * & -\mathcal{D}_i \end{bmatrix} < 0, \tag{11.31}$$

where Ξ_{11i}, Ξ_{12i}, Ξ_{22i}, Ξ_{23i}, Ξ_{33i} and $\mathcal{D}_i$ are given in Theorem 11.2.

In the case of time-invariant delays, that is, $d(t) \equiv d > 0$, system (11.1) reduces to system (10.9), and the corresponding Lyapunov functional candidate is

$$V(x_t, r_t) = V_1(x_t, r_t) + V_2(x_t, r_t)|_{d_1=d} + V_4(x_t, r_t)|_{d_1=d}, \tag{11.32}$$

where $V_1(x_t, r_t)$, $V_2(x_t, r_t)$, and $V_4(x_t, r_t)$ are given in (11.10). By using the similar proof method of Theorem 11.2, we can derive the following result.

Theorem 11.5. *For a given integer $m > 0$, system (10.9) is stochastically admissible and passive, if there exist matrices $P_i > 0$, $Q_i > 0$, $G_1 > 0$, $S_{li} > 0$, $U_l > 0$ $(l = 1, 2, \cdots, m)$, W_i, and a scalar $\gamma > 0$ such that for any $i \in \hat{\mathcal{S}}$, (11.4b), (11.4e) and (11.33) hold*

$$\begin{bmatrix} \hat{\Xi}_{11i} & \hat{\Xi}_{12i} & (g_1^{\mathrm{T}} A_i^{\mathrm{T}} + g_{m+1}^{\mathrm{T}} A_{di}^{\mathrm{T}})\hat{\mathcal{D}}_i \\ * & \hat{\Xi}_{22i} & B_{\omega i}^{\mathrm{T}} \hat{\mathcal{D}}_i \\ * & * & -\hat{\mathcal{D}}_i \end{bmatrix} < 0, \tag{11.33}$$

where $\hat{\mathcal{D}}_i = \sum_{l=1}^{m} \left[\left(\frac{d}{m}\right)^2 S_{li} + \frac{(2l-1)d^3}{2m^3} U_l \right]$ *and*

$$\begin{aligned} \hat{\Xi}_{11i} &= g_1^{\mathrm{T}}(E^{\mathrm{T}} P_i + W_i R^{\mathrm{T}}) A_i g_1 + g_1^{\mathrm{T}} A_i^{\mathrm{T}} (P_i E + R W_i^{\mathrm{T}}) g_1 \\ &\quad + g_1^{\mathrm{T}} \left[\sum_{j=1}^{s} \pi_{ij} E^{\mathrm{T}} P_j E \right] g_1 - \sum_{l=1}^{m} (g_l - g_{l+1})^{\mathrm{T}} E^{\mathrm{T}} S_{li} E (g_l - g_{l+1}) \\ &\quad + \hat{W}_1^{\mathrm{T}} (Q_i + \frac{d}{m} G_1) \hat{W}_1 - \hat{W}_2^{\mathrm{T}} Q_i \hat{W}_2 \\ &\quad + g_1^{\mathrm{T}} (E^{\mathrm{T}} P_i + W_i R^{\mathrm{T}}) A_{di} g_{m+1} + g_{m+1}^{\mathrm{T}} A_{di}^{\mathrm{T}} (P_i E + R W_i^{\mathrm{T}}) g_1, \\ \hat{\Xi}_{12i} &= g_1^{\mathrm{T}} (E^{\mathrm{T}} P_i + W_i R^{\mathrm{T}}) B_{\omega i} - g_1^{\mathrm{T}} C_i^{\mathrm{T}} - g_{m+1}^{\mathrm{T}} C_{di}^{\mathrm{T}}, \\ \hat{\Xi}_{22i} &= -\gamma I - D_{\omega i} - D_{\omega i}^{\mathrm{T}}. \end{aligned}$$

Based on Theorem 11.5, we can get the following stochastic stability criterion for system (10.5).

Corollary 11.6. *For a given integer $m > 0$, system (10.5) is stochastically admissible, if there exist matrices $P_i > 0$, $Q_i > 0$, $G_1 > 0$, $S_{li} > 0$, $U_l > 0$ $(l = 1, 2, \cdots, m)$ and W_i such that for any $i \in \hat{\mathcal{S}}$, (11.4b), (11.4e) and (11.34) hold*

$$\begin{bmatrix} \hat{\Xi}_{11i} & (g_1^{\mathrm{T}} A_i^{\mathrm{T}} + g_{m+1}^{\mathrm{T}} A_{di}^{\mathrm{T}})\hat{\mathcal{D}}_i \\ * & -\hat{\mathcal{D}}_i \end{bmatrix} < 0, \tag{11.34}$$

where $\hat{\Xi}_{11i}$ and $\hat{\mathcal{D}}_i$ are given in Theorem 11.5.

Remark 11.7. It is noted that when $m = 1$, $S_{1i} = S$, and $U_1 = \varepsilon I$ $(\varepsilon \to 0)$, the stability condition in Corollary 11.6 reduces to the one of [162]. Thus, Corollary 11.6 has theoretically less conservatism than that of [162] even for $m = 1$.

11.4 Numerical Examples

In this section, we will use four numerical examples to illustrate the usefulness and less conservatism of the proposed results in this chapter.

Example 11.8. Consider system (10.5) with[182]

$$A_1 = \begin{bmatrix} -3.4888 & 0.8057 \\ -0.6451 & -3.2684 \end{bmatrix}, A_{d1} = \begin{bmatrix} -0.8620 & -1.2919 \\ -0.6841 & -2.0729 \end{bmatrix},$$
$$A_2 = \begin{bmatrix} -2.4898 & 0.2895 \\ 1.3396 & -0.0211 \end{bmatrix}, A_{d2} = \begin{bmatrix} -2.8306 & 0.4978 \\ -0.8436 & -1.0115 \end{bmatrix}.$$

and

$$E = \begin{bmatrix} 1 & 0 \\ 0 & 1 \end{bmatrix},$$

that is, system (10.5) reduce to a regular system.

In this example, we choose the following transition probability matrix

$$\Pi = \begin{bmatrix} -a & a \\ 0.8 & -0.8 \end{bmatrix}.$$

Now we compare Corollary 11.6 with the results proposed in [31, 182], and the comparison under different a can be found in Table 11.1, from which it can be seen that our obtained results have less conservatism than those in [182] even for $m = 1$, and Corollary 11.6 gives the better result than [31] for the same m.

Fig. 11.1 shows the state responses of the considered system with $d = 1.3$ under the initial condition $x(t) = \begin{bmatrix} -12 \\ 4 \end{bmatrix}$, from which we find that the corresponding state responses converge to zero.

Table 11.1. Example 11.8: Comparison of the maximum time delay d for various a

a	0.1	0.5	0.8	1
[182]	0.6797	0.5794	0.5562	0.5465
[31]($m = 1$)	0.6797	0.5794	0.5562	0.5465
[31]($m = 2$)	0.7939	0.7007	0.6707	0.6581
[31]($m = 3$)	0.8130	0.7222	0.6926	0.6806
[31]($m = 5$)	0.8232	0.7327	0.7039	0.6934
Corollary 11.6 ($m = 1$)	1.1103	0.7718	0.6887	0.6563
Corollary 11.6 ($m = 2$)	1.2550	0.8816	0.8065	0.7783
Corollary 11.6 ($m = 3$)	1.2891	0.9121	0.8363	0.8098
Corollary 11.6 ($m = 5$)	1.3101	0.9339	0.8562	0.8306

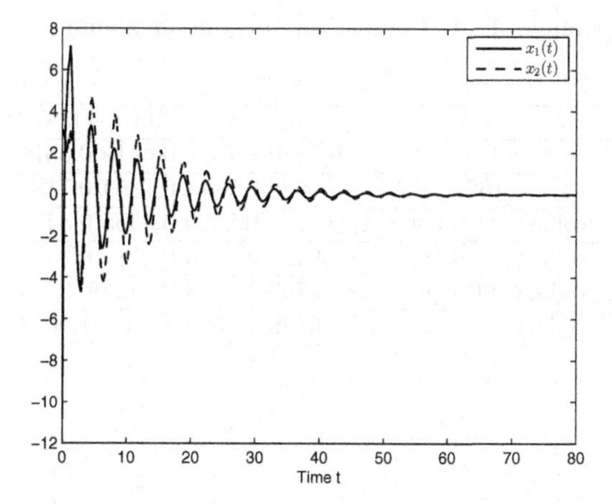

Fig. 11.1. Example 11.8: State responses of system

Example 11.9. Consider system (10.5) with

$$A_1 = \begin{bmatrix} -0.1793 & -0.7876 \\ 1.6790 & 1.6746 \end{bmatrix}, A_{d1} = \begin{bmatrix} -0.3649 & 0.6192 \\ -0.4381 & -0.0420 \end{bmatrix},$$

$$A_2 = \begin{bmatrix} -0.3946 & -2.3342 \\ -0.1439 & -1.3575 \end{bmatrix}, A_{d2} = \begin{bmatrix} -0.9503 & 1.1842 \\ -0.0672 & 0.3443 \end{bmatrix},$$

and

$$E = \begin{bmatrix} 1 & 0 \\ 0 & 0 \end{bmatrix}.$$

In this example, we choose

$$\Pi = \begin{bmatrix} -a & a \\ 0.3 & -0.3 \end{bmatrix}$$

and

$$R = \begin{bmatrix} 0 \\ 1 \end{bmatrix}.$$

Now we compare Corollary 11.6 with the results proposed in [162, 215], and Table 11.2 gives the comparison under different a. It can be found from Table 11.2 that our obtained results have less conservatism than those in [162, 215] even for $m = 1$.

Table 11.2. Example 11.9: Comparison of the maximum time delay d for various a

a	0.15	0.45	0.75	0.95
[215]	0.7280	0.7037	0.6919	0.6879
[162]	0.8686	0.7811	0.7364	0.7184
Corollary 11.6 ($m = 1$)	0.8799	0.8579	0.8455	0.8409
Corollary 11.6 ($m = 2$)	1.0454	0.9854	0.9517	0.9453
Corollary 11.6 ($m = 3$)	1.0703	1.0134	0.9824	0.9783
Corollary 11.6 ($m = 5$)	1.0849	1.0326	1.0056	1.0038

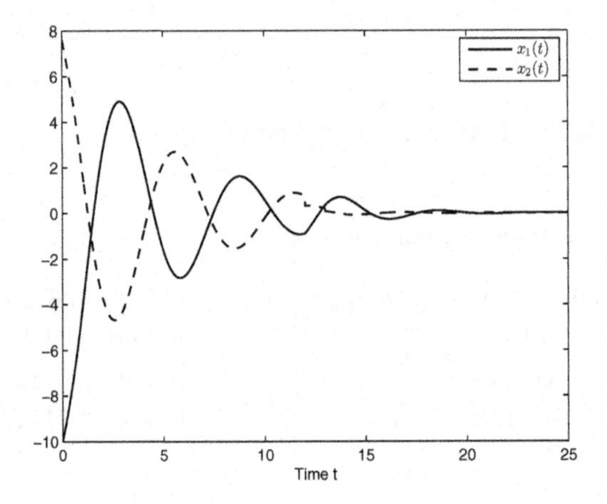

Fig. 11.2. Example 11.9: State responses of system

Fig. 11.2 shows the state responses of the considered system with $d = 1$ under the initial condition $x(t) = \begin{bmatrix} -10 \\ 7.6008 \end{bmatrix}$ and $r(0) = 1$, from which we find that the corresponding state responses converge to zero.

Example 11.10. Consider system (11.30) with two modes and E, R, A_1, A_2, A_{d1} and A_{d2} are given in Example 11.9.

In this example, we choose

$$\Pi = \begin{bmatrix} -a & a \\ 0.4 & -0.4 \end{bmatrix}.$$

We suppose $\mu = 0.2$ and $d_1 = 0.2$. For various a, Table 11.3 gives the allowable upper bound d_2 ensuring admissibility of the considered system by using Corollary 11.4 and [161]. It is clear that Corollary 11.4 in this chapter gives better results even for $m = 1$.

Table 11.3. Example 11.10: Comparison of the maximum upper bound d_2 for various a

a	0.1	0.3	0.5	0.7
[161]	0.7158	0.6985	0.6881	0.6814
Corollary 11.4 ($m = 1$)	0.8750	0.8284	0.8076	0.7918
Corollary 11.4 ($m = 2$)	0.8759	0.8292	0.8086	0.7930
Corollary 11.4 ($m = 3$)	0.8762	0.8294	0.8088	0.7933

Example 11.11. Consider system (11.1) with

$$A_1 = \begin{bmatrix} -13.1 & -13.7 \\ -15.4 & -23.8 \end{bmatrix}, A_{d1} = \begin{bmatrix} -18.6 & -10.4 \\ -25.2 & -16.8 \end{bmatrix},$$

$$B_{\omega 1} = \begin{bmatrix} 1.9 \\ 1.8 \end{bmatrix}, C_1 = \begin{bmatrix} 0.4 & -0.8 \end{bmatrix}, C_{d1} = \begin{bmatrix} 1.4 & -1.8 \end{bmatrix}, D_{\omega 1} = 2,$$

$$A_2 = \begin{bmatrix} -12 & -13.7 \\ -12.1 & -20 \end{bmatrix}, A_{d2} = \begin{bmatrix} -17 & -10.3 \\ -24 & -16 \end{bmatrix},$$

$$B_{\omega 2} = \begin{bmatrix} 1.7 \\ 1.5 \end{bmatrix}, C_2 = \begin{bmatrix} 0.5 & -0.9 \end{bmatrix}, C_{d2} = \begin{bmatrix} 1.5 & -1.3 \end{bmatrix}, D_{\omega 2} = 2,$$

and

$$E = \begin{bmatrix} 9 & 3 \\ 6 & 2 \end{bmatrix}.$$

In this example, we choose

$$\Pi = \begin{bmatrix} -a & a \\ 0.45 & -0.45 \end{bmatrix}$$

and

$$R = \begin{bmatrix} -2 \\ 3 \end{bmatrix}.$$

We suppose $\mu = 0.5$, $d_1 = 1$, and $d_2 = 1.45$. For various a, Table 11.4 gives the the optimal passivity performance $\gamma_{\min}$ by Theorem 11.2.

Table 11.4. Example 11.11: The optimal passivity performance $\gamma_{\min}$ for various a

a	0.1	0.3	0.5	0.7
Theorem 11.2 ($m = 1$)	1.1075	1.2155	1.2916	1.3505
Theorem 11.2 ($m = 2$)	0.3711	0.4603	0.5231	0.5571
Theorem 11.2 ($m = 3$)	0.2712	0.3568	0.4161	0.4475
Theorem 11.2 ($m = 4$)	0.2437	0.3145	0.3736	0.3996

11.5 Conclusions

In this chapter, the problem of delay-dependent passivity analysis has been discussed for SMJSs with time-delays. Based on the idea of delay partitioning, a delay-dependent passivity condition has been derived. The case of time-invariant delay has also been considered. The proposed Lyapunov functionals include some mode-dependent double integral terms, which make full use of the information of Markov process. Some numerical examples has been given to show the effectiveness and improvement of the proposed methods.

12

l_2-l_∞ Filtering for Discrete-Time SMJSs with Time-Varying Delays

12.1 Introduction

The robust H_∞ filtering problem for discrete-time SMJSs with parameter uncertainties and mode-dependent time delay has been considered in [100], where a filter has been designed such that the filtering error system is regular, causal, delay-dependent stochastically stable and satisfies a given H_∞ performance index. It should be pointed out that the filters designed in [100] are mode-dependent, that is, the Markov chain state, often called mode, is assumed to be available to the filter at any time. However, this assumption may sometimes be impossible to be satisfied, such as networked control systems without time stamp information [22, 87]. Thus, the ideal requirement inevitably limits the applications of the given results. In order to overcome the drawback of mode-dependent filter, the mode-independent filter has been utilized to solve the H_∞ filtering problem for MJSs with non-accessible jump mode information in [22]. The result proposed in [22] has been further improved by [87]. Although the importance of mode-independent filter has been widely recognized, only few results have been focused on SMJSs with time-delays.

The problem of mode-independent filter design for discrete-time SMJSs with time-varying delays is investigated in this chapter. Attention is focused on the design of state-space filter with mode-independent characterization, which guarantees a given l_2-l_∞ performance for the filtering error systems. Based on the delay partitioning technique, a delay-dependent condition is proposed to guarantee the existence of full-order and reduced-order filters in a unified framework, which is formulated in terms of LMIs. When these LMIs are feasible, a set of the parameters of a desired filter can be given. Finally, two numerical examples are given to show the effectiveness of the proposed methods.

Z.-G. Wu et al.: *Anal. & Synth. of Singular Syst. with Time-Delays*, LNCIS 443, pp. 165–181.
DOI: 10.1007/978-3-642-37497-5_12

12.2 Problem Formulation

Fix a probability space $(\Omega, \mathcal{F}, \mathcal{P})$ and consider the following discrete-time SMJS with time-varying delays:

$$\begin{cases} Ex(k+1) = A(r(k))x(k) + A_d(r(k))x(k-d(k)) + B(r(k))\omega(k), \\ y(k) = C(r(k))x(k) + C_d(r(k))x(k-d(k)) + D(r(k))\omega(k), \\ z(k) = L(r(k))x(k), \\ x(k) = \phi(k),\ k \in \mathbb{N}[-d_2,\ 0], \end{cases} \tag{12.1}$$

where $x(k) \in \mathbb{R}^n$ is the state, $y(k) \in \mathbb{R}^m$ is the measurement, $\omega(k) \in \mathbb{R}^q$ is the noise signal vector belonging to $l_2[0, \infty)$, $z(k) \in \mathbb{R}^p$ is the signal to be estimated, and $\phi(k)$ is the compatible initial condition. The matrix $E \in \mathbb{R}^{n\times n}$ is singular and it is assumed that $\operatorname{rank} E = r \leqslant n$. $A(r(k))$, $A_d(r(k))$, $B(r(k))$, $C(r(k))$, $C_d(r(k))$, $D(r(k))$ and $L(r(k))$ are known real constant matrices with appropriate dimensions. $d(k)$ is a time-varying delay satisfying $d_1 \leqslant d(k) \leqslant d_2$, where $d_1 > 0$ and $d_2 > 0$ are given integers representing the lower and upper bounds of time-varying delay $d(k)$, respectively. The parameter $r(k)$ represents a discrete-time homogeneous Markov chain and follows the same definition as that in Chapter 9.

In this chapter, we consider the following linear state-space filter

$$\begin{cases} \hat{x}(k+1) = A_f\hat{x}(k) + B_f y(k),\ \hat{x}(0) = 0, \\ \hat{z}(k) = C_f\hat{x}(k), \end{cases} \tag{12.2}$$

where $\hat{x} \in \mathbb{R}^{\hat{n}}$ is the filter state, $\hat{z}(t) \in \mathbb{R}^p$ is the estimation signal, and the constant matrices A_f, B_f and C_f are the filter parameters to be determined.

Remark 12.1. It should be pointed out that we not only consider the full-order filter (when $\hat{n} = n$), but also discuss the reduced-order filter (when $1 \leqslant \hat{n} < n$). As can be seen in Section 12.3, the two kinds of filters will be considered in a unified framework.

Remark 12.2. To the best of our knowledge, most results on the filtering problem of MJSs with or without time-delay require an indispensable assumption on the accessibility of the jump mode for the filter and belong to a kind of mode-dependent (stochastic) filters. However, this assumption may sometimes be impossible to be satisfied even for a simple system. It is clear that the filter (12.2) is a mode-independent (deterministic) filter. Therefore, it can be more powerful and has more wide applications when the jump mode is not available to the filter.

Augmenting the model of (12.1) to include the state of filter (12.2), we can obtain the following filtering error system:

$$\begin{cases} \bar{E}\bar{x}(k+1) = \bar{A}(r(k))\bar{x}(k) + \bar{A}_d(r(k))K\bar{x}(k-d(k)) + \bar{B}(r(k))\omega(k), \\ \quad \bar{z}(k) = \bar{C}(r(k))\bar{x}(k), \\ \quad \bar{x}(k) = \bar{\phi}(k) = \begin{bmatrix} \phi(k) \\ 0 \end{bmatrix}, \ k \in \mathbb{N}[-d_2, 0], \end{cases} \tag{12.3}$$

where $\bar{x}(k) = \begin{bmatrix} x(k) \\ \hat{x}(k) \end{bmatrix}$, $\bar{z}(k) = z(k) - \hat{z}(k)$, and

$$\begin{aligned} \bar{E} &= \begin{bmatrix} E & 0 \\ 0 & I \end{bmatrix}, \ \bar{A}(r(k)) = \begin{bmatrix} A(r(k)) & 0 \\ B_f C(r(k)) & A_f \end{bmatrix}, \\ \bar{A}_d(r(k)) &= \begin{bmatrix} A_d(r(k)) \\ B_f C_d(r(k)) \end{bmatrix}, \ \bar{B}(r(k)) = \begin{bmatrix} B(r(k)) \\ B_f D(r(k)) \end{bmatrix}, \\ \bar{C}(r(k)) &= \begin{bmatrix} L(r(k)) & -C_f \end{bmatrix}, \ K = \begin{bmatrix} I & 0 \end{bmatrix}. \end{aligned}$$

The main objective of this chapter is formulated as follows: for a given scalar $\gamma > 0$, develop a filter (12.2) such that the filtering error system (12.3) is stochastically admissible in the case of $\omega(k) = 0$, and satisfies

$$\sup_k \sqrt{\mathscr{E}\left\{\bar{z}(k)^{\mathrm{T}}\bar{z}(k)\right\}} < \gamma \sqrt{\sum_{k=0}^{\infty} \omega(k)^{\mathrm{T}}\omega(k)}, \tag{12.4}$$

under the zero-initial condition for any nonzero $\omega(k) \in l_2[0, \infty)$. In this case, the filtering error system (12.3) is said to be stochastically admissible with l_2-l_∞ performance γ.

12.3 Main Results

In this section, the delay partitioning technique will be developed to solve the l_2-l_∞ filtering problem. By applying the delay partitioning idea to the lower bound of time-varying $d(k)$, we can always describe $d_1 = m\tau$, where $m > 0$ and $\tau > 0$ are integers. For presentation convenience, we denote

$$\begin{aligned} \Upsilon(k) &= \begin{bmatrix} x(k)^{\mathrm{T}} & x(k-\tau)^{\mathrm{T}} & \cdots & x(k-(m-1)\tau)^{\mathrm{T}} \end{bmatrix}^{\mathrm{T}}, \\ \eta(k) &= \begin{bmatrix} \Upsilon(k-\tau)^{\mathrm{T}} & x(k-d(k))^{\mathrm{T}} & x(k-d_2)^{\mathrm{T}} \end{bmatrix}^{\mathrm{T}}, \\ \zeta(k) &= \begin{bmatrix} \bar{x}(k)^{\mathrm{T}} & \eta(k)^{\mathrm{T}} & \omega(k)^{\mathrm{T}} \end{bmatrix}^{\mathrm{T}}, \\ W_{Q_3} &= \begin{bmatrix} I_{(m-1)n} & 0_{(m-1)n \times 3n} \end{bmatrix}, \\ W_Q &= \begin{bmatrix} I_{mn} & 0_{mn \times 2n} \end{bmatrix}, \end{aligned}$$

$$\begin{aligned} W_{d_2} &= \left[0_{n\times(m+1)n}\ I_n\right], \\ W_d &= \left[0_{n\times mn}\ I_n\ 0_{n\times n}\right], \\ W_X &= \left[-E\ 0_{n\times(m+1)n}\right], \\ W_Y &= \left[0_{n\times(m-1)n}\ E\ -E\ 0_{n\times n}\right], \\ W_Z &= \left[0_{n\times mn}\ E\ -E\right]. \end{aligned}$$

We first present a performance analysis result for the filtering error system (12.3) and then give a representation of filter gains in terms of the feasible solutions to a set of LMIs.

Theorem 12.3. *For a given scalar $\gamma > 0$, system (12.3) is stochastically admissible with l_2-l_∞ performance γ, if there exist matrices $P_i > 0$, $Q = \begin{bmatrix} Q_1 & Q_2 \\ * & Q_3 \end{bmatrix} > 0$, $Z_1 > 0$, $Z_2 > 0$, $S_1 > 0$, $S_2 > 0$ and $\hat{S}_i$ such that for any $i \in \mathcal{I}$,*

$$\begin{bmatrix} \Xi_{1i} & \Xi_{2i} & \hat{S}_i\bar{R}^{\mathrm{T}}\bar{B}_i & \bar{A}_i^{\mathrm{T}}X_i & K^{\mathrm{T}}(A_i-E)^{\mathrm{T}}S \\ * & \Xi_{3i} & 0 & W_d^{\mathrm{T}}\bar{A}_{di}^{\mathrm{T}}X_i & W_d^{\mathrm{T}}A_{di}^{\mathrm{T}}S \\ * & * & -I & \bar{B}_i^{\mathrm{T}}X_i & B_i^{\mathrm{T}}S \\ * & * & * & -X_i & 0 \\ * & * & * & * & -S \end{bmatrix} < 0, \tag{12.5}$$

$$\begin{bmatrix} \bar{E}^{\mathrm{T}}P_i\bar{E} & \bar{C}_i^{\mathrm{T}} \\ * & \gamma^2 I \end{bmatrix} \geqslant 0, \tag{12.6}$$

where $X_i = \sum_{j=1}^{N}\pi_{ij}P_j$, $S = \tau^2 S_1 + d_{12}^2 S_2$, $d_{12} = d_2 - d_1$, $\bar{R} \in \mathbb{R}^{(n+\hat{n})\times(n-r)}$ is any matrix with full column satisfying $\bar{E}^{\mathrm{T}}\bar{R} = 0$, and

$$\begin{aligned} \Xi_{1i} &= -\bar{E}^{\mathrm{T}}P_i\bar{E} + K^{\mathrm{T}}(Q_1 + Z_1 + (d_{12}+1)Z_2 - E^{\mathrm{T}}S_1E)K + \mathrm{sym}\{\hat{S}_i\bar{R}^{\mathrm{T}}\bar{A}_i\}, \\ \Xi_{2i} &= K^{\mathrm{T}}Q_2W_{Q_3} - K^{\mathrm{T}}E^{\mathrm{T}}S_1W_X + \hat{S}_i\bar{R}^{\mathrm{T}}\bar{A}_{di}W_d, \\ \Xi_{3i} &= W_{Q_3}^{\mathrm{T}}Q_3W_{Q_3} - W_Q^{\mathrm{T}}QW_Q - W_{d_2}^{\mathrm{T}}Z_1W_{d_2} - W_d^{\mathrm{T}}Z_2W_d \\ &\quad - W_X^{\mathrm{T}}S_1W_X - W_Y^{\mathrm{T}}S_2W_Y - W_Z^{\mathrm{T}}S_2W_Z. \end{aligned}$$

Proof. Under the given condition, we first show that system (12.3) with $\omega(k) = 0$ is regular and causal. Since $\operatorname{rank}\bar{E} = \hat{n} + r$, we choose two nonsingular matrices M and G such that

$$M\bar{E}G = \begin{bmatrix} I_{\hat{n}+r} & 0 \\ 0 & 0 \end{bmatrix}. \tag{12.7}$$

Set

$$M\bar{A}_iG = \begin{bmatrix} A_{1i} & A_{2i} \\ A_{3i} & A_{4i} \end{bmatrix}, \ G^{\mathrm{T}}\hat{S}_i = \begin{bmatrix} S_{11i} \\ S_{12i} \end{bmatrix}, \ M^{-\mathrm{T}}\bar{R} = \begin{bmatrix} 0 \\ I \end{bmatrix} F, \tag{12.8}$$

where $F \in \mathbb{R}^{(n-r)\times(n-r)}$ is any nonsingular matrix. It can be seen that $\Xi_{1i} < 0$ implies

$$-\bar{E}^{\mathrm{T}}P_i\bar{E}+\mathrm{sym}\{\hat{S}_i\bar{R}^{\mathrm{T}}\bar{A}_i\}-\bar{E}^{\mathrm{T}}\bar{S}_1\bar{E}<0, \tag{12.9}$$

where $\bar{S}_1 = \mathrm{diag}\{S_1, 0\}$. Pre- and post-multiplying (12.9) by G^{T} and G, respectively, we have $S_{12i}F^{\mathrm{T}}A_{4i} + A_{4i}^{\mathrm{T}}FS_{12i}^{\mathrm{T}} < 0$, which implies A_{4i} is non-singular for any $i \in \mathcal{I}$. Thus, the pair $(\bar{E}, \bar{A}_i)$ is regular and causal for any $i \in \mathcal{I}$. According to Definition 9.2, system (12.3) with $\omega(k) = 0$ is regular and causal.

Next we will show that system (12.3) is stochastically stable with l_2-l_∞ performance γ. To the end, we define $\delta(k) = x(k+1) - x(k)$ and consider the following Lyapunov functional for system (12.3):

$$V(\bar{x}(k),k,r(k))=\sum_{l=1}^{4}V_l(\bar{x}(k),k,r(k)), \tag{12.10}$$

where

$$\begin{aligned}
V_1(\bar{x}(k),k,r(k)) &= \bar{x}(k)^{\mathrm{T}}\bar{E}^{\mathrm{T}}P(r(k))\bar{E}\bar{x}(k),\\
V_2(\bar{x}(k),k,r(k)) &= \sum_{s=k-\tau}^{k-1}\Upsilon(s)^{\mathrm{T}}Q\Upsilon(s),\\
V_3(\bar{x}(k),k,r(k)) &= \sum_{s=k-d_2}^{k-1}x(s)^{\mathrm{T}}Z_1x(s)+\sum_{j=-d_2+1}^{-d_1+1}\sum_{s=k-1+j}^{k-1}x(s)^{\mathrm{T}}Z_2x(s),\\
V_4(\bar{x}(k),k,r(k)) &= \tau\sum_{-\tau}^{-1}\sum_{s=k+g}^{k-1}\delta(s)^{\mathrm{T}}E^{\mathrm{T}}S_1E\delta(s)\\
&\quad+d_{12}\sum_{g=-d_2}^{-d_1-1}\sum_{s=k+g}^{k-1}\delta(s)^{\mathrm{T}}E^{\mathrm{T}}S_2E\delta(s).
\end{aligned}$$

Then, along the trajectory of system (12.3), we have that for any $i \in \mathcal{I}$,

$$\mathscr{E}\{\Delta V_1(k)\}=\bar{x}(k+1)^{\mathrm{T}}\bar{E}^{\mathrm{T}}X_i\bar{E}\bar{x}(k+1)-\bar{x}(k)^{\mathrm{T}}\bar{E}^{\mathrm{T}}P_i\bar{E}\bar{x}(k), \tag{12.11}$$

$$\begin{aligned}
\mathscr{E}\{\Delta V_2(k)\} &= \Upsilon(k)^{\mathrm{T}}Q\Upsilon(k)-\Upsilon(k-\tau)^{\mathrm{T}}Q\Upsilon(k-\tau)\\
&= \bar{x}(k)^{\mathrm{T}}K^{\mathrm{T}}Q_1K\bar{x}(k)+\mathrm{sym}\{\bar{x}(k)^{\mathrm{T}}K^{\mathrm{T}}Q_2W_{Q_3}\eta(k)\}\\
&\quad+\eta(k)^{\mathrm{T}}W_{Q_3}^{\mathrm{T}}Q_3W_{Q_3}\eta(k)-\eta(k)^{\mathrm{T}}W_Q^{\mathrm{T}}QW_Q\eta(k),
\end{aligned} \tag{12.12}$$

$$\begin{aligned}
\mathscr{E}\{\Delta V_3(k)\} &\leqslant \bar{x}(k)^{\mathrm{T}}K^{\mathrm{T}}(Z_1+(d_{12}+1)Z_2)K\bar{x}(k)-\eta(k)^{\mathrm{T}}W_{d_2}^{\mathrm{T}}Z_1W_{d_2}\eta(k)\\
&\quad-\eta(k)^{\mathrm{T}}W_d^{\mathrm{T}}Z_2W_d\eta(k),
\end{aligned} \tag{12.13}$$

$$
\begin{aligned}
\mathscr{E}\{\Delta V_4(k)\} &= \tau^2\delta(k)^{\mathrm{T}}E^{\mathrm{T}}S_1E\delta(k) - \tau\sum_{s=k-\tau}^{k-1}\delta(s)^{\mathrm{T}}E^{\mathrm{T}}S_1E\delta(s)\\
&\quad + d_{12}^2\delta(k)^{\mathrm{T}}E^{\mathrm{T}}S_2E\delta(k)\\
&\quad - d_{12}\sum_{s=k-d(k)}^{k-d_1-1}\delta(s)^{\mathrm{T}}E^{\mathrm{T}}S_2E\delta(s) - d_{12}\sum_{s=k-d_2}^{k-d(k)-1}\delta(s)^{\mathrm{T}}E^{\mathrm{T}}S_2E\delta(s)\\
&\leqslant \delta(k)^{\mathrm{T}}E^{\mathrm{T}}SE\delta(k) - \bar{x}(k)^{\mathrm{T}}K^{\mathrm{T}}E^{\mathrm{T}}S_1EK\bar{x}(k)\\
&\quad - \mathrm{sym}\{\bar{x}(k)^{\mathrm{T}}K^{\mathrm{T}}E^{\mathrm{T}}S_1W_X\eta(k)\}\\
&\quad - \eta(k)^{\mathrm{T}}\left(W_X^{\mathrm{T}}S_1W_X + W_Y^{\mathrm{T}}S_2W_Y + W_Z^{\mathrm{T}}S_2W_Z\right)\eta(k),
\end{aligned}
\tag{12.14}
$$

On the other hand, it can be found that

$$
\bar{E}\bar{x}(k+1) = \bar{A}_i\bar{x}(k) + \bar{A}_{di}W_d\eta(k) + \bar{B}_i\omega(k), \tag{12.15}
$$

$$
E\delta(k) = (A_i - E)K\bar{x}(k) + A_{di}W_d\eta(k) + B_i\omega(k) \tag{12.16}
$$

and $2\bar{x}(k)^{\mathrm{T}}\hat{S}_i\bar{R}^{\mathrm{T}}\bar{E}\bar{x}(k+1) \equiv 0$, that is,

$$
\begin{aligned}
f(k) = \mathrm{sym}\{&\bar{x}(k)^{\mathrm{T}}\hat{S}_i\bar{R}^{\mathrm{T}}\bar{A}_i\bar{x}(k) + \bar{x}(k)^{\mathrm{T}}\hat{S}_i\bar{R}^{\mathrm{T}}\bar{A}_{di}W_d\eta(k)\\
&+ \bar{x}(k)^{\mathrm{T}}\hat{S}_i\bar{R}^{\mathrm{T}}\bar{B}_i\omega(k)\} \equiv 0.
\end{aligned}
\tag{12.17}
$$

Then, for any nonzero $\omega(k) \in l_2[0,\infty)$ and under zero initial condition, we get from (12.11)-(12.17) that

$$
\begin{aligned}
J &= \mathscr{E}\left\{V(k) - \sum_{s=0}^{k-1}\omega(s)^{\mathrm{T}}\omega(s)\right\}\\
&= \mathscr{E}\left\{\sum_{s=0}^{k-1}\left(\Delta V(s) + f(s) - \omega(s)^{\mathrm{T}}\omega(s)\right)\right\}\\
&\leqslant \mathscr{E}\left\{\sum_{s=0}^{k-1}\zeta(s)^{\mathrm{T}}\Sigma_i\zeta(s)\right\},
\end{aligned}
\tag{12.18}
$$

where

$$
\begin{aligned}
\Sigma_i &= \begin{bmatrix}\Xi_{1i} & \Xi_{2i} & \hat{S}_i\bar{R}^{\mathrm{T}}\bar{B}_i\\ * & \Xi_{3i} & 0\\ * & * & -I\end{bmatrix} + \begin{bmatrix}\bar{A}_i^{\mathrm{T}}\\ W_d^{\mathrm{T}}\bar{A}_{di}^{\mathrm{T}}\\ \bar{B}_i^{\mathrm{T}}\end{bmatrix} X_i \begin{bmatrix}\bar{A}_i^{\mathrm{T}}\\ W_d^{\mathrm{T}}\bar{A}_{di}^{\mathrm{T}}\\ \bar{B}_i^{\mathrm{T}}\end{bmatrix}^{\mathrm{T}}\\
&\quad + \begin{bmatrix}K^{\mathrm{T}}(A_i - E)^{\mathrm{T}}\\ W_d^{\mathrm{T}}A_{di}^{\mathrm{T}}\\ B_i^{\mathrm{T}}\end{bmatrix} S \begin{bmatrix}K^{\mathrm{T}}(A_i - E)^{\mathrm{T}}\\ W_d^{\mathrm{T}}A_{di}^{\mathrm{T}}\\ B_i^{\mathrm{T}}\end{bmatrix}^{\mathrm{T}}.
\end{aligned}
$$

According to Schur complement, we get from (12.5) that $\Sigma_i < 0$ for any $i \in \mathcal{I}$, that is, $J < 0$ for any nonzero $\omega(k) \in l_2[0, \infty)$, which implies that

$$\mathscr{E}\left\{\bar{x}(k)^{\mathrm{T}}\bar{E}^{\mathrm{T}}P_i\bar{E}\bar{x}(k)\right\} \leqslant \mathscr{E}\left\{V(k)\right\} < \sum_{s=0}^{k-1}\omega(s)^{\mathrm{T}}\omega(s). \tag{12.19}$$

On the other hand, applying Schur complement, we obtain from (12.6) that $\bar{C}_i^{\mathrm{T}}\bar{C}_i \leqslant \gamma^2\bar{E}^{\mathrm{T}}P_i\bar{E}$. Then we can conclude that for all $k > 0$,

$$\begin{aligned}\mathscr{E}\left\{\bar{z}(k)^{\mathrm{T}}\bar{z}(k)\right\} &= \mathscr{E}\left\{\bar{x}(k)^{\mathrm{T}}\bar{C}_i^{\mathrm{T}}\bar{C}_i\bar{x}(k)\right\}\\ &\leqslant \gamma^2\mathscr{E}\left\{\bar{x}(k)^{\mathrm{T}}\bar{E}^{\mathrm{T}}P_i\bar{E}\bar{x}(k)\right\}\\ &< \gamma^2\sum_{s=0}^{k-1}\omega(s)^{\mathrm{T}}\omega(s)\\ &\leqslant \gamma^2\sum_{s=0}^{\infty}\omega(s)^{\mathrm{T}}\omega(s),\end{aligned} \tag{12.20}$$

which implies (12.4) holds under the zero-initial condition for any nonzero $\omega(k) \in l_2[0, \infty)$. On the other hand, applying Schur complement, we get from (12.5) that

$$\begin{aligned}&\begin{bmatrix}\Xi_{1i} & \Xi_{2i}\\ * & \Xi_{3i}\end{bmatrix} + \begin{bmatrix}\bar{A}_i^{\mathrm{T}}\\ W_d^{\mathrm{T}}\bar{A}_{di}^{\mathrm{T}}\end{bmatrix}X_i\begin{bmatrix}\bar{A}_i^{\mathrm{T}}\\ W_d^{\mathrm{T}}\bar{A}_{di}^{\mathrm{T}}\end{bmatrix}^{\mathrm{T}}\\ &\qquad + \begin{bmatrix}K^{\mathrm{T}}(A_i - E)^{\mathrm{T}}\\ W_d^{\mathrm{T}}A_{di}^{\mathrm{T}}\end{bmatrix}S\begin{bmatrix}K^{\mathrm{T}}(A_i - E)^{\mathrm{T}}\\ W_d^{\mathrm{T}}A_{di}^{\mathrm{T}}\end{bmatrix}^{\mathrm{T}} < 0,\end{aligned} \tag{12.21}$$

which guarantees there exists a scalar $\rho > 0$ such that

$$\mathscr{E}\left\{\Delta V(k)\right\} < -\rho||\bar{x}(k)||^2 \tag{12.22}$$

in case of $\omega(k) \equiv 0$. Using the similar method of [6], it can be seen from (12.10) and (12.22) that system (12.3) with $\omega(k) \equiv 0$ is stochastically stable. This completes the proof.

Remark 12.4. In terms of the delay partitioning technique, Theorem 12.3 provides a filtering analysis result for the l_2-l_∞ performance of discrete-time SMJSs with time-varying delays. When there are no time-delays in system, that is, $A_{di} = 0$ and $C_{di} = 0$ for any $i \in \mathcal{I}$, (12.5) reduces to

$$\begin{bmatrix}\Xi_{1i} & \hat{S}_i\bar{R}^{\mathrm{T}}\bar{B}_i & \bar{A}_i^{\mathrm{T}}X_i\\ * & -I & \bar{B}_i^{\mathrm{T}}X_i\\ * & * & -X_i\end{bmatrix} < 0. \tag{12.23}$$

It can be seen that (12.6) and (12.23) are the l_2-l_∞ performance conditions for discrete-time SMJSs with delay free.

Remark 12.5. In the proof of Theorem 12.3, the matrix $\hat{S}_i$ is introduced via the null term (12.17) and plays a key role in proving the regularity and causality of discrete-time SMJSs with time-varying delays. It is worth pointing out that for state-space systems the matrix $\hat{S}_i$ is not necessary and the null term (12.17) will vanish because of the non-singularity of E.

Based on Theorem 12.3, we are now ready to deal with the l_2-l_∞ filtering problem, and a sufficient condition for the existence of a suitable filter is presented as follows.

Theorem 12.6. *For a given scalar $\gamma > 0$, system (12.3) is stochastically admissible with l_2-l_∞ performance γ, if there exist matrices $\begin{bmatrix} P_{1i} & P_{2i} \\ * & P_{3i} \end{bmatrix} > 0$, $\begin{bmatrix} Q_1 & Q_2 \\ * & Q_3 \end{bmatrix} > 0$, $Z_1 > 0$, $Z_2 > 0$, $S_1 > 0$, $S_2 > 0$, S_{1i}, S_{2i}, T_{1i}, T_{2i}, V, $\bar{A}_f$, $\bar{B}_f$ and $\bar{C}_f$ such that for any $i \in \mathcal{I}$,*

$$\begin{bmatrix} \Xi_{11i} & \Xi_{12i} & \Xi_{13i} & S_{1i}R^{\mathrm{T}}B_i & \Xi_{15i} & \Xi_{16i} & (A_i - E)^{\mathrm{T}}S \\ * & -P_{3i} & S_{2i}R^{\mathrm{T}}A_{di}W_d & S_{2i}R^{\mathrm{T}}B_i & \bar{A}_f^{\mathrm{T}}\hat{M}^{\mathrm{T}} & \bar{A}_f^{\mathrm{T}} & 0 \\ * & * & \Xi_{3i} & 0 & \Xi_{35i} & \Xi_{36i} & W_d^{\mathrm{T}}A_{di}^{\mathrm{T}}S \\ * & * & * & -I & \Xi_{45i} & \Xi_{46i} & B_i^{\mathrm{T}}S \\ * & * & * & * & \Xi_{55i} & \Xi_{56i} & 0 \\ * & * & * & * & * & \Xi_{66i} & 0 \\ * & * & * & * & * & * & -S \end{bmatrix} < 0, \quad (12.24)$$

$$\begin{bmatrix} E^{\mathrm{T}}P_{1i}E & E^{\mathrm{T}}P_{2i} & L_i^{\mathrm{T}} \\ * & P_{3i} & -C_f^{\mathrm{T}} \\ * & * & \gamma^2 I \end{bmatrix} \geqslant 0, \quad (12.25)$$

where Ξ_{3i}, S and d_{12} are given in Theorem 12.3, $\hat{M} = \left[I_{\hat{n}} \; 0_{\hat{n}\times(n-\hat{n})}\right]^{\mathrm{T}}$, $X_{li} = \sum_{j=1}^{N} \pi_{ij}P_{lj}$ $(l = 1, 2, 3)$, and

$$\begin{aligned}
\Xi_{11i} &= -E^{\mathrm{T}}P_{1i}E + \mathrm{sym}\{S_{1i}R^{\mathrm{T}}A_i\} + Q_1 + Z_1 + (d_{12}+1)Z_2 - E^{\mathrm{T}}S_1E, \\
\Xi_{12i} &= -E^{\mathrm{T}}P_{2i} + A_i^{\mathrm{T}}RS_{2i}^{\mathrm{T}}, \\
\Xi_{13i} &= Q_2W_{Q_3} - E^{\mathrm{T}}S_1W_X + S_{1i}R^{\mathrm{T}}A_{di}W_d, \\
\Xi_{15i} &= A_i^{\mathrm{T}}T_{1i}^{\mathrm{T}} + C_i^{\mathrm{T}}\bar{B}_f^{\mathrm{T}}\hat{M}^{\mathrm{T}}, \\
\Xi_{16i} &= A_i^{\mathrm{T}}T_{2i}^{\mathrm{T}} + C_i^{\mathrm{T}}\bar{B}_f^{\mathrm{T}}, \\
\Xi_{35i} &= W_d^{\mathrm{T}}\left(A_{di}^{\mathrm{T}}T_{1i}^{\mathrm{T}} + C_{di}^{\mathrm{T}}\bar{B}_f^{\mathrm{T}}\hat{M}^{\mathrm{T}}\right), \\
\Xi_{36i} &= W_d^{\mathrm{T}}\left(A_{di}^{\mathrm{T}}T_{2i}^{\mathrm{T}} + C_{di}^{\mathrm{T}}\bar{B}_f^{\mathrm{T}}\right), \\
\Xi_{45i} &= B_i^{\mathrm{T}}T_{1i}^{\mathrm{T}} + D_i^{\mathrm{T}}\bar{B}_f^{\mathrm{T}}\hat{M}^{\mathrm{T}}, \\
\Xi_{46i} &= B_i^{\mathrm{T}}T_{2i}^{\mathrm{T}} + D_i^{\mathrm{T}}\bar{B}_f^{\mathrm{T}}, \\
\Xi_{55i} &= X_{1i} - T_{1i} - T_{1i}^{\mathrm{T}},
\end{aligned}$$

$$\Xi_{56i} = X_{2i} - T_{2i}^{\mathrm{T}} - \hat{M}V,$$
$$\Xi_{66i} = X_{3i} - V - V^{\mathrm{T}}.$$

Furthermore, if (12.24) *and* (12.25) *are solvable, the desired filter* (12.2) *can be chosen with parameters as*

$$A_f = V^{-1}\bar{A}_f,\ B_f = V^{-1}\bar{B}_f,\ C_f = \bar{C}_f. \tag{12.26}$$

Proof. By introducing the slack variable T_i, the equivalent form of (12.5) can be given in the following

$$\begin{bmatrix} \Xi_{1i} & \Xi_{2i} & \hat{S}_i\bar{R}^{\mathrm{T}}\bar{B}_i & \bar{A}_i^{\mathrm{T}}T_i^{\mathrm{T}} & H^{\mathrm{T}}(A_i - E)^{\mathrm{T}}S \\ * & \Xi_{3i} & 0 & W_d^{\mathrm{T}}\bar{A}_{di}^{\mathrm{T}}T_i^{\mathrm{T}} & W_d^{\mathrm{T}}A_{di}^{\mathrm{T}}S \\ * & * & -I & \bar{B}_i^{\mathrm{T}}T_i^{\mathrm{T}} & B_i^{\mathrm{T}}S \\ * & * & * & X_i - T_i - T_i^{\mathrm{T}} & 0 \\ * & * & * & * & -S \end{bmatrix} < 0. \tag{12.27}$$

Choosing $\bar{R} = \begin{bmatrix} R \\ 0 \end{bmatrix}$, where $R \in \mathbb{R}^{n\times(n-r)}$ is any matrix with full column satisfying $E^{\mathrm{T}}R = 0$, we can see that $\bar{R}$ is a matrix with full column and satisfies $\bar{E}^{\mathrm{T}}\bar{R} = 0$. Partition T_i as

$$T_i = \begin{bmatrix} T_{1i} & \hat{M}T_3 \\ \bar{T}_{2i} & T_4 \end{bmatrix}. \tag{12.28}$$

If condition (12.27) is true, then $T_i + T_i^{\mathrm{T}} > 0$, which implies that T_4 is invertible. Define

$$J_1 = \begin{bmatrix} I & 0 \\ 0 & T_3T_4^{-1} \end{bmatrix},\ J_2 = \mathrm{diag}\{J_1, I, I, J_1, I\}. \tag{12.29}$$

Pre- and post-multiplying the left hand side of (12.27) by J_2 and J_2^{T}, respectively, and introducing the following new variables

$$T_{2i} = T_3T_4^{-1}\bar{T}_{2i},\ V = T_3T_4^{-\mathrm{T}}T_3^{\mathrm{T}},\ \begin{bmatrix} P_{1i} & P_{2i} \\ * & P_{3i} \end{bmatrix} = J_1P_iJ_1^{\mathrm{T}},\ J_1\hat{S}_i = \begin{bmatrix} S_{1i} \\ S_{2i} \end{bmatrix},$$
$$\bar{A}_f = T_3A_fT_4^{-\mathrm{T}}T_3^{\mathrm{T}},\ \bar{B}_f = T_3B_f,\ \bar{C}_f = C_fT_4^{-\mathrm{T}}T_3^{\mathrm{T}}, \tag{12.30}$$

we can find (12.24) is equivalent to (12.27). Similarly, pre- and post-multiplying the left hand side of (12.6) by $\mathrm{diag}\{J_1, I\}$ and $\mathrm{diag}\{J_1, I\}^{\mathrm{T}}$, respectively, we can find (12.25) is equivalent to (12.6). On the other hand, since V is nonsingular, we can get that T_3 is also nonsingular. Then we can find that

$$A_f = T_3^{-1}\bar{A}_fT_3^{-\mathrm{T}}T_4^{\mathrm{T}},\ B_f = T_3^{-1}\bar{B}_f,\ C_f = \bar{C}_fT_3^{-\mathrm{T}}T_4^{\mathrm{T}}. \tag{12.31}$$

Therefore, the following filter

$$\begin{cases} x_f(k+1) = T_3^{-1}\bar{A}_f T_3^{-\mathrm{T}} T_4^{\mathrm{T}} x_f(k) + T_3^{-1}\bar{B}_f y(k), \\ \hat{z}(k) = \bar{C}_f T_3^{-\mathrm{T}} T_4^{\mathrm{T}} x_f(k) \end{cases} \tag{12.32}$$

guarantees the filtering error system (12.3) to be stochastically admissible with l_2-l_∞ performance γ. Next, performing an linear transformation $\hat{x}(k) = T_3^{-\mathrm{T}} T_4^{\mathrm{T}} x_f(k)$ on the state in (12.32) yields the filter (12.2), where A_f, B_f, and C_f are defined in (12.26). This completes the proof.

Remark 12.7. It is noted that Theorem 12.6 presents a sufficient condition for the existence of l_2-l_∞ filter for discrete-time SMJSs with time-varying delays. It should be pointed out that the conservatism of Theorem 12.6 lies in the parameter m, which refers to the number of delay partitioning, that is, the conservatism is reduced as the partitions grow. On the other hand, the computational complexity is also dependent on the partition number m, that is, the computational complexity is increased as the partitioning becomes thinner. Therefore, there is a tradeoff between the computational complexity and the l_2-l_∞ performance of system.

Remark 12.8. The robust H_∞ filtering problem for discrete-time SMJSs with parameter uncertainties and mode-dependent time-delay has been discussed in [100]. But the result of [100] requires an indispensable assumption on the accessibility of the jump mode for the filter, and is thus invalid when the jump mode $r(k)$ is not available to the filter. Therefore our result is much more desirable and applicable than the one in [100].

When $A_{di} = 0$ and $C_{di} = 0$ for any $i \in \mathcal{I}$, that is, system (12.1) reduces to a discrete-time SMJS without time delays, we have the following corollary based on Theorem 12.6.

Corollary 12.9. *For a given scalar $\gamma > 0$, system*

$$\begin{cases} \bar{E}\bar{x}(k+1) = \bar{A}(r(k))\bar{x}(k) + \bar{B}(r(k))\omega(k), \\ \bar{z}(k) = \bar{C}(r(k))\bar{x}(k) \end{cases} \tag{12.33}$$

is stochastically admissible with l_2-l_∞ performance γ, if there exist matrices $\begin{bmatrix} P_{1i} & P_{2i} \\ * & P_{3i} \end{bmatrix} > 0$, S_{1i}, S_{2i}, T_{1i}, T_{2i}, V, $\bar{A}_f$, $\bar{B}_f$ and $\bar{C}_f$ such that (12.25) *and* (12.34) *hold for any $i \in \mathcal{I}$,*

$$\begin{bmatrix} \hat{\Xi}_{11i} & \Xi_{12i} & S_{1i}R^{\mathrm{T}}B_i & \Xi_{15i} & \Xi_{16i} \\ * & -P_{3i} & S_{2i}R^{\mathrm{T}}B_i & \bar{A}_f^{\mathrm{T}}\hat{M}^{\mathrm{T}} & \bar{A}_f^{\mathrm{T}} \\ * & * & -I & \Xi_{45i} & \Xi_{46i} \\ * & * & * & \Xi_{55i} & \Xi_{56i} \\ * & * & * & * & \Xi_{66i} \end{bmatrix} < 0, \tag{12.34}$$

where Ξ_{12i}, Ξ_{15i}, Ξ_{16i}, Ξ_{45i}, Ξ_{46i}, Ξ_{55i}, Ξ_{56i}, Ξ_{66i}, $\hat{M}$ *and* X_{li} $(l = 1, 2, 3)$ *are given in Theorem 12.6, and* $\hat{\Xi}_{11i} = -E^{\mathrm{T}} P_{1i} E + \mathrm{sym}\{S_{1i} R^{\mathrm{T}} A_i\}$. *Furthermore, if* (12.25) *and* (12.34) *are solvable, the desired filter* (12.2) *can be chosen with parameters defined in* (12.26).

Remark 12.10. It is noted that the matrix $\hat{M}$ defined in Theorem 12.6 and Corollary 12.9 plays an important role in formulating the full-order ($\hat{n} = n$ and $\hat{M} = I_n$) and reduced-order filters ($\hat{n} < n$ and $\hat{M} = \begin{bmatrix} I_{\hat{n}} & 0_{\hat{n}\times(n-\hat{n})} \end{bmatrix}^{\mathrm{T}}$) in a unified framework.

Remark 12.11. It should be pointed out that the LMIs in (12.24) and (12.25) (or (12.25) and (12.34)) are not only over the matrix variables, but also over the scalar γ^2. This implies that, by setting $\gamma^2 = \delta$ and minimizing δ subject to (12.24) and (12.25) (or (12.25) and (12.34)), we can obtain the optimal l_2-l_∞ performance γ by $\gamma = \sqrt{\delta^*}$, where δ^* is the optimal value of δ, and the corresponding filter as well.

12.4 Numerical Examples

In this section, two numerical examples will be given to demonstrate the applicability of the proposed approaches in the previous section.

Example 12.12. Consider system (12.1) with

$$\begin{aligned}
A_1 &= \begin{bmatrix} 0.8 & 2.4 \\ 1.7 & 3.3 \end{bmatrix}, \; A_2 = \begin{bmatrix} 0.7 & 2.1 \\ 1.85 & 3.75 \end{bmatrix}, \\
A_{d1} &= \begin{bmatrix} 0.1 & 0.1 \\ -0.39 & -0.87 \end{bmatrix}, \; A_{d2} = \begin{bmatrix} 0.1 & 0 \\ -0.61 & -1.33 \end{bmatrix}, \\
B_1 &= \begin{bmatrix} 0.1 \\ 1 \end{bmatrix}, \; B_2 = \begin{bmatrix} 0 \\ -1 \end{bmatrix}, \; C_1 = \begin{bmatrix} 2 \\ 5 \end{bmatrix}^{\mathrm{T}}, \; C_2 = \begin{bmatrix} 1 \\ 1 \end{bmatrix}^{\mathrm{T}}, \\
C_{d1} &= \begin{bmatrix} 1.5 & 3.3 \end{bmatrix}, \; C_{d2} = \begin{bmatrix} 1.6 & 3.6 \end{bmatrix}, \; D_1 = 0.9, \; D_2 = -1, \\
L_1 &= \begin{bmatrix} -3 & -9 \end{bmatrix}, \; L_2 = \begin{bmatrix} 5 & 15 \end{bmatrix}.
\end{aligned}$$

The singular matrix

$$E = \begin{bmatrix} 1 & 3 \\ 0 & 0 \end{bmatrix}$$

and the transition probability matrix

$$\Pi = \begin{bmatrix} 0.3 & 0.7 \\ 0.8 & 0.2 \end{bmatrix}.$$

It is assumed that the time-varying delay $d(k)$ satisfies $6 \leqslant d(k) \leqslant 9$, that is, the lower bound of the time-varying delay is 6 and the upper bound of the time-varying delay is 9.

Our purpose is to determine the minimum l_2-l_∞ performances of the filtering error system (12.3) for different values of the partition number m and the filter order $\hat{n}$. When the partition number $m = 2$ and the filter order $\hat{n} = 1$, by solving the convex optimization problem in (12.24) and (12.25), the minimum l_2-l_∞ performance obtained is $\gamma = 12.0968$. However, if we assume $\hat{n} = 2$, the minimum l_2-l_∞ performance obtained is $\gamma = 12.0803$ for the same partition number m. This shows that for a fixed partition number m, the higher order $\hat{n}$ corresponds to the smaller l_2-l_∞ performance γ. When the partition number $m = 3$ and the filter order $\hat{n} = 2$, by solving the convex optimization problem in (12.24) and (12.25), the minimum l_2-l_∞ performance obtained is $\gamma = 12.0225$. When the partition number $m = 6$ and the filter order $\hat{n} = 2$, the minimum l_2-l_∞ performance obtained is $\gamma = 11.9905$. Therefore it can be found that the larger partition number m corresponds to the smaller l_2-l_∞ performance γ when the filter order $\hat{n}$ is fixed, that is, for the same filter order $\hat{n}$, the minimum l_2-l_∞ performance γ becomes smaller as the partitioning becomes thinner. A more detailed comparison for different m and $\hat{n}$ is provided in Table 12.1.

Table 12.1. Example 12.12: Minimum l_2-l_∞ performance γ with different (m, τ) and $\hat{n}$

(m, τ)	$(1, 6)$	$(2, 3)$	$(3, 2)$	$(6, 1)$
Theorem 12.6 ($\hat{n} = 1$)	12.5193	12.0968	12.0383	12.0058
Theorem 12.6 ($\hat{n} = 2$)	12.5021	12.0803	12.0225	11.9905

The second task in this example is for the given scalars $\hat{n}$, γ and m, to find a filter (12.2) such that the filtering error system (12.3) is stochastically admissible with l_2-l_∞ performance γ. When $\hat{n} = 1$, $\gamma = 15$ and $m = 6$, by using Matlab LMI control Toolbox to solve the LMIs in (12.24) and (12.25), we can get the following reduced-order filter:

$$\begin{cases} \hat{x}(k+1) = 0.6674\hat{x}(k) + 0.0933y(k), \\ \hat{z}(k) = -0.0387\hat{x}(k). \end{cases} \tag{12.35}$$

When $\hat{n} = 2$, $\gamma = 14$ and $m = 3$, we can get the following full-order filter:

$$\begin{cases} \hat{x}(k+1) = \begin{bmatrix} 0.6913 & 0.0077 \\ 0.0017 & 0.0047 \end{bmatrix} \hat{x}(k) + \begin{bmatrix} 0.1013 \\ 0.0035 \end{bmatrix} y(k), \\ \hat{z}(k) = \begin{bmatrix} -0.0183 & -1.3993 \end{bmatrix} \hat{x}(k). \end{cases} \tag{12.36}$$

With the filter (12.36), the simulation results are shown in Figs. 12.1-12.5, where the external disturbance

$$\omega(k) = \begin{cases} 4, & 10 \leqslant k \leqslant 15 \\ -5, & 30 \leqslant k \leqslant 70 \\ 0, & \text{elsewhere} \end{cases}.$$

In the simulation, the time-varying delay $d(k)$ and the Markov chain $r(k)$ are generated randomly and are shown in Figs. 12.1 and 12.2, respectively. The state responses of the considered system and the filter (12.36) are depicted in Figs. 12.3 and 12.4, respectively. Fig. 12.5 is the simulation result of the error response of $z(k) - \hat{z}(k)$. It is clearly observed from the simulation results that the designed l_2-l_∞ filter (12.36) satisfies the specified requirements.

Example 12.13. : Consider system (12.1) with delay free and

$$A_1 = \begin{bmatrix} -1 & 0.5 & 1 \\ -1 & -0.3 & 1 \\ 0.5 & 0 & 1 \end{bmatrix}, \ A_2 = \begin{bmatrix} -0.9 & 0.45 & 0.7 \\ -0.9 & -0.2 & 0.9 \\ -0.5 & 0.1 & 1 \end{bmatrix},$$

$$B_1 = \begin{bmatrix} 0.1 \\ 0.1 \\ 0.5 \end{bmatrix}, \ B_2 = \begin{bmatrix} -0.5 \\ 0.1 \\ 0.1 \end{bmatrix}, \ C_1 = \begin{bmatrix} 0.9 \\ 0.3 \\ 0.2 \end{bmatrix}^{\mathrm{T}}, \ C_2 = \begin{bmatrix} 0.9 \\ -0.3 \\ -0.2 \end{bmatrix}^{\mathrm{T}},$$

$$D_1 = 0.7, \ D_2 = -0.7, \ L_1 = \begin{bmatrix} 0.4 \\ 1 \\ -0.3 \end{bmatrix}^{\mathrm{T}}, \ L_2 = \begin{bmatrix} -0.5 \\ -0.9 \\ 0.2 \end{bmatrix}^{\mathrm{T}}.$$

The singular matrix

$$E = \begin{bmatrix} 1 & 1 & 0 \\ 1 & -1 & 1 \\ 2 & 0 & 1 \end{bmatrix}$$

and the transition probability matrix

$$\Pi = \begin{bmatrix} a & 1-a \\ 0.2 & 0.8 \end{bmatrix},$$

where the parameter $a \geqslant 0$ in matrix Π can take different values for extensive discussion purpose.

Our purpose here is for given scalars a, $\hat{n}$ and γ, to find a filter (12.2) such that the filtering error system (12.33) is stochastically admissible with l_2-l_∞ performance γ. Given $a = 0.5$, $\hat{n} = 1$ and $\gamma = 0.5$, and by using Matlab LMI control Tool box to solve the LMIs in (12.25) and (12.34), we can get the following reduced-order filter:

$$\begin{cases} \hat{x}(k+1) = 0.1404\hat{x}(k) - 0.1687y(k), \\ \hat{z}(k) = 0.1733\hat{x}(k). \end{cases} \tag{12.37}$$

When $a = 0.7$, $\hat{n} = 3$ and $\gamma = 0.6$, using Matlab LMI control Tool box to solve the LMIs in (12.34), we can get the following full-order filter:

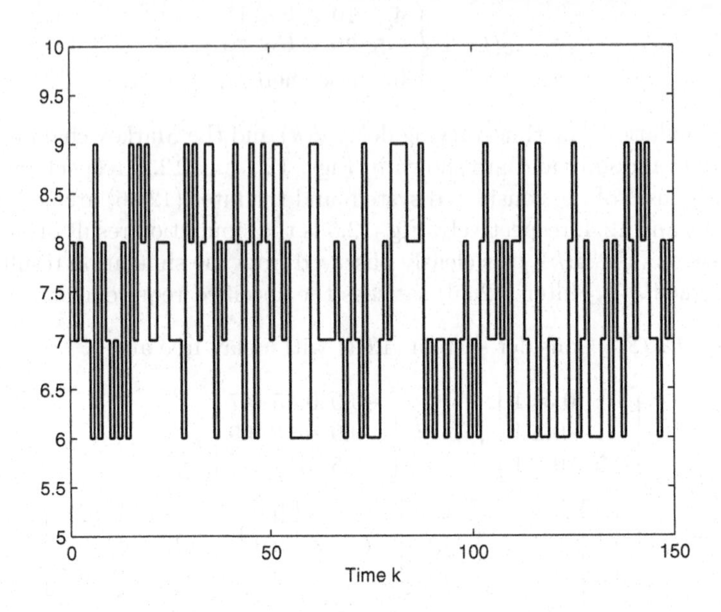

Fig. 12.1. Example 12.12: Time-varying delay $d(k)$

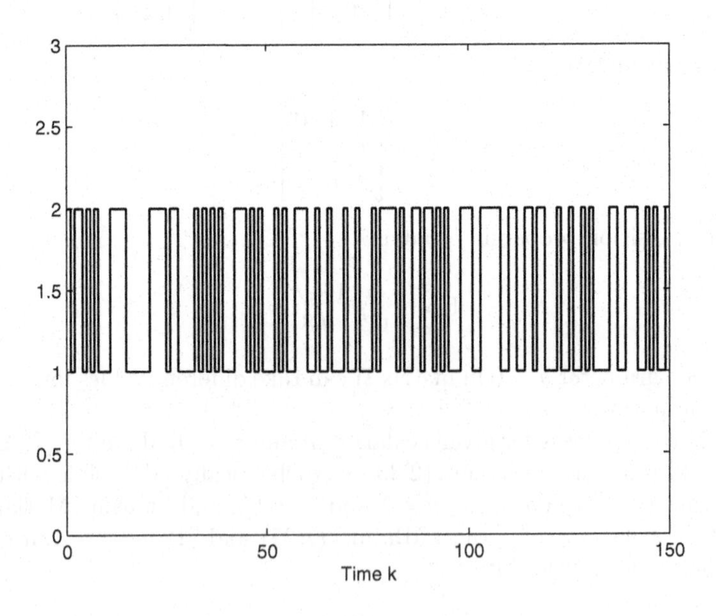

Fig. 12.2. Example 12.12: Markov chain $r(k)$

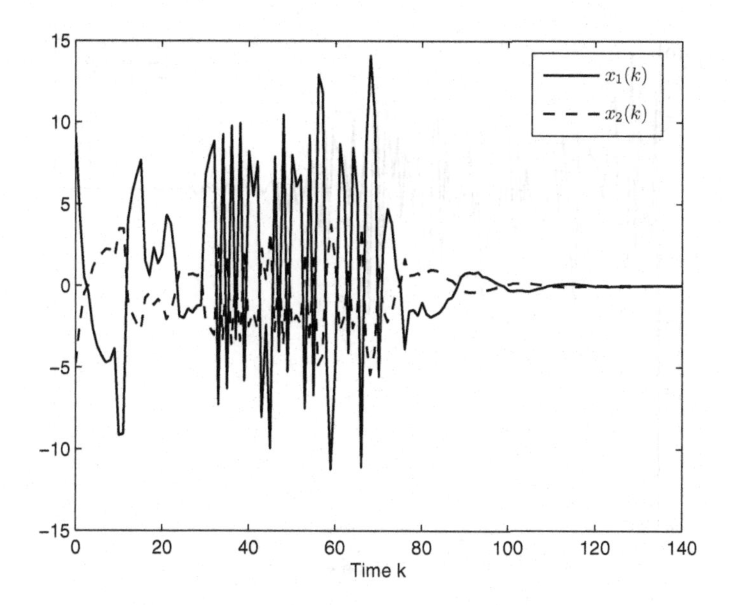

Fig. 12.3. Example 12.12: State trajectories of system

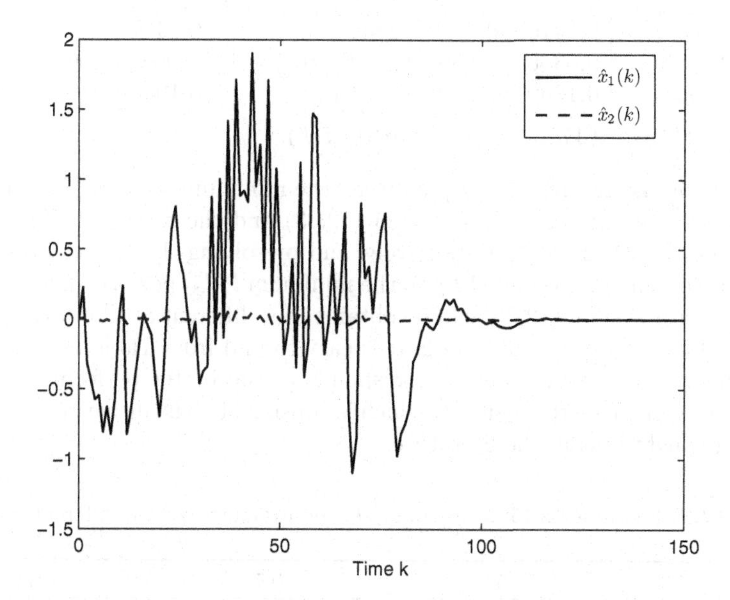

Fig. 12.4. Example 12.12: State trajectories of filter

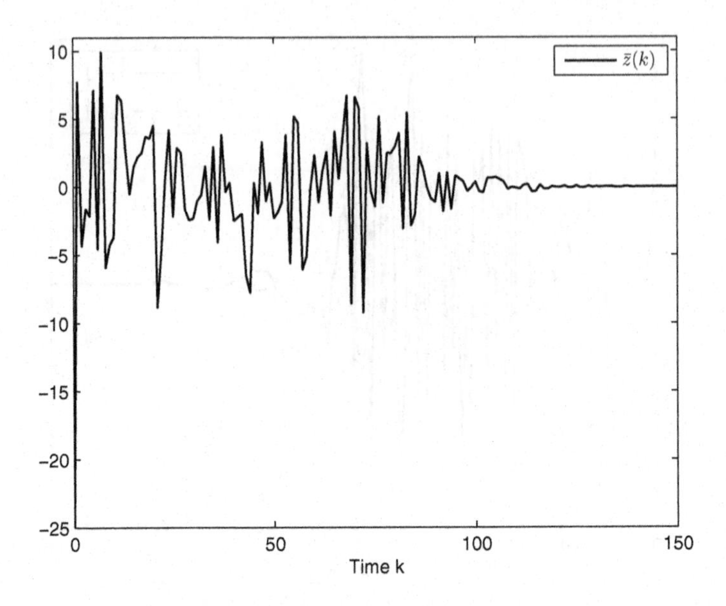

Fig. 12.5. Example 12.12: Estimation error

$$\begin{cases} \hat{x}(k+1) = \begin{bmatrix} 0.1410 & -0.0090 & 0.1747 \\ 0.0356 & 0.1289 & 0.2104 \\ 0.1650 & 0.1253 & 0.3804 \end{bmatrix} \hat{x}(k) + \begin{bmatrix} -0.2447 \\ 0.1008 \\ -0.1047 \end{bmatrix} y(k), \\ \hat{z}(k) = \begin{bmatrix} 0.1725 & 0.0054 & -0.0881 \end{bmatrix} \hat{x}(k). \end{cases} \tag{12.38}$$

Now, we discuss the relationship between the minimum l_2-l_∞ performance γ achieved by the filtering error system (12.33) and the filter order $\hat{n}$. Given filter order $\hat{n} = 1$, 2 and 3, respectively, and by solving the convex optimization problem in (12.25) and (12.34), the minimum l_2-l_∞ performance γ of the filtering error system (12.33) can be obtained for different a. The corresponding results are listed in Table 12.2. It is easy to find from Table 12.2 that for the same a, the higher order $\hat{n}$ corresponds to the better performance that the filtering error system can get, which is consistent to intuitive analysis and existing conclusions in the literature.

Table 12.2. Example 12.13: Minimum l_2-l_∞ performance γ with different a and $\hat{n}$

a	0.5	0.7	0.9
Corollary 12.9 ($\hat{n} = 1$)	0.3120	0.3278	0.3420
Corollary 12.9 ($\hat{n} = 2$)	0.3066	0.3212	0.3329
Corollary 12.9 ($\hat{n} = 3$)	0.3061	0.3189	0.3258

12.5 Conclusion

In this chapter, the problem of l_2-l_∞ mode-independent filter design has been investigated for discrete-time SMJSs with state delays. A sufficient condition has been derived for the filtering error systems to be regular, causal and delay-dependent stochastically stable with a given l_2-l_∞ performance γ. Based on the obtained analysis result, the full-order and reduced-order filters have been designed in a unified framework. The addressed filter design problem has been converted into the feasibility problem of a set of LMIs, which can be checked efficiently via the Matlab LMI Toolbox. The obtained delay-dependent results rely upon the partitioning size. The numerical examples have been given to show the effectiveness and the potential of the proposed techniques.

13

Mode-Independent H_∞ Filter Design for SMJSs with Constant Delays

13.1 Introduction

In the previous chapter, we have investigated the problem of mode-independent filter design for discrete-time SMJSs with time-varying delays. The following two chapters are concerned with the H_∞ filtering problem for continuous-time SMJSs with time delays. The purpose of this chapter is to design a mode-independent filter such that the error system is stochastically admissible with H_∞ performance index. The conditions for the solvability of this problem are proposed by strict LMIs. In order to enlarge the application scope of the obtained results, the assumption that all the transition rates have to be precisely known is not necessary and the designed filter is deterministic. These strict LMI-based conditions include both delay-dependent and delay-independent results and can be applied to systems with unknown transition jump rates or completely known transition jump rates. Some numerical examples will be finally given to show the effectiveness of the methods.

13.2 Problem Formulation

Let $\{r_t, t \geqslant 0\}$ be a continuous-time Markov process with right continuous trajectories and follow the same definition as that in Chapter 10. Fix a probability space $(\Omega, \mathcal{F}, \mathcal{P})$ and consider the following SMJS with time delays:

$$\begin{cases} E\dot{x}(t) = A(r_t)x(t) + A_d(r_t)x(t-d) + B_\omega(r_t)\omega(t), \\ y(t) = C(r_t)x(t) + C_d(r_t)x(t-d) + D_\omega(r_t)\omega(t), \\ z(t) = L(r_t)x(t), \\ x(t) = \phi(t),\ t \in [-d, 0], \end{cases} \tag{13.1}$$

where $x(t) \in \mathbb{R}^n$ is the state, $y(t) \in \mathbb{R}^s$ is the measurement, $z(t) \in \mathbb{R}^q$ is the signal to be estimated, $\omega(t) \in \mathbb{R}^p$ is the disturbance input that belongs to $\mathscr{L}_2[0,\infty)$, d is the constant time-delay, and $\phi(t) \in C_{n,d}$ is a compatible

Z.-G. Wu et al.: *Anal. & Synth. of Singular Syst. with Time-Delays*, LNCIS 443, pp. 183–198.
DOI: 10.1007/978-3-642-37497-5_13

vector valued initial function. The matrix $E \in \mathbb{R}^{n\times n}$ may be singular and it is assumed that rank $E = r \leqslant n$. $A(r_t)$, $A_d(r_t)$, $B_\omega(r_t)$, $C(r_t)$, $C_d(r_t)$, $D_\omega(r_t)$ and $L(r_t)$ are known real constant matrices with appropriate dimensions for each $r_t \in \mathcal{S}$.

Generally speaking, for some practical systems, the transition rates we get will never be precise, and always will have some errors that may restrict applications of the established results. Therefore, in this chapter, we will assume that the transition rates are unknown but satisfy

$$0 \leqslant \underline{\pi}_i \leqslant \pi_{ij} \leqslant \bar{\pi}_i, \forall i,j \in \mathcal{S}, j \neq i, \tag{13.2}$$

where $\bar{\pi}_i$ and $\underline{\pi}_i$ are known scalars.

The purpose of this chapter is that for system (13.1) to design a linear filter as

$$\begin{cases} E\dot{\hat{x}}(t) = A_f\hat{x}(t) + B_f y(t), \\ \hat{z}(t) = C_f\hat{x}(t), \end{cases} \tag{13.3}$$

where $\hat{x} \in \mathbb{R}^n$ is the filter state, $\hat{z}(t) \in \mathbb{R}^q$ is the estimation signal, and the constant matrices A_f, B_f and C_f are the filter matrices to be designed.

Define

$$\begin{aligned} \bar{z}(t) &= z(t) - \hat{z}(t), \\ \bar{x}(t) &= \left[x(t)^{\mathrm{T}}\ \hat{x}(t)^{\mathrm{T}}\right]^{\mathrm{T}}, \end{aligned}$$

then we obtain the filter error system as follows

$$\begin{cases} \bar{E}\dot{\bar{x}}(t) = \bar{A}(r_t)\bar{x}(t) + \bar{A}_d(r_t)\bar{x}(t-d) + \bar{B}_\omega(r_t)\omega(t), \\ \bar{z}(t) = \bar{L}(r_t)\bar{x}(t), \end{cases} \tag{13.4}$$

where

$$\begin{aligned} \bar{E} &= \begin{bmatrix} E & 0 \\ 0 & E \end{bmatrix}, \ \bar{A}(r_t) = \begin{bmatrix} A(r_t) & 0 \\ B_f C(r_t) & A_f \end{bmatrix}, \\ \bar{A}_d(r_t) &= \begin{bmatrix} A_d(r_t) & 0 \\ B_f C_d(r_t) & 0 \end{bmatrix}, \ \bar{B}_\omega(r_t) = \begin{bmatrix} B_\omega(r_t) \\ B_f D_\omega(r_t) \end{bmatrix}, \\ \bar{L}(r_t) &= \begin{bmatrix} L(r_t) & -C_f \end{bmatrix}. \end{aligned}$$

The H_∞ filtering problem to be addressed in this chapter is formulated as follows: for a given scalar $\gamma > 0$ and system (13.1), design a linear filter (13.3) such that the filtering error system (13.4) is delay-dependent or delay-independent stochastically admissible with H_∞ performance γ.

13.3 Main Results

In this section, the H_∞ filter problem will be investigated for system (13.1) by LMI approach.

13.3.1 BRL

We first provide the following delay-dependent BRL for system

$$\begin{cases} E\dot{x}(t) = A(r_t)x(t) + A_d(r_t)x(t-d) + B_\omega(r_t)\omega(t), \\ z(t) = L(r_t)x(t), \\ x(t) = \phi(t), \ t \in [-d, 0], \end{cases} \tag{13.5}$$

which will play a key role in the derivation of our main results.

Theorem 13.1. *For a given scalar $\gamma > 0$, system (13.5) is stochastically admissible with H_∞ performance γ, if there exist matrices $P_i > 0$, $Q_i > 0$, $Q > 0$, $Z > 0$ and S_i such that for every $i \in \mathcal{S}$,*

$$\begin{bmatrix} \Xi_{1i} & \Xi_{2i} & \Xi_{4i} & L_i^{\mathrm{T}} \\ * & \Xi_{3i} & 0 & 0 \\ * & * & -\gamma^2 I & \\ * & * & * & -I \end{bmatrix} + d^2 \begin{bmatrix} A_i^{\mathrm{T}} \\ A_{di}^{\mathrm{T}} \\ B_{\omega i}^{\mathrm{T}} \\ 0 \end{bmatrix} Z \begin{bmatrix} A_i^{\mathrm{T}} \\ A_{di}^{\mathrm{T}} \\ B_{\omega i}^{\mathrm{T}} \\ 0 \end{bmatrix}^{\mathrm{T}} < 0, \tag{13.6}$$

$$\sum_{j=1}^{s} \pi_{ij} Q_j < Q, \tag{13.7}$$

where $R \in \mathbb{R}^{n\times(n-r)}$ is any matrix with full column satisfying $E^{\mathrm{T}}R = 0$ and

$$\begin{aligned} \Xi_{1i} = & \sum_{j=1}^{s} \pi_{ij} E^{\mathrm{T}} P_j E + E^{\mathrm{T}} P_i A_i + S_i R^{\mathrm{T}} A_i \\ & + A_i^{\mathrm{T}} R S_i^{\mathrm{T}} + A_i^{\mathrm{T}} P_i E - E^{\mathrm{T}} Z E + Q_i + dQ, \\ \Xi_{2i} = & E^{\mathrm{T}} P_i A_{di} + S_i R^{\mathrm{T}} A_{di} + E^{\mathrm{T}} Z E, \\ \Xi_{3i} = & -Q_i - E^{\mathrm{T}} Z E, \\ \Xi_{4i} = & E^{\mathrm{T}} P_i B_{\omega i} + S_i R^{\mathrm{T}} B_{\omega i}. \end{aligned}$$

Proof. Firstly, we prove the regularity and the absence of impulses of system (13.5). Since $\operatorname{rank} E = r \leqslant n$, there exist nonsingular matrices G and H such that

$$\bar{E} = GEH = \begin{bmatrix} I_r & 0 \\ 0 & 0 \end{bmatrix}. \tag{13.8}$$

Then

$$R = G^{\mathrm{T}} \begin{bmatrix} 0 \\ I \end{bmatrix} M. \tag{13.9}$$

where $M \in \mathbb{R}^{(n-r)\times(n-r)}$ is any nonsingular matrix. Denote

$$GA_iH = \begin{bmatrix} A_{i1} & A_{i2} \\ A_{i3} & A_{i4} \end{bmatrix}, \ G^{-\mathrm{T}} P_i G^{-1} = \begin{bmatrix} P_{i1} & P_{i2} \\ * & P_{i3} \end{bmatrix}, \ H^{\mathrm{T}} S_i = \begin{bmatrix} S_{i1} \\ S_{i2} \end{bmatrix}, \tag{13.10}$$

for every $i \in \mathcal{S}$. Pre- and post-multiplying $\Xi_{1i} < 0$ by H^{T} and H, respectively, we have

$$A_{i4}^{\mathrm{T}} M S_{i2}^{\mathrm{T}} + S_{i2} M^{\mathrm{T}} A_{i4} < 0, \tag{13.11}$$

which implies A_{i4} is nonsingular for every $i \in \mathcal{S}$, and thus the pair (E, A_i) is regular and impulse free for every $i \in \mathcal{S}$. Thus, by Definition 10.1, system (13.5) with $\omega(t) = 0$ is regular and impulse free.

Next, we will show the stochastic stability of system (13.5). Define a new process $\{(x_t, r_t), t \geqslant 0\}$ by $\{x_t = x(t+\theta), -2d \leqslant \theta \leqslant 0\}$, then $\{(x_t, r_t), t \geqslant d\}$ is a Markov process with initial state $(\phi(\cdot), r_0)$. Now, for $t \geqslant d$, define the following Lyapunov functional for system (13.5) with $\omega(t) \equiv 0$,

$$V(x_t, r_t, t) = V_1(x_t, r_t, t) + V_2(x_t, r_t, t) + V_3(x_t, r_t, t) + V_4(x_t, r_t, t), \tag{13.12}$$

where

$$\begin{aligned}
V_1(x_t, r_t, t) &= x(t)^{\mathrm{T}} E^{\mathrm{T}} P(r_t) E x(t), \\
V_2(x_t, r_t, t) &= \int_{t-d}^{t} x(\alpha)^{\mathrm{T}} Q(r_t) x(\alpha)\,\mathrm{d}\alpha, \\
V_3(x_t, r_t, t) &= d \int_{-d}^{0} \int_{t+\beta}^{t} \dot{x}(\alpha)^{\mathrm{T}} E^{\mathrm{T}} Z E \dot{x}(\alpha)\,\mathrm{d}\alpha \mathrm{d}\beta, \\
V_4(x_t, r_t, t) &= \int_{-d}^{0} \int_{t+\beta}^{t} x(\alpha)^{\mathrm{T}} Q x(\alpha)\,\mathrm{d}\alpha \mathrm{d}\beta.
\end{aligned}$$

Then, for each $i \in \mathcal{S}$ and $t \geqslant d$, we have

$$\begin{aligned}
\mathcal{A}V(x_t, i, t) \leqslant{}& x(t)^{\mathrm{T}} (E^{\mathrm{T}} P_i + S_i R^{\mathrm{T}}) E \dot{x}(t) + x(t)^{\mathrm{T}} \left[\sum_{j=1}^{s} \pi_{ij} E^{\mathrm{T}} P_j E \right] x(t) \\
&+ x(t)^{\mathrm{T}} Q_i x(t) - x(t-d)^{\mathrm{T}} Q_i x(t-d) \\
&+ \int_{t-d}^{t} x(\alpha)^{\mathrm{T}} \left[\sum_{j=1}^{s} \pi_{ij} Q_j \right] x(\alpha)\,\mathrm{d}\alpha \\
&+ d^2 \dot{x}(t)^{\mathrm{T}} E^{\mathrm{T}} Z E \dot{x}(t) - d \int_{t-d}^{t} \dot{x}(\alpha)^{\mathrm{T}} E^{\mathrm{T}} Z E \dot{x}(\alpha)\,\mathrm{d}\alpha \\
&+ d x(t)^{\mathrm{T}} Q x(t) - \int_{t-d}^{t} x(\alpha)^{\mathrm{T}} Q x(\alpha)\,\mathrm{d}\alpha.
\end{aligned} \tag{13.13}$$

According to Jensen inequality, and using this and (13.7), we have that, for every $i \in \mathcal{S}$,

$$\mathcal{A}V(x_t, i, t) \leqslant \begin{bmatrix} x(t) \\ x(t-d) \end{bmatrix}^{\mathrm{T}} \left(\begin{bmatrix} \Xi_{1i} & \Xi_{2i} \\ * & \Xi_{3i} \end{bmatrix} + \begin{bmatrix} A_i^{\mathrm{T}} \\ A_{di}^{\mathrm{T}} \end{bmatrix} d^2 Z \begin{bmatrix} A_i^{\mathrm{T}} \\ A_{di}^{\mathrm{T}} \end{bmatrix}^{\mathrm{T}} \right) \begin{bmatrix} x(t) \\ x(t-d) \end{bmatrix}. \tag{13.14}$$

Using (13.6), it is easy to see that there exits a scalar $\lambda > 0$ such that for every $i \in \mathcal{S}$,

$$\mathcal{A}V(x_t, i, t) \leqslant -\lambda \|x(t)\|^2. \tag{13.15}$$

Using the similar method in the proof of Theorem 10.2, we can find that system (13.5) with $\omega(t) \equiv 0$ is stochastically stable.

In the following, we establish the H_∞ performance of system (13.5). For this purpose, we consider the stochastic Lyapunov functional (13.12) and the following index for system (13.5):

$$J_{z\omega}(t) = \mathscr{E}\left\{\int_0^t \left[z(s)^{\mathrm{T}} z(s) - \gamma^2 \omega(s)^{\mathrm{T}} \omega(s)\right] \mathrm{d}s\right\}. \tag{13.16}$$

Under zero initial condition, it easy to see that

$$\begin{aligned} J_{z\omega}(t) &\leqslant \mathscr{E}\left\{\int_0^t \left[z(s)^{\mathrm{T}} z(s) - \gamma^2 \omega(s)^{\mathrm{T}} \omega(s) + \mathcal{A}V(x_s, i, s)\right] \mathrm{d}s\right\} \\ &\leqslant \mathscr{E}\left\{\int_0^t \eta(s)^{\mathrm{T}} \Theta_i \eta(s) \mathrm{d}s\right\}, \end{aligned} \tag{13.17}$$

where

$$\eta(t)^{\mathrm{T}} = \left[x(t)^{\mathrm{T}} \; x(t-d)^{\mathrm{T}} \; \omega(t)^{\mathrm{T}}\right],$$

$$\Theta_i = \begin{bmatrix} \Xi_{1i} + L_i^{\mathrm{T}} L_i & \Xi_{2i} & \Xi_{4i} \\ * & \Xi_{3i} & 0 \\ * & * & -\gamma^2 I \end{bmatrix} + \begin{bmatrix} A_i^{\mathrm{T}} \\ A_{di}^{\mathrm{T}} \\ B_{\omega i}^{\mathrm{T}} \end{bmatrix} d^2 Z \begin{bmatrix} A_i^{\mathrm{T}} \\ A_{di}^{\mathrm{T}} \\ B_{\omega i}^{\mathrm{T}} \end{bmatrix}^{\mathrm{T}}.$$

Hence, by Schur complement, we can obtain from (13.6) that for all $t > 0$, $J_{z\omega}(t) < 0$. Therefore, we arrive at (10.8) for any nonzero $\omega(t) \in \mathscr{L}_2[0, \infty)$. This completes the proof.

Remark 13.2. It is noted that a delay-dependent BRL is given in Theorem 13.1 for system (13.5) to be stochastically admissible with H_∞ performance γ, which is formulated by strict LMIs with coefficient matrices of the original SS, and thus can be tested easily by the LMI toolbox.

13.3.2 Mode-Independent H_∞ Filter Design

We are now ready to deal with the delay-dependent H_∞ filtering problem of system (13.1). For this purpose, applying Theorem 13.1 to the filter error system (13.4), we have that system (13.4) is stochastically admissible with H_∞ performance γ, if there exist matrices $P_i > 0$, $Q_i > 0$, $Q > 0$, $Z > 0$ and S_i such that for every $i \in \mathcal{S}$,

$$\Xi_i = \begin{bmatrix} \tilde{\Xi}_{1i} & \tilde{\Xi}_{2i} & \tilde{\Xi}_{4i} & \bar{L}_i^{\mathrm{T}} \\ * & \tilde{\Xi}_{3i} & 0 & 0 \\ * & * & -\gamma^2 I & 0 \\ * & * & * & -I \end{bmatrix} + d^2 \begin{bmatrix} \bar{A}_i^{\mathrm{T}} \\ \bar{A}_{di}^{\mathrm{T}} \\ \bar{B}_{\omega i}^{\mathrm{T}} \\ 0 \end{bmatrix} Z \begin{bmatrix} \bar{A}_i^{\mathrm{T}} \\ \bar{A}_{di}^{\mathrm{T}} \\ \bar{B}_{\omega i}^{\mathrm{T}} \\ 0 \end{bmatrix}^{\mathrm{T}} < 0, \tag{13.18}$$

$$\sum_{j=1}^{s} \pi_{ij} Q_j < Q, \tag{13.19}$$

where $R \in \mathbb{R}^{2n\times 2(n-r)}$ is any matrix with full column satisfying $\bar{E}^{\mathrm{T}} R = 0$ and

$$\begin{aligned}
\tilde{\Xi}_{1i} =& \sum_{j=1}^{s} \pi_{ij} \bar{E}^{\mathrm{T}} P_j \bar{E} + \bar{E}^{\mathrm{T}} P_i \bar{A}_i + S_i R^{\mathrm{T}} \bar{A}_i + \bar{A}_i^{\mathrm{T}} R S_i^{\mathrm{T}} \\
&+ \bar{A}_i^{\mathrm{T}} P_i \bar{E} - \bar{E}^{\mathrm{T}} Z \bar{E} + Q_i + dQ, \\
\tilde{\Xi}_{2i} =& \bar{E}^{\mathrm{T}} P_i \bar{A}_{di} + S_i R^{\mathrm{T}} \bar{A}_{di} + \bar{E}^{\mathrm{T}} Z \bar{E}, \\
\tilde{\Xi}_{3i} =& -Q_i - \bar{E}^{\mathrm{T}} Z \bar{E}, \\
\tilde{\Xi}_{4i} =& \bar{E}^{\mathrm{T}} P_i \bar{B}_{\omega i} + S_i R^{\mathrm{T}} \bar{B}_{\omega i}.
\end{aligned}$$

It is easy to find that every $i \in \mathcal{S}$

$$\Xi_i = \Pi_i \Psi_i \Pi_i^{\mathrm{T}} < 0, \tag{13.20}$$

where

$$\Psi_i = \begin{bmatrix} \Psi_{1i} & \Psi_{2i} & \Psi_{4i} & M_i^{\mathrm{T}} \bar{B}_{\omega i} & \bar{L}_i^{\mathrm{T}} \\ * & \Psi_{3i} & N_i^{\mathrm{T}} \bar{A}_{di} & N_i^{\mathrm{T}} \bar{B}_{\omega i} & 0 \\ * & * & \tilde{\Xi}_{3i} & 0 & 0 \\ * & * & * & -\gamma^2 I & 0 \\ * & * & * & * & -I \end{bmatrix},$$

$$\begin{aligned}
\Psi_{1i} =& \sum_{j=1}^{s} \pi_{ij} \bar{E}^{\mathrm{T}} P_j \bar{E} + M_i^{\mathrm{T}} \bar{A}_i + \bar{A}_i^{\mathrm{T}} M_i - \bar{E}^{\mathrm{T}} Z \bar{E} + Q_i + dQ, \\
\Psi_{2i} =& \bar{E}^{\mathrm{T}} P_i + S_i R^{\mathrm{T}} - M_i^{\mathrm{T}} + \bar{A}_i^{\mathrm{T}} N_i, \\
\Psi_{3i} =& -N_i - N_i^{\mathrm{T}} + d^2 Z, \\
\Psi_{4i} =& M_i^{\mathrm{T}} \bar{A}_{di} + \bar{E}^{\mathrm{T}} Z \bar{E}, \\
\Pi_i =& \begin{bmatrix} I & \bar{A}_i^{\mathrm{T}} & 0 & 0 & 0 \\ 0 & \bar{A}_{di}^{\mathrm{T}} & I & 0 & 0 \\ 0 & \bar{B}_{\omega i}^{\mathrm{T}} & 0 & I & 0 \\ 0 & 0 & 0 & 0 & I \end{bmatrix}.
\end{aligned}$$

Clearly, $\Psi_i < 0$ implies $\Xi_i < 0$. Choose the matrices in Ψ_i and (13.19) as

$$\begin{aligned}
&P_i = \begin{bmatrix} P_{1i} & P_{2i} \\ * & P_{3i} \end{bmatrix} > 0,\ M_i^{\mathrm{T}} = \begin{bmatrix} Y_{1i} & Y \\ Y_{2i} & Y \end{bmatrix},\ N_i^{\mathrm{T}} = \begin{bmatrix} Y_{3i} & Y \\ Y_{4i} & Y \end{bmatrix},\ S_i = \begin{bmatrix} S_{1i} & S_{2i} \\ S_{3i} & S_{4i} \end{bmatrix} \\
&Q_i = \begin{bmatrix} \tilde{Q}_i & 0 \\ 0 & \varepsilon I \end{bmatrix} > 0,\ Q = \begin{bmatrix} \tilde{Q} & 0 \\ 0 & 2\varepsilon I \end{bmatrix} > 0,\ Z = \begin{bmatrix} \tilde{Z} & 0 \\ 0 & \varepsilon I \end{bmatrix} > 0,\ R = \begin{bmatrix} \hat{R} & 0 \\ 0 & \hat{R} \end{bmatrix},
\end{aligned} \tag{13.21}$$

where $\varepsilon \to 0$ and $\hat{R} \in \mathbb{R}^{n\times(n-r)}$ is any matrix with full column satisfying $E^{\mathrm{T}}\hat{R} = 0$. Then, letting $\tilde{A} = YA_f$, $\tilde{B} = YB_f$ and $\tilde{C} = C_f$, we get the following delay-dependent result.

Theorem 13.3. *For a given scalar $\gamma > 0$, system (13.4) is stochastically admissible with H_∞ performance γ, if there exist matrices $\begin{bmatrix} P_{1i} & P_{2i} \\ * & P_{3i} \end{bmatrix} > 0$, $\tilde{Q}_i > 0$, $\tilde{Q} > 0$, $\tilde{Z} > 0$, S_{1i}, S_{2i}, S_{3i}, S_{4i}, Y_{1i}, Y_{2i}, Y_{3i}, Y_{4i}, Y, $\tilde{A}$, $\tilde{B}$ and $\tilde{C}$ such that for every $i \in \mathcal{S}$,*

$$\begin{bmatrix} \Omega_{1i} & \Omega_{2i} & \Omega_{4i} & \Omega_{7i} & \Omega_{11i} & \Omega_{16i} & L_i^{\mathrm{T}} \\ * & \Omega_{3i} & \Omega_{5i} & \Omega_{8i} & \Omega_{12i} & \Omega_{17i} & -\tilde{C}^{\mathrm{T}} \\ * & * & \Omega_{6i} & \Omega_{9i} & \Omega_{13i} & \Omega_{18i} & 0 \\ * & * & * & \Omega_{10i} & \Omega_{14i} & \Omega_{19i} & 0 \\ * & * & * & * & \Omega_{15i} & 0 & 0 \\ * & * & * & * & * & -\gamma^2 I & 0 \\ * & * & * & * & * & * & -I \end{bmatrix} < 0, \tag{13.22}$$

$$\bar{\pi}_i \sum_{j=1, j\neq i}^{s} \tilde{Q}_j - (s-1)\underline{\pi}_i \tilde{Q}_i < \tilde{Q}, \tag{13.23}$$

where $\hat{R} \in \mathbb{R}^{n\times(n-r)}$ is any matrix with full column satisfying $E^{\mathrm{T}}\hat{R} = 0$ and

$$\begin{aligned}
\Omega_{1i} &= \bar{\pi}_i \sum_{j=1, j\neq i}^{s} E^{\mathrm{T}} P_{1j} E - (s-1)\underline{\pi}_i E^{\mathrm{T}} P_{1i} E + Y_{1i} A_i + A_i^{\mathrm{T}} Y_{1i}^{\mathrm{T}} \\
&\quad + \tilde{B} C_i + C_i^{\mathrm{T}} \tilde{B}^{\mathrm{T}} - E^{\mathrm{T}} \tilde{Z} E + \tilde{Q}_i + d\tilde{Q}, \\
\Omega_{2i} &= \bar{\pi}_i \sum_{j=1, j\neq i}^{s} E^{\mathrm{T}} P_{2j} E - (s-1)\underline{\pi}_i E^{\mathrm{T}} P_{2i} E + A_i^{\mathrm{T}} Y_{2i}^{\mathrm{T}} + \tilde{A} + C_i^{\mathrm{T}} \tilde{B}^{\mathrm{T}}, \\
\Omega_{3i} &= \bar{\pi}_i \sum_{j=1, j\neq i}^{s} E^{\mathrm{T}} P_{3j} E - (s-1)\underline{\pi}_i E^{\mathrm{T}} P_{3i} E + \tilde{A} + \tilde{A}^{\mathrm{T}}, \\
\Omega_{4i} &= E^{\mathrm{T}} P_{1i} + S_{1i} \hat{R}^{\mathrm{T}} + A_i^{\mathrm{T}} Y_{3i}^{\mathrm{T}} + C_i^{\mathrm{T}} \tilde{B}^{\mathrm{T}} - Y_{1i}, \\
\Omega_{5i} &= E^{\mathrm{T}} P_{2i}^{\mathrm{T}} + S_{3i} \hat{R}^{\mathrm{T}} + \tilde{A}^{\mathrm{T}} - Y_{2i}, \\
\Omega_{6i} &= -Y_{3i}^{\mathrm{T}} - Y_{3i} + d^2 \tilde{Z}, \\
\Omega_{7i} &= E^{\mathrm{T}} P_{2i} + S_{2i} \hat{R}^{\mathrm{T}} + A_i^{\mathrm{T}} Y_{4i}^{\mathrm{T}} + C_i^{\mathrm{T}} \tilde{B}^{\mathrm{T}} - Y, \\
\Omega_{8i} &= E^{\mathrm{T}} P_{3i} + S_{4i} \hat{R}^{\mathrm{T}} + \tilde{A}^{\mathrm{T}} - Y, \\
\Omega_{9i} &= -Y - Y_{4i}^{\mathrm{T}}, \\
\Omega_{10i} &= -Y - Y^{\mathrm{T}}, \\
\Omega_{11i} &= Y_{1i} A_{di} + \tilde{B} C_{di} + E^{\mathrm{T}} \tilde{Z} E, \\
\Omega_{12i} &= Y_{2i} A_{di} + \tilde{B} C_{di},
\end{aligned}$$

$$\begin{aligned}
\Omega_{13i} &= Y_{3i}A_{di} + \tilde{B}C_{di}, \\
\Omega_{14i} &= Y_{4i}A_{di} + \tilde{B}C_{di}, \\
\Omega_{15i} &= -\tilde{Q}_i - E^{\mathrm{T}}\tilde{Z}E, \\
\Omega_{16i} &= Y_{1i}B_{\omega i} + \tilde{B}D_{\omega i}, \\
\Omega_{17i} &= Y_{2i}B_{\omega i} + \tilde{B}D_{\omega i}, \\
\Omega_{18i} &= Y_{3i}B_{\omega i} + \tilde{B}D_{\omega i}, \\
\Omega_{19i} &= Y_{4i}B_{\omega i} + \tilde{B}D_{\omega i}.
\end{aligned}$$

Furthermore, if (13.22) *and* (13.23) *are solvable, the parameters of the desired filter can be chosen by*

$$A_f = Y^{-1}\tilde{A},\ B_f = Y^{-1}\tilde{B},\ C_f = \tilde{C}. \tag{13.24}$$

Furthermore, if the matrices, in (13.21), $\tilde{Q}_i = \hat{Q}$, $\tilde{Z} = \varepsilon I$ and $\tilde{Q} = \varepsilon I/d$ ($\varepsilon \to 0$), we can get the following delay-independent result.

Theorem 13.4. *For a given scalar $\gamma > 0$, system* (13.4) *is stochastically admissible with H_∞ performance γ, if there exist matrices $\begin{bmatrix} P_{1i} & P_{2i} \\ * & P_{3i} \end{bmatrix} > 0$, $\hat{Q} > 0$, S_{1i}, S_{2i}, S_{3i}, S_{4i}, Y_{1i}, Y_{2i}, Y_{3i}, Y_{4i}, Y, $\tilde{A}$, $\tilde{B}$ and $\tilde{C}$ such that for every $i \in \mathcal{S}$,*

$$\begin{bmatrix}
\tilde{\Omega}_{1i} & \Omega_{2i} & \Omega_{4i} & \Omega_{7i} & \tilde{\Omega}_{11i} & \Omega_{16i} & L_i^{\mathrm{T}} \\
* & \Omega_{3i} & \Omega_{5i} & \Omega_{8i} & \Omega_{12i} & \Omega_{17i} & -\tilde{C}^{\mathrm{T}} \\
* & * & \tilde{\Omega}_{6i} & \Omega_{9i} & \Omega_{13i} & \Omega_{18i} & 0 \\
* & * & * & \Omega_{10i} & \Omega_{14i} & \Omega_{19i} & 0 \\
* & * & * & * & -\hat{Q} & 0 & 0 \\
* & * & * & * & * & -\gamma^2 I & 0 \\
* & * & * & * & * & * & -I
\end{bmatrix} < 0, \tag{13.25}$$

where $\hat{R}$, Ω_{2i}, Ω_{3i}, Ω_{4i}, Ω_{5i}, Ω_{7i}, Ω_{8i}, Ω_{9i}, Ω_{10i}, Ω_{12i}, Ω_{13i}, Ω_{14i}, Ω_{16i}, Ω_{17i}, Ω_{18i} and Ω_{19i} are given in Theorem 13.3, and

$$\begin{aligned}
\tilde{\Omega}_{1i} &= \bar{\pi}_i \sum_{j=1, j\neq i}^{s} E^{\mathrm{T}}P_{1j}E - (s-1)\underline{\pi}_i E^{\mathrm{T}}P_{1i}E + Y_{1i}A_i + A_i^{\mathrm{T}}Y_{1i}^{\mathrm{T}} \\
&\quad + \tilde{B}C_i + C_i^{\mathrm{T}}\tilde{B}^{\mathrm{T}} + \hat{Q}, \\
\tilde{\Omega}_{6i} &= -Y_{3i}^{\mathrm{T}} - Y_{3i}, \\
\tilde{\Omega}_{11i} &= Y_{1i}A_{di} + \tilde{B}C_{di}.
\end{aligned}$$

Furthermore, if (13.25) *is solvable, the parameters of the desired filter can be chosen by* (13.24).

When $A_{di} = 0$ and $C_{di} = 0$ for any $i \in \mathcal{S}$, that is, system (13.1) reduces to a SMJS without time delay, we have the following result from Theorem 13.4.

Corollary 13.5. *For a given scalar $\gamma > 0$, system*

$$\begin{cases} \bar{E}\dot{\bar{x}}(t) = \bar{A}(r_t)\bar{x}(t) + \bar{B}_\omega(r_t)\omega(t), \\ \bar{z}(t) = \bar{L}(r_t)\bar{x}(t), \end{cases} \tag{13.26}$$

is stochastically admissible with H_∞ performance γ, if there exist matrices $\begin{bmatrix} P_{1i} & P_{2i} \\ * & P_{3i} \end{bmatrix} > 0$, S_{1i}, S_{2i}, S_{3i}, S_{4i}, Y_{1i}, Y_{2i}, Y_{3i}, Y_{4i}, Y, $\tilde{A}$, $\tilde{B}$ and $\tilde{C}$ such that for every $i \in \mathcal{S}$,

$$\begin{bmatrix} \hat{\Omega}_{1i} & \Omega_{2i} & \Omega_{4i} & \Omega_{7i} & \Omega_{16i} & L_i^{\mathrm{T}} \\ * & \Omega_{3i} & \Omega_{5i} & \Omega_{8i} & \Omega_{17i} & -\tilde{C}^{\mathrm{T}} \\ * & * & \tilde{\Omega}_{6i} & \Omega_{9i} & \Omega_{18i} & 0 \\ * & * & * & \Omega_{10i} & \Omega_{19i} & 0 \\ * & * & * & * & -\gamma^2 I & 0 \\ * & * & * & * & * & -I \end{bmatrix} < 0, \tag{13.27}$$

where $\hat{R}$, Ω_{2i}, Ω_{3i}, Ω_{4i}, Ω_{5i}, $\tilde{\Omega}_{6i}$, Ω_{7i}, Ω_{8i}, Ω_{9i}, Ω_{10i}, Ω_{16i}, Ω_{17i}, Ω_{18i} and Ω_{19i} are given in Theorem 13.4, and

$$\begin{aligned} \hat{\Omega}_{1i} = \bar{\pi}_i \sum_{j=1, j\neq i}^{s} E^{\mathrm{T}} P_{1j} E - (s-1)\underline{\pi}_i E^{\mathrm{T}} P_{1i} E \\ + Y_{1i} A_i + A_i^{\mathrm{T}} Y_{1i}^{\mathrm{T}} + \tilde{B} C_i + C_i^{\mathrm{T}} \tilde{B}^{\mathrm{T}}. \end{aligned}$$

Furthermore, if (13.27) is solvable, the parameters of the desired filter can be chosen by (13.24).

Remark 13.6. The delay-dependent and delay-independent H_∞ filter design problems for the addressed SSs with Markov jump parameters are discussed in Theorem 13.3 and Theorem 13.4, respectively. While the H_∞ filter design problem for SMJSs without time delay is investigated in Corollary 13.5. The sufficient conditions of the existence of the desired filters are established by strict LMIs, which are in contrast with [4, 167], where the nonstrict LMI conditions have been reported. Moreover, different with the results of [61, 139, 167, 182], all the established conditions do not require the precise knowledge of transition rates and the system modes to be accessible for the filter. Thus, our results are more powerful and desirable than [4, 61, 139, 167, 182].

Remark 13.7. When all the transition rates we get are precisely known, we can choose $\bar{\pi}_i$ and $\underline{\pi}_i$ represent the lower and upper bounds, respectively, that is, $\underline{\pi}_i = \min\{\pi_{ij}, j \neq i\}$ and $\bar{\pi}_i = \max\{\pi_{ij}, j \neq i\}$. Thus, the results on the design methods of H_∞ filter reported here can also be used to deal with system with known transition rates and cover the traditional MJSs as a special case.

13.4 Numerical Examples

In this section, some numerical examples are presented to illustrate the usefulness and flexibility of the results developed in this chapter.

Example 13.8. Consider system (13.5) with $E = I$, two modes and the following parameters:

$$
\begin{aligned}
A_1 &= \begin{bmatrix} -3.5 & 0.8 \\ -0.6 & -3.3 \end{bmatrix}, \; A_2 = \begin{bmatrix} -2.5 & 0.3 \\ 1.4 & -0.1 \end{bmatrix}, \\
A_{d1} &= \begin{bmatrix} -0.9 & -1.3 \\ -0.7 & -2.1 \end{bmatrix}, \; A_{d2} = \begin{bmatrix} -2.8 & 0.5 \\ -0.8 & -1.0 \end{bmatrix}, \\
B_{\omega 1} &= \begin{bmatrix} 0.5 \\ 0.4 \end{bmatrix}, \; B_{\omega 2} = \begin{bmatrix} 0.3 \\ 0.2 \end{bmatrix}, \\
L_1 &= \begin{bmatrix} 0.1 & 0.3 \end{bmatrix}, \; L_2 = \begin{bmatrix} 0.2 & 0.15 \end{bmatrix},
\end{aligned}
$$

that is, system (13.5) reduces to a regular system.

To compare our delay-dependent stochastic stability conditions with those in [16, 61, 144], we consider system (13.5) with $\omega(t) = 0$ and suppose $\pi_{22} = -0.8$. Table 13.1 presents the comparison results with various π_{11}, which shows that the delay-dependent stochastic stability condition in Theorem 13.1 gives better results than those in [16, 61, 144].

Now, let us consider system (13.5) and choose $\pi_{11} = -0.2$ and $\pi_{22} = -0.8$. Table 13.2 gives the comparison results on minimum allowed γ for various d by different methods. It is clear that the results of Theorem 13.1 are less conservative than those of [61, 144]. In particular, when $d = 0.6$, the methods of [61, 144] fail, but the minimum allowed $\gamma = 0.4332$ by using Theorem 13.1.

Example 13.9. Consider system (10.5) with two modes, that is, $\mathcal{S} = \{1, 2\}$. The mode switching is governed by the transition rate matrix

$$
\Pi = \begin{bmatrix} -0.45 & 0.45 \\ 0.3 & -0.3 \end{bmatrix}.
$$

System parameters are described as follows

$$
\begin{aligned}
A_1 &= \begin{bmatrix} 0.4972 & 0 \\ 0 & -0.9541 \end{bmatrix}, \; A_2 = \begin{bmatrix} 0.5121 & 0 \\ 0 & -0.7215 \end{bmatrix}, \\
A_{d1} &= \begin{bmatrix} -1.010 & 1.5415 \\ 0 & 0.5449 \end{bmatrix}, \; A_{d2} = \begin{bmatrix} -0.8521 & 1.9721 \\ 0 & 0.4321 \end{bmatrix}.
\end{aligned}
$$

and the singular matrix

$$
E = \begin{bmatrix} 1 & 0 \\ 0 & 0 \end{bmatrix}.
$$

Table 13.1. Example 13.9: Comparisons of maximum allowed d

π_{11}	-0.40	-0.55	-0.70	-0.85	-1.00
[16, 61, 144]	0.5044	0.5025	0.5010	0.4998	0.4987
Theorem 13.1	0.6302	0.6152	0.6051	0.5974	0.5910

Table 13.2. Example 13.9: Comparisons of minimum allowed γ

d	0.2	0.3	0.4	0.5	0.6
[61, 144]	0.0642	0.0971	0.2060	3.2465	—
Theorem 13.1	0.0510	0.0704	0.1152	0.1999	0.4332

Applying Theorem 13.1, it can be shown that system is regular, impulse free and stochastically stable for any constant time delay d satisfying $0 \leqslant d \leqslant 1.1152$. But the result of [7] fails to determine the stochastic stability of the above system.

Example 13.10. Consider system (13.1) with the following parameters:

$$A_1 = \begin{bmatrix} -3 & 1 & 0 \\ 0.3 & -2.5 & a \\ -0.1 & 0.3 & -3.8 \end{bmatrix}, \; A_2 = \begin{bmatrix} -2.5 & 0.5 & -0.1 \\ 0.1 & -3.5 & 0.3 \\ -0.1 & 1 & -2 \end{bmatrix},$$

$$B_{\omega 1} = \begin{bmatrix} 1 \\ 0 \\ 1 \end{bmatrix}, \; B_{\omega 2} = \begin{bmatrix} -0.6 \\ 0.5 \\ 0 \end{bmatrix}, \; C_1 = \begin{bmatrix} 0.8 \\ 0.3 \\ 0 \end{bmatrix}^{\mathrm{T}}, \; C_2 = \begin{bmatrix} -0.5 \\ 0.2 \\ 0.3 \end{bmatrix}^{\mathrm{T}},$$

$$L_1 = \begin{bmatrix} 0.5 \\ -0.1 \\ 1 \end{bmatrix}^{\mathrm{T}}, \; L_2 = \begin{bmatrix} 0 \\ 1 \\ 0.6 \end{bmatrix}^{\mathrm{T}}, \; D_{\omega 1} = 0.2, \; D_{\omega 2} = 0.5.$$

(1) We choose $E = I$, $A_{d1} = A_{d2} = 0$ and $C_{d1} = C_{d2} = 0$, that is, system (13.1) reduces to a regular system without time delay, which has been considered in [87]. When $a = 1$, and the transition rate matrix is exactly known and given by

$$\Pi = \begin{bmatrix} -0.5 & 0.5 \\ 0.3 & -0.3 \end{bmatrix}.$$

By using the methods of[22] and [87], the achieved minimum H_∞ performances γ of the resulting filtering error system are 0.74038 and 0.3028, respectively, while applying Corollary 13.5 proposed here can obtain the minimal attenuation level $\gamma = 0.2019$, which is 72.73% and 33.32% smaller than those in [22] and [87], respectively, with the filter parameters

$$A_f = \begin{bmatrix} -2.4614 & -3.6782 & 0.5636 \\ 0.1029 & -2.5787 & 0.7094 \\ -0.4577 & -3.8497 & -2.8816 \end{bmatrix},$$

$$B_f = \begin{bmatrix} -1.0703 \\ -0.5197 \\ -1.4948 \end{bmatrix}, \; C_f = \begin{bmatrix} 0.0195 & -1.6458 & -0.5525 \end{bmatrix}.$$

Moreover, when $a > 25$, [22, 87] can not conclude whether there exists a desired H_∞ filtering, but our approach result is still valid even for $a = 37$.

(2) When $a = 2$, $A_{d1} = A_{d2} = 0$, $C_{d1} = C_{d2} = 0$ and

$$E = \begin{bmatrix} 1 & 0.9 & 0 \\ 0.9 & 1 & 0 \\ 0 & 0 & 0 \end{bmatrix}.$$

Regarding the transition probability matrix Π, two cases will be considered.

Case 1: Π is perfectly known and given by

$$\Pi = \begin{bmatrix} -1 & 1 \\ 1.1 & -1.1 \end{bmatrix}.$$

The minimum attenuation level γ of the resulting filtering error system obtained from Corollary 13.5 is 0.2339, and the corresponding filter is given by

$$A_f = \begin{bmatrix} -2.6067 & 0.7678 & 0.1555 \\ 0.2249 & -3.9845 & 2.2317 \\ -0.3709 & 0.5784 & -4.9699 \end{bmatrix},$$

$$B_f = \begin{bmatrix} -0.5721 \\ -0.7144 \\ -1.4287 \end{bmatrix}, \; C_f = \begin{bmatrix} 0.1506 & -0.5548 & -1.1822 \end{bmatrix}.$$

Case 2: Π is unknown and $\bar{\pi}_1 = \bar{\pi}_2 = 1.3$, $\underline{\pi}_1 = \underline{\pi}_2 = 0.7$.

The minimum attenuation level γ obtained with H_∞ filter design of Corollary 13.5 is $\gamma = 0.2750$, and the corresponding filter matrices are

$$A_f = \begin{bmatrix} -1.1537 & -1.4273 & 1.7353 \\ 3.4426 & -7.0047 & 2.6217 \\ -2.8670 & 3.3906 & -5.7857 \end{bmatrix},$$

$$B_f = \begin{bmatrix} -0.5765 \\ -0.7229 \\ -1.4255 \end{bmatrix}, \; C_f = \begin{bmatrix} 0.0343 & -0.3195 & -1.3153 \end{bmatrix}.$$

Although the result of [4] fails to determine whether there exist the desired H_∞ filters for both above two cases.

(3) Now choose

$$A_{d1} = \begin{bmatrix} -0.1669 & 0.0802 & 1.6820 \\ -0.8162 & -0.9373 & 0.5936 \\ 2.0941 & 0.6357 & 0.7902 \end{bmatrix},$$

$$A_{d2} = \begin{bmatrix} 0.1053 & -0.1948 & -0.6855 \\ -0.1586 & 0.0755 & -0.2684 \\ 0.8709 & -0.5266 & -1.1883 \end{bmatrix},$$

$$C_{d1} = \begin{bmatrix} 0.2486 \\ 0.1025 \\ -0.0410 \end{bmatrix}^{\mathrm{T}}, \quad C_{d2} = \begin{bmatrix} -2.2476 \\ -0.5108 \\ 0.2492 \end{bmatrix}^{\mathrm{T}}, \quad E = \begin{bmatrix} 1 & 0 & 0 \\ 0 & 1 & 0 \\ 0 & 0 & 0 \end{bmatrix}.$$

Regarding the transition rate matrix Π, two cases will also be considered.

Case 1: Π is perfectly known and given by

$$\Pi = \begin{bmatrix} -0.5 & 0.5 \\ 0.6 & -0.6 \end{bmatrix}.$$

Case 2: Π is unknown and $\bar{\pi}_1 = \bar{\pi}_2 = 0.9$, $\underline{\pi}_1 = \underline{\pi}_2 = 0.7$.

When $a = 2$, applying the delay-independent result of Theorem 13.4 in this chapter, we can obtain the minimum guaranteed H_∞ error $\gamma = 1.2937$ for Case 1 and $\gamma = 1.3962$ for Case 2. While the condition in [167] fails to conclude whether or not there exists H_∞ filter for both cases.

When $a = -4$, it can be verified that the conditions in [167] and Theorem 13.4 reported in this chapter are not valid any more, which implies that the conditions in [167] and Theorem 13.4 here fail to conclude whether or not there exist filters for this system with perfectly known transition rates or unknown transition rates. However, by Theorem 13.3 in this chapter, it can be calculated that for all $0 \leqslant d \leqslant 1.8515$ (for Case 1) or $0 \leqslant d \leqslant 1.1892$ (for Case 2), there exist H_∞ filters. Thus, for this case, the delay-dependent condition in Theorem 13.3 has less conservativeness than the delay-independent conditions of [167] and Theorem 13.4. For example, we assume $d = 0.8$. For Case 1, the minimum attenuation level γ of the resulting filtering error system obtained from Theorem 13.3 is 0.6893. While, for Case 2, $\gamma = 0.9933$ and the corresponding filter is given by

$$A_f = \begin{bmatrix} -1.6463 & 1.4952 & -1.0499 \\ -1.1719 & -1.9662 & -2.0601 \\ -0.0322 & 0.5762 & -2.0600 \end{bmatrix},$$

$$B_f = \begin{bmatrix} -1.9720 \\ 0.8883 \\ -0.8963 \end{bmatrix}, \quad C_f = \begin{bmatrix} -0.3683 & 0.2372 & -1.0489 \end{bmatrix}.$$

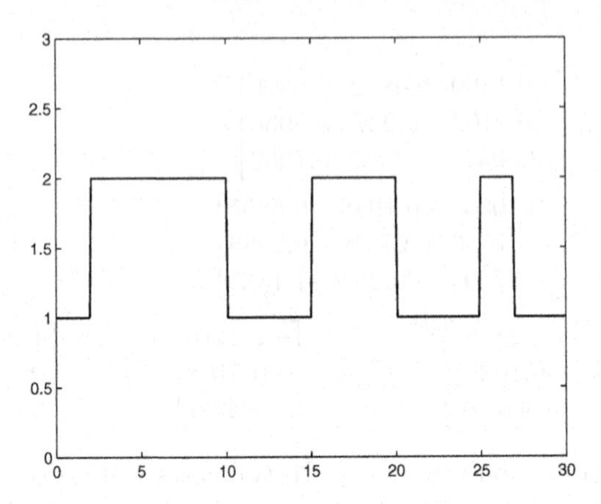

Fig. 13.1. Example 13.10: Markov mode

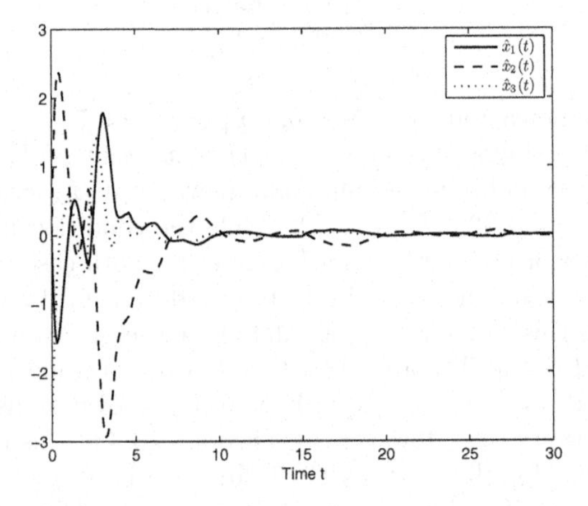

Fig. 13.2. Example 13.10: State responses for the designed filter

With this filter, one of the possible realizations of the Markov mode is plotted in Fig. 13.1, where the initial mode is assumed to be $r_0 = 1$. Fig. 13.2 shows the state responses of the designed filter with $\phi(t) = [5\ 3\ 4.2453]^{\mathrm{T}}$, $t \in [-0.8, 0]$ and the disturbance $\omega(t) = \sin(t)e^{-0.1t}$, and the error estimation signal $\bar{z}(t) = z(t) - \hat{z}(t)$ is depicted in Fig. 13.3.

In addition, the obtained minimum attenuation levels versus different d are shown in Fig. 13.4 for Case 1 and Case 2, respectively. From Fig. 13.4, we can easily find that the minimal disturbance rejection γ for the above mentioned system depends on the knowledge of transition rates. More specifically, for the same d, the results with unknown transition rates are larger than the ones we get for the case of perfectly known transition rates. This is reasonable since the latter uses more information on the transition rates. However, when the

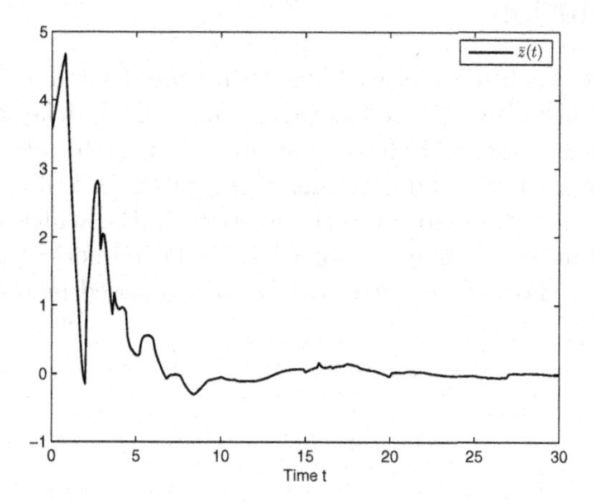

Fig. 13.3. Example 13.10: Error estimation

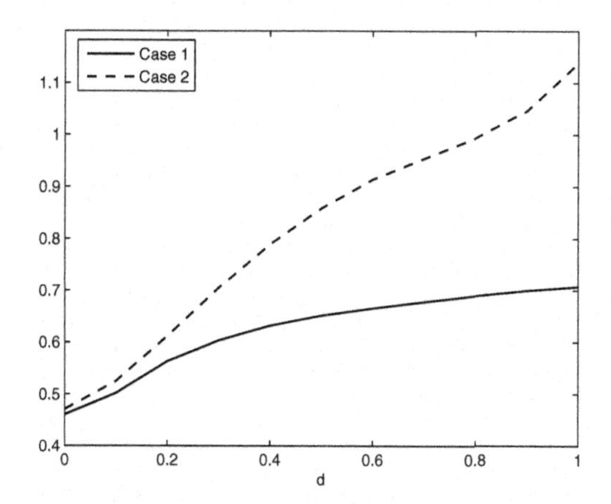

Fig. 13.4. Example 13.10: H_∞ performance comparison between Case 1 and Case 2

transition rates cannot be precisely known, the former will be more useful and meaningful.

Through this example, we find that our results not only can be applied to system with totally known transition rates, but also can be applied to system with unknown transition rates. This proves their power and the correctness of Remark 13.7.

13.5 Conclusion

The problem of H_∞ filtering for SMJSs with time delay and unknown transition rates has been investigated in this chapter. Both delay-dependent and delay-independent approaches have been presented, with sufficient conditions obtained for the existence of deterministic H_∞ filters. All the results reported in this chapter have been formulated by strict LMIs, which can be readily solved using standard numerical software. Some numerical examples have been provided to show the effectiveness of the proposed methods.

14

Mode-Dependent H_∞ Filtering for SMJSs with Time-Varying Delays

14.1 Introduction

In the above chapter, we have investigated the problem of mode-independent H_∞ filter design for continuous-time SMJSs with constant delays. In this chapter, we will deal with the problem of mode-dependent H_∞ filtering for SSs with both time-varying delays and Markov jump parameters. A BRL is proposed to guarantee the considered system to be mean-square exponentially admissible and satisfy a prescribed H_∞ performance level. Based on this, a LMI-based approach is proposed to design the desired filter. Finally, some illustrative examples are provided to demonstrate the effectiveness of the proposed methods.

14.2 Problem Formulation

Let $\{r_t, t \geqslant 0\}$ be a continuous-time Markov process with right continuous trajectories and follow the same definition as that in Chapter 10. Fix a probability space $(\Omega, \mathcal{F}, \mathcal{P})$ and consider the following SS with Markov jump parameters and time-varying delays:

$$\begin{cases} E\dot{x}(t) = A(r_t)x(t) + A_d(r_t)x(t-d(t)) + B_\omega(r_t)\omega(t), \\ \quad y(t) = C(r_t)x(t) + C_d(r_t)x(t-d(t)) + D_\omega(r_t)\omega(t), \\ \quad z(t) = L(r_t)x(t), \\ \quad x(t) = \phi(t), \ t \in [-d_2, 0], \end{cases} \tag{14.1}$$

where $x(t) \in \mathbb{R}^n$ is the state, $y(t) \in \mathbb{R}^s$ is the measurement, $z(t) \in \mathbb{R}^q$ is the signal to be estimated, $\omega(t) \in \mathbb{R}^p$ is the disturbance input that belongs to $\mathscr{L}_2[0,\infty)$, and $\phi(t) \in C_{n,d_2}$ is a compatible vector valued initial function. The matrix $E \in \mathbb{R}^{n\times n}$ may be singular and it is assumed that $\operatorname{rank} E = r \leqslant n$. $A(r_t)$, $A_d(r_t)$, $B_\omega(r_t)$, $C(r_t)$, $C_d(r_t)$, $D_\omega(r_t)$ and $L(r_t)$ are known real constant matrices with appropriate dimensions for each $r_t \in \mathcal{S}$. $d(t)$ is a

Z.-G. Wu et al.: *Anal. & Synth. of Singular Syst. with Time-Delays*, LNCIS 443, pp. 199–213.
DOI: 10.1007/978-3-642-37497-5_14

time-varying continuous function that satisfies $0 < d_1 \leqslant d(t) \leqslant d_2$, $\dot{d}(t) \leqslant \mu$, where d_1 and d_2 are the time-delay lower and upper bounds, respectively, and $0 \leqslant \mu < 1$ is the time-delay variation rate.

Definition 14.1

1. System (11.30) is said to be mean-square exponentially stable, if there exist scalars $\alpha > 0$ and $\beta > 0$ such that $\mathscr{E}\{\|x(t)\|^2\} \leqslant \alpha e^{-\beta t}\|\phi(t)\|^2_{d_2}$, $t > 0$.
2. System (11.30) is said to be mean-square exponentially admissible, if it is regular, impulse free and mean-square exponentially stable.

Definition 14.2. *For a given scalar $\gamma > 0$, system (14.1) is said to be mean-square exponentially admissible with H_∞ performance γ, if it with $\omega(t) \equiv 0$ is mean-square exponentially admissible, and under zero initial condition, it satisfies (10.8) for any non-zero $\omega(t) \in \mathscr{L}_2[0,\infty)$.*

In this chapter, in order to estimate $z(t)$, we are interested in designing a filter of the following structure:

$$\begin{cases} E_f\dot{\hat{x}}(t) = A_f(r_t)\hat{x}(t) + B_f(r_t)y(t), \\ \hat{z}(t) = C_f(r_t)\hat{x}(t), \end{cases} \tag{14.2}$$

where $\hat{x} \in \mathbb{R}^n$, $\hat{z}(t) \in \mathbb{R}^q$, and the constant matrices E_f, $A_f(r_t)$, $B_f(r_t)$ and $C_f(r_t)$ are the filter matrices with appropriate dimensions, which are to be designed.

Defining

$$\begin{aligned} \bar{z}(t) &= z(t) - \hat{z}(t), \\ \bar{x}(t) &= \left[x(t)^{\mathrm{T}} \ \hat{x}(t)^{\mathrm{T}}\right]^{\mathrm{T}}, \end{aligned}$$

and combining (14.1) and (14.2), we obtain the filtering error dynamics as follows:

$$\begin{cases} \bar{E}\dot{\bar{x}}(t) = \bar{A}(r_t)\bar{x}(t) + \bar{A}_d(r_t)\bar{x}(t-d(t)) + \bar{B}_\omega(r_t)\omega(t), \\ \bar{z}(t) = \bar{L}(r_t)\bar{x}(t), \end{cases} \tag{14.3}$$

where

$$\begin{aligned} \bar{E} &= \begin{bmatrix} E & 0 \\ 0 & E_f \end{bmatrix}, \\ \bar{A}(r_t) &= \begin{bmatrix} A(r_t) & 0 \\ B_f(r_t)C(r_t) & A_f(r_t) \end{bmatrix}, \\ \bar{A}_d(r_t) &= \begin{bmatrix} A_d(r_t) & 0 \\ B_f(r_t)C_d(r_t) & 0 \end{bmatrix}, \\ \bar{B}_\omega(r_t) &= \begin{bmatrix} B_\omega(r_t) \\ B_f(r_t)D_\omega(r_t) \end{bmatrix}, \\ \bar{L}(r_t) &= \begin{bmatrix} L(r_t) & -C_f(r_t) \end{bmatrix}. \end{aligned}$$

The purpose of this chapter is to design H_∞ filter (14.2) for system (14.1) such that the filtering error system (14.3) is mean-square exponentially admissible with H_∞ performance γ, where the criteria are dependent on the length of time-delay.

14.3 Main Results

In this section, the delay-dependent H_∞ filtering problem will be discussed by LMI approach.

14.3.1 BRL

We initially present the following BRL, which will play a key role in solving the aforementioned problem.

Theorem 14.3. *For a given scalar $\gamma > 0$, system (14.1) is mean-square exponentially admissible with H_∞ performance γ, if there exist matrices $Q_1 > 0$, $Q_2 > 0$, $Q_3 > 0$, $Z_1 > 0$, $Z_2 > 0$ and P_i such that for every $i \in \mathcal{S}$,*

$$E^{\mathrm{T}} P_i = P_i^{\mathrm{T}} E \geqslant 0, \tag{14.4}$$

$$\Xi_i = \begin{bmatrix} \Xi_{11i} & \Xi_{12i} & 0 & 0 & P_i^{\mathrm{T}} B_{\omega i} & A_i^{\mathrm{T}} W \\ * & \Xi_{22i} & E^{\mathrm{T}} Z_2 E & \Xi_{24i} & 0 & A_{di}^{\mathrm{T}} W \\ * & * & \Xi_{33i} & 0 & 0 & 0 \\ * & * & * & \Xi_{44i} & 0 & 0 \\ * & * & * & * & -\gamma^2 I & B_{\omega i}^{\mathrm{T}} W \\ * & * & * & * & * & -W \end{bmatrix} < 0, \tag{14.5}$$

where $d_{12} = d_2 - d_1$, $W = d_{12} d_2^2 Z_1 + d_{12}^2 Z_2$, and

$$\begin{aligned} \Xi_{11i} &= \sum_{j=1}^{s} \pi_{ij} E^{\mathrm{T}} P_j + P_i^{\mathrm{T}} A_i + A_i^{\mathrm{T}} P_i + \sum_{k=1}^{3} Q_k - d_{12} E^{\mathrm{T}} Z_1 E + L_i^{\mathrm{T}} L_i, \\ \Xi_{12i} &= P_i^{\mathrm{T}} A_{di} + d_{12} E^{\mathrm{T}} Z_1 E, \\ \Xi_{22i} &= -(1-\mu) Q_3 - E^{\mathrm{T}}((d_{12} + d_2) Z_1 + 2 Z_2) E, \\ \Xi_{33i} &= -Q_1 - E^{\mathrm{T}} Z_2 E, \\ \Xi_{24i} &= E^{\mathrm{T}}(d_2 Z_1 + Z_2) E, \\ \Xi_{44i} &= -Q_2 - E^{\mathrm{T}}(d_2 Z_1 + Z_2) E. \end{aligned}$$

Proof. Firstly, we prove the mean-square exponential admissibility of system (11.30). Since $\operatorname{rank} E = r \leqslant n$, there exist nonsingular matrices G and H such that

$$GEH = \begin{bmatrix} I_r & 0 \\ 0 & 0 \end{bmatrix}. \tag{14.6}$$

Denote

$$GA_iH = \begin{bmatrix} A_{i1} & A_{i2} \\ A_{i3} & A_{i4} \end{bmatrix}, \; G^{-T}P_iH = \begin{bmatrix} \bar{P}_{i1} & \bar{P}_{i2} \\ \bar{P}_{i3} & \bar{P}_{i4} \end{bmatrix}. \tag{14.7}$$

From (14.4) and using the expressions in (14.6) and (14.7), it is easy to obtain that $\bar{P}_{i2} = 0$ for every $i \in \mathcal{S}$. Pre- and post-multiplying $\Xi_{11i} < 0$ by H^{T} and H, respectively, we have

$$A_{i4}^{\mathrm{T}}\bar{P}_{i4} + \bar{P}_{i4}^{\mathrm{T}}A_{i4} < 0, \tag{14.8}$$

which implies A_{i4} are nonsingular for every $i \in \mathcal{S}$ and thus the pairs (E, A_i) are regular and impulse free for every $i \in \mathcal{S}$. Thus, by Definition 10.1, system (11.30) is regular and impulse free

Next, we will show the mean-square exponential stability of system (11.30). Define a new process $\{(x_t, r_t), t \geqslant 0\}$ by $\{x_t = x(t+\theta), -2d_2 \leqslant \theta \leqslant 0\}$, then $\{(x_t, r_t), t \geqslant 0\}$ is a Markov process with initial state $(\phi(\cdot), r_0)$. Now, define the following Lyapunov functional for system (11.30),

$$\begin{aligned} V(x_t, r_t, t) &= x(t)^{\mathrm{T}}E^{\mathrm{T}}P(r_t)x(t) + \sum_{k=1}^{2}\int_{t-d_k}^{t} x(\alpha)^{\mathrm{T}}Q_kx(\alpha)\,\mathrm{d}\alpha \\ &\quad + \int_{t-d(t)}^{t} x(\alpha)^{\mathrm{T}}Q_3x(\alpha)\,\mathrm{d}\alpha \\ &\quad + d_{12}d_2\int_{-d_2}^{0}\int_{t+\beta}^{t} \dot{x}(\alpha)^{\mathrm{T}}E^{\mathrm{T}}Z_1E\dot{x}(\alpha)\,\mathrm{d}\alpha\mathrm{d}\beta \\ &\quad + d_{12}\int_{-d_2}^{-d_1}\int_{t+\beta}^{t} \dot{x}(\alpha)^{\mathrm{T}}E^{\mathrm{T}}Z_2E\dot{x}(\alpha)\,\mathrm{d}\alpha\mathrm{d}\beta. \end{aligned} \tag{14.9}$$

Then, for each $i \in \mathcal{S}$, we have

$$\begin{aligned} \mathcal{A}V(x_t, i, t) &\leqslant 2x(t)^{\mathrm{T}}E^{\mathrm{T}}P_i\dot{x}(t) + x(t)^{\mathrm{T}}\left[\sum_{j=1}^{s}\pi_{ij}E^{\mathrm{T}}P_j\right]x(t) \\ &\quad + \sum_{k=1}^{3}x(t)^{\mathrm{T}}Q_kx(t) - \sum_{k=1}^{2}x(t-d_k)^{\mathrm{T}}Q_kx(t-d_k) \\ &\quad - (1-\mu)x(t-d(t))^{\mathrm{T}}Q_3x(t-d(t)) + \dot{x}(t)^{\mathrm{T}}E^{\mathrm{T}}WE\dot{x}(t) \\ &\quad - d_{12}d_2\int_{t-d_2}^{t} \dot{x}(\alpha)^{\mathrm{T}}E^{\mathrm{T}}Z_1E\dot{x}(\alpha)\,\mathrm{d}\alpha \\ &\quad - d_{12}\int_{t-d_2}^{t-d_1} \dot{x}(\alpha)^{\mathrm{T}}E^{\mathrm{T}}Z_2E\dot{x}(\alpha)\,\mathrm{d}\alpha. \end{aligned} \tag{14.10}$$

According to Jensen inequality, the following inequality is true:

$$
\begin{aligned}
&- d_{12}d_2 \int_{t-d_2}^{t} \dot{x}(\alpha)^{\mathrm{T}} E^{\mathrm{T}} Z_1 E \dot{x}(\alpha)\, \mathrm{d}\alpha \\
=& - d_{12}d_2 \int_{t-d(t)}^{t} \dot{x}(\alpha)^{\mathrm{T}} E^{\mathrm{T}} Z_1 E \dot{x}(\alpha)\, \mathrm{d}\alpha - d_{12}d_2 \int_{t-d_2}^{t-d(t)} \dot{x}(\alpha)^{\mathrm{T}} E^{\mathrm{T}} Z_1 E \dot{x}(\alpha)\, \mathrm{d}\alpha \\
\leqslant& - d(t)d_{12} \int_{t-d(t)}^{t} \dot{x}(\alpha)^{\mathrm{T}} E^{\mathrm{T}} Z_1 E \dot{x}(\alpha)\, \mathrm{d}\alpha \\
&- (d_2 - d(t))d_2 \int_{t-d_2}^{t-d(t)} \dot{x}(\alpha)^{\mathrm{T}} E^{\mathrm{T}} Z_1 E \dot{x}(\alpha)\, \mathrm{d}\alpha \\
\leqslant& - d_{12} \left[\int_{t-d(t)}^{t} \dot{x}(\alpha)^{\mathrm{T}} E^{\mathrm{T}}\, \mathrm{d}\alpha \right] Z_1 \left[\int_{t-d(t)}^{t} E \dot{x}(\alpha)\, \mathrm{d}\alpha \right] \\
&- d_2 \left[\int_{t-d_2}^{t-d(t)} \dot{x}(\alpha)^{\mathrm{T}} E^{\mathrm{T}}\, \mathrm{d}\alpha \right] Z_1 \left[\int_{t-d_2}^{t-d(t)} E \dot{x}(\alpha)\, \mathrm{d}\alpha \right] \\
=& \begin{bmatrix} x(t) \\ x(t-d(t)) \\ x(t-d_2) \end{bmatrix}^{\mathrm{T}} \begin{bmatrix} -d_{12}E^{\mathrm{T}} Z_1 E & d_{12}E^{\mathrm{T}} Z_1 E & 0 \\ * & -(d_{12}+d_2)E^{\mathrm{T}} Z_1 E & d_2 E^{\mathrm{T}} Z_1 E \\ * & * & -d_2 E^{\mathrm{T}} Z_1 E \end{bmatrix} \\
&\times \begin{bmatrix} x(t) \\ x(t-d(t)) \\ x(t-d_2) \end{bmatrix}.
\end{aligned} \tag{14.11}
$$

Similarly, we can also get

$$
\begin{aligned}
&- d_{12} \int_{t-d_2}^{t-d_1} \dot{x}(\alpha)^{\mathrm{T}} E^{\mathrm{T}} Z_1 E \dot{x}(\alpha)\, \mathrm{d}\alpha \\
\leqslant& \begin{bmatrix} x(t-d(t)) \\ x(t-d_1) \\ x(t-d_2) \end{bmatrix}^{\mathrm{T}} \begin{bmatrix} -2E^{\mathrm{T}} Z_2 E & E^{\mathrm{T}} Z_2 E & E^{\mathrm{T}} Z_2 E \\ * & -E^{\mathrm{T}} Z_2 E & 0 \\ * & * & -E^{\mathrm{T}} Z_2 E \end{bmatrix} \begin{bmatrix} x(t-d(t)) \\ x(t-d_1) \\ x(t-d_2) \end{bmatrix}.
\end{aligned} \tag{14.12}
$$

Thus, we have that, for every $i \in \mathcal{S}$,

$$
\mathcal{A}V(x_t, i, t) \leqslant \xi(t)^{\mathrm{T}} \Psi_i \xi(t), \tag{14.13}
$$

where

$$
\Psi_i = \begin{bmatrix} \Xi_{11i} & \Xi_{12i} & 0 & 0 \\ * & \Xi_{22i} & E^{\mathrm{T}} Z_2 E & \Xi_{24i} \\ * & * & \Xi_{33i} & 0 \\ * & * & * & \Xi_{44i} \end{bmatrix} + \begin{bmatrix} A_i^{\mathrm{T}} \\ A_{di}^{\mathrm{T}} \\ 0 \\ 0 \end{bmatrix} W \begin{bmatrix} A_i^{\mathrm{T}} \\ A_{di}^{\mathrm{T}} \\ 0 \\ 0 \end{bmatrix}^{\mathrm{T}},
$$

$$
\xi(t) = \left[x(t)^{\mathrm{T}}\ x(t-d(t))^{\mathrm{T}}\ x(t-d_1)^{\mathrm{T}}\ x(t-d_2)^{\mathrm{T}} \right]^{\mathrm{T}}.
$$

Using (14.5) and Schur complement, it is easy to see that there exits a scalar $\lambda > 0$ such that for every $i \in \mathcal{S}$,

$$\mathcal{A}V(x_t, i, t) \leqslant -\lambda \|x(t)\|^2. \tag{14.14}$$

Since A_{i4} is nonsingular for every $i \in \mathcal{S}$, we set

$$\hat{G} = \begin{bmatrix} I_r & -A_{i2}A_{i4}^{-1} \\ 0 & A_{i4}^{-1} \end{bmatrix} G. \tag{14.15}$$

It is easy to get

$$\hat{G}EH = \begin{bmatrix} I_r & 0 \\ 0 & 0 \end{bmatrix}, \ \hat{G}A_iH = \begin{bmatrix} \hat{A}_{i1} & 0 \\ \hat{A}_{i3} & I \end{bmatrix}, \tag{14.16}$$

where $\hat{A}_{i1} = A_{i1} - A_{i2}A_{i4}^{-1}A_{i3}$ and $\hat{A}_{i3} = A_{i4}^{-1}A_{i3}$. Denote

$$\hat{G}A_{di}H = \begin{bmatrix} A_{id1} & A_{id2} \\ A_{id3} & A_{id4} \end{bmatrix}, \ \hat{G}^{-\mathrm{T}}P_iH = \begin{bmatrix} P_{i1} & P_{i2} \\ P_{i3} & P_{i4} \end{bmatrix}, \ H^{\mathrm{T}}Q_3H = \begin{bmatrix} Q_{11} & Q_{21} \\ * & Q_{22} \end{bmatrix}. \tag{14.17}$$

On the other hand, it can be seen from (14.5) that

$$\begin{bmatrix} I \\ I \\ I \\ I \end{bmatrix}^{\mathrm{T}} \begin{bmatrix} \Xi_{11i} & \Xi_{12i} & 0 & 0 \\ * & \Xi_{22i} & E^{\mathrm{T}}Z_2E & \Xi_{24i} \\ * & * & \Xi_{33i} & 0 \\ * & * & * & \Xi_{44i} \end{bmatrix} \begin{bmatrix} I \\ I \\ I \\ I \end{bmatrix} < 0, \tag{14.18}$$

which implies

$$\sum_{j=1}^{s} \pi_{ij}E^{\mathrm{T}}P_j + P_i^{\mathrm{T}}(A_i + A_{di}) + (A_i + A_{di})^{\mathrm{T}}P_i < 0. \tag{14.19}$$

It is easy to find the matrices P_i satisfying (14.19) are nonsingular for every $i \in \mathcal{S}$. If not, then there exist some $i_0 \in \mathcal{S}$ and a vector $\eta_0 \neq 0$ such that $P_{i_0}\eta_0 = 0$. Therefore, pre- and post-multiplying (14.19) by η_0^{T} and η_0, respectively, we get $\eta_0^{\mathrm{T}}(\sum_{j=1}^{s} \pi_{i_0j}E^{\mathrm{T}}P_j)\eta_0 < 0$. This is a contradiction since $E^{\mathrm{T}}P_i \geqslant 0$ implies $\eta_0^{\mathrm{T}}(\sum_{j=1}^{s} \pi_{i_0j}E^{\mathrm{T}}P_j)\eta_0 \geqslant 0$. Thus, considering (14.4), we can deduce that $P_{i1} > 0$ and $P_{i2} = 0$ for every $i \in \mathcal{S}$. Define

$$\zeta(t) = \begin{bmatrix} \zeta_1(t) \\ \zeta_2(t) \end{bmatrix} = H^{-1}x(t). \tag{14.20}$$

Then, system (11.30) is equivalent to

$$\begin{cases} \dot{\zeta}_1(t) = \hat{A}_{i1}\zeta_1(t) + A_{id1}\zeta_1(t-d(t)) + A_{id2}\zeta_2(t-d(t)), \\ -\zeta_2(t) = \hat{A}_{i13}\zeta_1(t) + A_{id3}\zeta_1(t-d(t)) + A_{id4}\zeta_2(t-d(t)), \\ \zeta(t) = \psi(t) = H^{-1}\phi(t), \ t \in [-d_2, 0]. \end{cases} \tag{14.21}$$

To prove the exponential stability in mean square, we define a new function as

$$W(x_t, r_t, t) = e^{\varepsilon t} V(x_t, r_t, t), \tag{14.22}$$

where $\varepsilon > 0$ and then, by Dynkin's formula [109], we find that for every $i \in \mathcal{S}$,

$$\mathscr{E}\{W(x_t, r_t, t)\} \leqslant \mathscr{E}\{W(x_0, r_0, 0)\} + \mathscr{E}\left\{\int_0^t e^{\varepsilon s}\left[\varepsilon V(x_s, r_s, s) - \lambda \|x(s)\|^2\right] \mathrm{d}s\right\}. \tag{14.23}$$

By using the similar analysis method of [108], it can be seen from (14.9), (14.22) and (14.23) that, if ε is chosen small enough, a constant $k > 0$ can be found such that for $t > 0$,

$$\begin{aligned}\min_{i \in \mathcal{S}}\{\lambda_{\min}(P_{i1})\}\mathscr{E}\left\{\|\zeta_1\|^2\right\} &\leqslant \mathscr{E}\left\{x(t)^{\mathrm{T}} E^{\mathrm{T}} P_i x(t)\right\} \\ &\leqslant \mathscr{E}\{V(x_t, r_t, t)\} \leqslant k e^{-\varepsilon t}\|\phi(t)\|_{d_2}^2. \end{aligned}\tag{14.24}$$

Hence, for any $t > 0$

$$\mathscr{E}\left\{\|\zeta_1\|^2\right\} \leqslant \alpha e^{-\varepsilon t}\|\phi(t)\|_{d_2}^2, \tag{14.25}$$

where $\alpha = (\min_{i \in \mathcal{S}}\{\lambda_{\min}(P_{i1})\})^{-1} k$. Defining

$$e(t) = \hat{A}_{i13}\zeta_1(t) + A_{id3}\zeta_1(t - d(t)) \tag{14.26}$$

and from (14.25), we have that a constant $m > 0$ can be found such that when $t > 0$,

$$\mathscr{E}\left\{\|e(t)\|^2\right\} \leqslant m e^{-\varepsilon t}\|\phi(t)\|_{d_2}^2. \tag{14.27}$$

To study the mean-square exponential stability of $\zeta_2(t)$, we construct a function as

$$J(t) = \zeta_2(t)^{\mathrm{T}} Q_{22}\zeta_2(t) - \zeta_2(t - d(t))^{\mathrm{T}} Q_{22}\zeta_2(t - d(t)). \tag{14.28}$$

By pre-multiplying the second equation of (14.21) with $\zeta_2(t)^{\mathrm{T}} P_{i4}^{\mathrm{T}}$, we obtain that

$$0 = \zeta_2(t)^{\mathrm{T}} P_{i4}^{\mathrm{T}}\zeta_2(t) + \zeta_2(t)^{\mathrm{T}} P_{i4}^{\mathrm{T}} A_{id4}\zeta_2(t - d(t)) + \zeta_2(t)^{\mathrm{T}} P_{i4}^{\mathrm{T}} e(t). \tag{14.29}$$

Adding (14.29) to (14.28) yields that

$$\begin{aligned} J(t) &= \zeta_2(t)^{\mathrm{T}}(P_{i4}^{\mathrm{T}} + P_{i4} + Q_{22})\zeta_2(t) + 2\zeta_2(t)^{\mathrm{T}} P_{i4}^{\mathrm{T}} A_{id4}\zeta_2(t - d(t)) \\ &\quad - \zeta_2(t - d(t))^{\mathrm{T}} Q_{22}\zeta_2(t - d(t)) + 2\zeta_2(t)^{\mathrm{T}} P_{i4}^{\mathrm{T}} e(t) \\ &\leqslant \begin{bmatrix} \zeta_2(t) \\ \zeta_2(t - d(t)) \end{bmatrix}^{\mathrm{T}} \begin{bmatrix} P_{i4}^{\mathrm{T}} + P_{i4} + Q_{22} & P_{i4}^{\mathrm{T}} A_{id4} \\ * & -Q_{22} \end{bmatrix} \begin{bmatrix} \zeta_2(t) \\ \zeta_2(t - d(t)) \end{bmatrix} \\ &\quad + \eta_1\zeta_2(t)^{\mathrm{T}}\zeta_2(t) + \eta_1^{-1} e(t)^{\mathrm{T}} P_{i4} P_{i4}^{\mathrm{T}} e(t), \end{aligned}\tag{14.30}$$

where η_1 is any positive scalar. Pre- and post-multiplying

$$\begin{bmatrix} \Xi_{11i} & \Xi_{12i} \\ * & \Xi_{22i} \end{bmatrix} < 0 \tag{14.31}$$

by $\begin{bmatrix} H & 0 \\ 0 & H \end{bmatrix}^{\mathrm{T}}$ and $\begin{bmatrix} H & 0 \\ 0 & H \end{bmatrix}$, respectively, we get a constant $\eta_2 > 0$ can be found such that

$$\begin{bmatrix} P_{i4}^{\mathrm{T}} + P_{i4} + Q_{22} & P_{i4}^{\mathrm{T}} A_{id4} \\ * & -Q_{22} \end{bmatrix} \leqslant - \begin{bmatrix} \eta_2 I & 0 \\ 0 & 0 \end{bmatrix}. \tag{14.32}$$

On the other hand, since η_1 can be can be chosen arbitrarily, η_1 can be chosen small enough such that $\eta_2 - \eta_1 > 0$. Then we can always find a scalar $\eta_3 > 1$ such that

$$Q_{22} - (\eta_1 - \eta_2) I \geqslant \eta_3 Q_{22}. \tag{14.33}$$

It follows from (14.28), (14.30) and (14.33) that

$$\zeta_2(t)^{\mathrm{T}} Q_{22} \zeta_2(t) \leqslant \eta_3^{-1} \zeta_2(t - d(t))^{\mathrm{T}} Q_{22} \zeta_2(t - d(t)) + (\eta_1 \eta_3)^{-1} e(t)^{\mathrm{T}} P_{i4} P_{i4}^{\mathrm{T}} e(t), \tag{14.34}$$

which infers

$$\mathscr{E}\{f(t)\} \leqslant \eta_3^{-1} \mathscr{E}\{ \sup_{t-d_2 \leqslant s \leqslant t} f(s)\} + \xi e^{-\delta t}, \tag{14.35}$$

where $0 < \delta < \min\{\varepsilon, d_2^{-1} \ln \eta_3\}$, $f(t) = \zeta_2(t)^{\mathrm{T}} Q_{22} \zeta_2(t)$ and

$$\xi = (\eta_1 \eta_3)^{-1} m \max_{i \in \mathcal{S}} \|P_{i4}\|^2 \|\phi(t)\|_{d_2}^2.$$

Therefore, applying Lemma 1.11 to the above inequality yields that

$$\mathscr{E}\{\|\zeta_2(s)\|^2\} \leqslant \lambda_{\min}^{-1}(Q_{22}) \lambda_{\max}(Q_{22}) e^{-\delta t} \|\zeta_2(s)\|_{d_2}^2 + \frac{\lambda_{\min}^{-1}(Q_{22}) \xi e^{-\delta t}}{1 - \eta_3^{-1} e^{\delta d_2}}, \tag{14.36}$$

which means that by combining (14.25) that system (11.30) is mean-square exponentially stable.

In the sequel, we shall deal with the H_∞ performance of system (14.1). For this purpose, we consider the Lyapunov functional (14.9) and the following index for system (14.1):

$$J_{zw}(t) = \mathscr{E}\left\{ \int_0^t \left[z(s)^{\mathrm{T}} z(s) - \gamma^2 \omega(s)^{\mathrm{T}} \omega(s) \right] \mathrm{d}s \right\}. \tag{14.37}$$

Under zero initial condition, it easy to see that

$$J_{zw}(t) \leqslant \mathscr{E}\left\{ \int_0^t \left[z(s)^{\mathrm{T}} z(s) - \gamma^2 \omega(s)^{\mathrm{T}} \omega(s) + \mathcal{A} V(x_s, i, s) \right] \mathrm{d}s \right\}$$

$$\leqslant \mathscr{E}\left\{\int_0^t \begin{bmatrix} \xi(s) \\ \omega(s) \end{bmatrix}^{\mathrm{T}} \Theta_i \begin{bmatrix} \xi(s) \\ \omega(s) \end{bmatrix} \mathrm{d}s \right\}, \tag{14.38}$$

where

$$\Theta_i = \begin{bmatrix} \Xi_{11i} & \Xi_{12i} & 0 & 0 & P_i^{\mathrm{T}} B_{\omega i} \\ * & \Xi_{22i} & E^{\mathrm{T}} Z_2 E & \Xi_{24i} & 0 \\ * & * & \Xi_{33i} & 0 & 0 \\ * & * & * & \Xi_{44i} & 0 \\ * & * & * & * & -\gamma^2 I \end{bmatrix} + \begin{bmatrix} A_i^{\mathrm{T}} \\ A_{di}^{\mathrm{T}} \\ 0 \\ 0 \\ B_{\omega i}^{\mathrm{T}} \end{bmatrix} W \begin{bmatrix} A_i^{\mathrm{T}} \\ A_{di}^{\mathrm{T}} \\ 0 \\ 0 \\ B_{\omega i}^{\mathrm{T}} \end{bmatrix}^{\mathrm{T}}.$$

Hence, by Schur complement, we can obtain from (14.5) that for all $t > 0$, $J_{z\omega}(t) < 0$, which implies (10.8) holds for any nonzero $\omega(t) \in \mathscr{L}_2[0, \infty)$. This completes the proof.

14.3.2 Mode-Dependent H_∞ Filter Design

Next, we will design H_∞ filter (14.2). The following theorem presents a sufficient condition of the existence of the filter for system (14.1).

Theorem 14.4. *For a given scalar $\gamma > 0$, system (14.3) is mean-square exponentially admissible with H_∞ performance γ, if there exist matrices $S_1 > 0$, $S_2 > 0$, $S_3 > 0$, $Z_1 > 0$, $Z_2 > 0$, X_i, U_i, $\bar{A}_{fi}$, $\bar{B}_{fi}$ and $\bar{C}_{fi}$ such that for every $i \in \mathcal{S}$,*

$$E^{\mathrm{T}} X_i = X_i^{\mathrm{T}} E \geqslant 0, \tag{14.39}$$

$$E^{\mathrm{T}} U_i = U_i^{\mathrm{T}} E \geqslant 0, \tag{14.40}$$

$$E^{\mathrm{T}}(X_i - U_i) = (X_i - U_i)^{\mathrm{T}} E \geqslant 0, \tag{14.41}$$

$$\Delta_i = \begin{bmatrix} \Delta_{1i} & \Delta_{2i} & \Delta_{4i} & 0 & 0 & (X_i - U_i)^{\mathrm{T}} B_{\omega i} & A_i^{\mathrm{T}} U & L_i^{\mathrm{T}} - \bar{C}_{fi}^{\mathrm{T}} \\ * & \Delta_{3i} & \Delta_{5i} & 0 & 0 & X_i^{\mathrm{T}} B_{\omega i} - \bar{B}_{fi} D_{\omega i} & A_i^{\mathrm{T}} U & L_i^{\mathrm{T}} \\ * & * & \Delta_{6i} & E^{\mathrm{T}} Z_2 E & \Delta_{8i} & 0 & A_{di}^{\mathrm{T}} U & 0 \\ * & * & * & \Delta_{7i} & 0 & 0 & 0 & 0 \\ * & * & * & * & \Delta_{9i} & 0 & 0 & 0 \\ * & * & * & * & * & -\gamma^2 I & B_{\omega i}^{\mathrm{T}} U & 0 \\ * & * & * & * & * & * & -U & 0 \\ * & * & * & * & * & * & * & -I \end{bmatrix} < 0, \tag{14.42}$$

where $U = d_{12} d_2^2 Z_1 + d_{12}^2 Z_2$ and

$$\Delta_{1i} = A_i^{\mathrm{T}}(X_i - U_i) + (X_i - U_i)^{\mathrm{T}} A_i + \sum_{j=1}^{s} \pi_{ij} E^{\mathrm{T}}(X_j - U_j)$$

$$
\begin{aligned}
&- d_{12}E^{\mathrm{T}}Z_1E + \sum_{k=1}^{3} S_k,\\
\Delta_{2i} &= A_i^{\mathrm{T}}X_i - C_i^{\mathrm{T}}\bar{B}_{fi}^{\mathrm{T}} - \bar{A}_{fi}^{\mathrm{T}} + (X_i - U_i)^{\mathrm{T}}A_i\\
&\quad + \sum_{j=1}^{s}\pi_{ij}E^{\mathrm{T}}(X_j - U_j) - d_{12}E^{\mathrm{T}}Z_1E + \sum_{k=1}^{3} S_k,\\
\Delta_{3i} &= A_i^{\mathrm{T}}X_i - C_i^{\mathrm{T}}\bar{B}_{fi}^{\mathrm{T}} + X_i^{\mathrm{T}}A_i\\
&\quad - \bar{B}_{fi}C_i + \sum_{j=1}^{s}\pi_{ij}E^{\mathrm{T}}X_j - d_{12}E^{\mathrm{T}}Z_1E + \sum_{k=1}^{3} S_k,\\
\Delta_{4i} &= (X_i - U_i)^{\mathrm{T}}A_{di} + d_{12}E^{\mathrm{T}}Z_1E,\\
\Delta_{5i} &= X_i^{\mathrm{T}}A_{di} - \bar{B}_{fi}C_{di} + d_{12}E^{\mathrm{T}}Z_1E,\\
\Delta_{6i} &= -(1-\mu)S_3 - E^{\mathrm{T}}((d_{12}+d_2)Z_1 + 2Z_2)E,\\
\Delta_{7i} &= -S_1 - E^{\mathrm{T}}Z_2E,\\
\Delta_{8i} &= E^{\mathrm{T}}(d_2Z_1 + Z_2)E,\\
\Delta_{9i} &= -S_2 - E^{\mathrm{T}}(d_2Z_1 + Z_2)E.
\end{aligned}
$$

Furthermore, if (14.39)-(14.42) are solvable, the parameters of the desired filter can be chosen by

$$
A_{fi} = U_i^{-\mathrm{T}}\bar{A}_{fi},\ B_{fi} = U_i^{-\mathrm{T}}\bar{B}_{fi},\ C_{fi} = \bar{C}_{fi},\ E_f = E. \tag{14.43}
$$

Proof. Considering (14.42), we have that

$$
\begin{bmatrix} I\\ I\\ I\\ I\end{bmatrix}^{\mathrm{T}}
\begin{bmatrix} \Delta_{1i} & \Delta_{4i} & 0 & 0\\ * & \Delta_{6i} & E^{\mathrm{T}}Z_2E & \Delta_{8i}\\ * & * & \Delta_{7i} & 0\\ * & * & * & \Delta_{9i}\end{bmatrix}
\begin{bmatrix} I\\ I\\ I\\ I\end{bmatrix} < 0, \tag{14.44}
$$

which implies

$$
(A_i+A_{di})^{\mathrm{T}}(X_i-U_i)+(X_i-U_i)^{\mathrm{T}}(A_i+A_{di})+\sum_{j=1}^{s}\pi_{ij}E^{\mathrm{T}}(X_j-U_j) < 0. \tag{14.45}
$$

Using the same approach in Theorem 14.3, we can find from (14.41) and (14.45) that $X_j - U_j$ are nonsingular for every $i \in \mathcal{S}$. Define

$$
P_i = \begin{bmatrix} X_i & -U_i\\ -U_i & U_i\end{bmatrix},\ J_i = \begin{bmatrix} (X_i - U_i)^{-1} & I\\ (X_i - U_i)^{-1} & 0\end{bmatrix},\ \hat{E} = \begin{bmatrix} E & 0\\ 0 & E\end{bmatrix},\ H = \begin{bmatrix} I\\ 0\end{bmatrix}^{\mathrm{T}}. \tag{14.46}
$$

Then noting (14.39)-(14.41), we have

$$
\hat{E}^{\mathrm{T}}P_i = \begin{bmatrix} E^{\mathrm{T}}X_i & -E^{\mathrm{T}}U_i\\ -E^{\mathrm{T}}U_i & E^{\mathrm{T}}U_i\end{bmatrix} = P_i^{\mathrm{T}}\hat{E}. \tag{14.47}
$$

It can be deduced from (14.41) that

$$E^{\mathrm{T}}X_i - (-E^{\mathrm{T}}U_i)(E^{\mathrm{T}}U_i)^{+}(-E^{\mathrm{T}}U_i) = E^{\mathrm{T}}(X_i - U_i) \geqslant 0, \tag{14.48}$$

and

$$\begin{aligned}-E^{\mathrm{T}}U_i(I - (E^{\mathrm{T}}U_i)(E^{\mathrm{T}}U_i)^{+}) &= -E^{\mathrm{T}}U_i + (E^{\mathrm{T}}U_i)[(E^{\mathrm{T}}U_i)]^{\mathrm{T}}[(E^{\mathrm{T}}U_i)^{+}]^{\mathrm{T}} \\ &= -E^{\mathrm{T}}U_i + (E^{\mathrm{T}}U_i)[(E^{\mathrm{T}}U_i)^{+}(E^{\mathrm{T}}U_i)]^{\mathrm{T}} \\ &= -E^{\mathrm{T}}U_i + E^{\mathrm{T}}U_i \\ &= 0. \end{aligned} \tag{14.49}$$

Considering (14.40), (14.48) and (14.49), and using Lemma 1.3, we have

$$\hat{E}^{\mathrm{T}}P_i = P_i^{\mathrm{T}}\hat{E} \geqslant 0. \tag{14.50}$$

Now, pre- and post-multiplying (14.42) by $\mathrm{diag}\{(X_i - U_i)^{-\mathrm{T}}, I, I, I, I, I, I, I\}$ and its transpose, respectively, we obtain

$$\begin{bmatrix} \Sigma_{1i} & \Sigma_{2i} & 0 & 0 & J_i^{\mathrm{T}}P_i^{\mathrm{T}}\bar{B}_{\omega i} & J_i^{\mathrm{T}}\bar{A}_i^{\mathrm{T}}H^{\mathrm{T}}U & J_i^{\mathrm{T}}\bar{L}_i^{\mathrm{T}} \\ * & \Delta_{6i} & E^{\mathrm{T}}Z_2E & \Delta_{8i} & 0 & H\bar{A}_{di}^{\mathrm{T}}H^{\mathrm{T}}U & 0 \\ * & * & \Delta_{7i} & 0 & 0 & 0 & 0 \\ * & * & * & \Delta_{9i} & 0 & 0 & 0 \\ * & * & * & * & -\gamma^2 I & \bar{B}_{\omega i}^{\mathrm{T}}H^{\mathrm{T}}U & 0 \\ * & * & * & * & * & -U & 0 \\ * & * & * & * & * & * & -I \end{bmatrix} < 0, \tag{14.51}$$

where

$$\Sigma_{1i} = J_i^{\mathrm{T}}\Big(\sum_{j=1}^{s}\pi_{ij}\hat{E}^{\mathrm{T}}P_j + P_i^{\mathrm{T}}\bar{A}_i + \bar{A}_i^{\mathrm{T}}P_i + \sum_{k=1}^{3}H^{\mathrm{T}}S_kH - d_{12}\hat{E}^{\mathrm{T}}H^{\mathrm{T}}Z_1H\hat{E}\Big)J_i,$$

$$\Sigma_{2i} = J_i^{\mathrm{T}}(P_i^{\mathrm{T}}\bar{A}_{di}H^{\mathrm{T}} + d_{12}\hat{E}^{\mathrm{T}}H^{\mathrm{T}}Z_1E),$$

and the matrices $\bar{A}_i$, $\bar{A}_{di}$, $\bar{B}_{\omega i}$ and $\bar{L}_i$ are given in (14.3) with the parameters E_f, A_{fi}, B_{fi} and C_{fi} are given in (14.43), respectively. Then pre- and post-multiplying (14.51) by $\mathrm{diag}\{J_i^{-\mathrm{T}}, I, I, I, I, I, I\}$ and its transpose, respectively, we obtain

$$\begin{bmatrix} \Upsilon_{1i} & \Upsilon_{2i} & 0 & 0 & P_i^{\mathrm{T}}\bar{B}_{\omega i} & \bar{A}_i^{\mathrm{T}}H^{\mathrm{T}}U & \bar{L}_i^{\mathrm{T}} \\ * & \Delta_{6i} & E^{\mathrm{T}}Z_2E & \Delta_{8i} & 0 & H\bar{A}_{di}^{\mathrm{T}}H^{\mathrm{T}}U & 0 \\ * & * & \Delta_{7i} & 0 & 0 & 0 & 0 \\ * & * & * & \Delta_{9i} & 0 & 0 & 0 \\ * & * & * & * & -\gamma^2 I & \bar{B}_{\omega i}^{\mathrm{T}}H^{\mathrm{T}}U & 0 \\ * & * & * & * & * & -U & 0 \\ * & * & * & * & * & * & -I \end{bmatrix} < 0, \tag{14.52}$$

where

$$\Upsilon_{1i} = \sum_{j=1}^{s} \pi_{ij} \hat{E}^{\mathrm{T}} P_j + P_i^{\mathrm{T}} \bar{A}_i + \bar{A}_i^{\mathrm{T}} P_i + \sum_{k=1}^{3} H^{\mathrm{T}} S_k H - d_{12} \hat{E}^{\mathrm{T}} H^{\mathrm{T}} Z_1 H \hat{E},$$
$$\Upsilon_{2i} = P_i^{\mathrm{T}} \bar{A}_{di} H^{\mathrm{T}} + d_{12} \hat{E}^{\mathrm{T}} H^{\mathrm{T}} Z_1 E.$$

Then, we can always find a small enough scalar $\sigma > 0$ such that

$$\begin{bmatrix} \Lambda_{1i} & \Lambda_{2i} & 0 & 0 & P_i^{\mathrm{T}} \bar{B}_{\omega i} & \bar{A}_i^{\mathrm{T}} \bar{U} & \bar{L}_i^{\mathrm{T}} \\ * & \Lambda_{3i} & \hat{E}^{\mathrm{T}} \bar{Z}_2 \hat{E} & \hat{E}^{\mathrm{T}} (d_2 \bar{Z}_1 + \bar{Z}_2) \hat{E} & 0 & \bar{A}_{di}^{\mathrm{T}} \bar{U} & 0 \\ * & * & -\bar{S}_1 - \hat{E}^{\mathrm{T}} \bar{Z}_2 \hat{E} & 0 & 0 & 0 & 0 \\ * & * & * & -\bar{S}_2 - \hat{E}^{\mathrm{T}} (d_2 \bar{Z}_1 + \bar{Z}_2) \hat{E} & 0 & 0 & 0 \\ * & * & * & * & -\gamma^2 I & \bar{B}_{\omega i}^{\mathrm{T}} \bar{U} & 0 \\ * & * & * & * & * & -\bar{U} & 0 \\ * & * & * & * & * & * & -I \end{bmatrix} < 0, \tag{14.53}$$

where

$$\Lambda_{1i} = \sum_{j=1}^{s} \pi_{ij} \hat{E}^{\mathrm{T}} P_j + P_i^{\mathrm{T}} \bar{A}_i + \bar{A}_i^{\mathrm{T}} P_i + \sum_{k=1}^{3} \bar{S}_k - d_{12} \hat{E}^{\mathrm{T}} \bar{Z}_1 \hat{E},$$
$$\Lambda_{2i} = P_i^{\mathrm{T}} \bar{A}_{di} + d_{12} \hat{E}^{\mathrm{T}} \bar{Z}_1 \hat{E},$$
$$\Lambda_{3i} = -(1-\mu)\bar{S}_3 - \hat{E}^{\mathrm{T}}((d_{12} + d_2)\bar{Z}_1 + 2\bar{Z}_2)\hat{E},$$
$$\bar{S}_k = \begin{bmatrix} S_k & 0 \\ 0 & \sigma I \end{bmatrix}, k = 1, 2, 3,$$
$$\bar{Z}_l = \begin{bmatrix} Z_l & 0 \\ 0 & \sigma I \end{bmatrix}, l = 1, 2,$$
$$\bar{U} = d_{12} d_2^2 \bar{Z}_1 + d_{12}^2 \bar{Z}_2.$$

Therefore, by Schur complement and Theorem 14.3, we have that system (14.3) is mean square exponentially admissible with H_∞ performance γ. This completes the proof.

Remark 14.5. Note that Theorem 14.4 provides a delay-range-dependent sufficient condition for the solvability of the H_∞ filter problem of SMJSs with time-delays. The gain matrices of the desired filter (14.2) can be constructed through the solutions of (14.39)-(14.42). It is worth mentioning that the conditions of Theorem 14.3 and Theorem 14.4 are formulated by the original system matrices. Thus, the computational problems arising from decomposition of the original SS can be avoided, which makes the analysis and design procedures simple and reliable. Although, equality constraints, which may lead to numerical problems when checking the conditions, appear in Theorem 14.3 and Theorem 14.4, this computational difficulty can be overcome by employing some transformation applied in Chapter 4.

14.4 Numerical Examples

In this section, we shall give some numerical examples to demonstrate the applicability of the proposed approaches.

Example 14.6. Consider system (11.30) with $E = I$, two modes and the following parameters, which is borrowed from [182]:

$$A_1 = \begin{bmatrix} -3.4888 & 0.8057 \\ -0.6451 & -3.2684 \end{bmatrix}, \; A_{d1} = \begin{bmatrix} -0.8620 & -1.2919 \\ -0.6841 & -2.0729 \end{bmatrix},$$

$$A_2 = \begin{bmatrix} -2.4898 & 0.2895 \\ 1.3396 & -0.0211 \end{bmatrix}, \; A_{d2} = \begin{bmatrix} -2.8306 & 0.4978 \\ -0.8436 & -1.0115 \end{bmatrix}.$$

To compare the stochastic stability condition in Theorem 14.3 with that in [182], we choose $\pi_{22} = -0.8$ and $\mu = 0.9$. For given $d_1 \to 0^+$, the computed upper bound d_2 by different methods can be found in Table 14.1, from which we get that our result is an improvement over that in [182]. Next, we consider system with $d_1 > 0$. The allowable upper bound d_2 for different d_1 and π_{11} can be found in Table 14.2. It can be seen from Tables 14.1 and 14.2 that for Theorem 14.3 of our chapter, the allowable upper bound d_2 is dependent on the lower bound d_1 when π_{11} is fixed.

Table 14.1. Example 14.6: Allowable upper bound d_2 for different π_{11}

π_{11}	−0.1	−0.3	−0.5	−0.7	−0.9
[182]	0.4021	0.4010	0.4001	0.3993	0.3987
Theorem 14.3	0.4252	0.4250	0.4248	0.4246	0.4244

Table 14.2. Example 14.6: Allowable upper bound d_2 for different π_{11} and d_1

	$\pi_{11} = -0.1$	$\pi_{11} = -0.3$	$\pi_{11} = -0.5$	$\pi_{11} = -0.7$	$\pi_{11} = -0.9$
$d_1 = 0.1$	0.4336	0.4332	0.4328	0.4324	0.4321
$d_1 = 0.2$	0.4459	0.4450	0.4442	0.4434	0.4428

Example 14.7. Consider system (14.1) with two modes. The system parameters are described as follows:

$$A_1 = \begin{bmatrix} -0.1793 & -0.7876 \\ 1.6790 & 1.6746 \end{bmatrix}, \; A_2 = \begin{bmatrix} -0.3946 & -2.3342 \\ -0.1439 & -1.3575 \end{bmatrix},$$

$$A_{d1} = \begin{bmatrix} -0.3649 & 0.6192 \\ -0.4381 & -0.0420 \end{bmatrix}, \; A_{d2} = \begin{bmatrix} -0.9503 & 1.1842 \\ -0.0672 & 0.3443 \end{bmatrix},$$

$$B_{\omega 1} = \begin{bmatrix} -0.2 \\ 0.5 \end{bmatrix}, \; B_{\omega 2} = \begin{bmatrix} 0.5 \\ 0.2 \end{bmatrix}, \; L_1 = L_2 = \begin{bmatrix} 0.9 \\ 0.5 \end{bmatrix}^{\mathrm{T}}.$$

The singular matrix E and the transition rate matrix Π are given by the following expression:

$$E = \begin{bmatrix} 1 & 0 \\ 0 & 0 \end{bmatrix},$$
$$\Pi = \begin{bmatrix} -a & a \\ 0.4 & -0.4 \end{bmatrix}.$$

We firstly suppose $\mu = 0.2$, $\omega(t) = 0$ and $d_1 = 0.2$. For various a, Table 14.3 gives the allowable upper bound d_2 ensuring the mean-square exponentially admissibility of the considered system by using Theorem 14.3. Next, we take disturbance $\omega(t)$ into account and let $\mu = 0.3$, $d_1 = 0.3$ and $a = 0.2$. The minimum H_∞ performance γ for various d_2 can be obtained in Table 14.4. While the results of [7, 9, 159] cannot be applied to deal with the above system, because only time-invariant delay has been discussed in these papers.

Table 14.3. Example 14.7: Allowable upper bound d_2 for different a

a	0.1	0.3	0.5	0.7	0.9
Theorem 14.3	0.7158	0.6985	0.6881	0.6814	0.6773

Table 14.4. Example 14.7: Minimum H_∞ performance γ for different d_2

d_2	0.35	0.4	0.45	0.5	0.55
Theorem 14.3	0.4802	0.5388	0.6184	0.7248	0.8638

Example 14.8. Consider system (14.1) with two modes. For mode 1, the system is described as

$$A_1 = \begin{bmatrix} -1.1423 & 1.4521 \\ -2.5 & -3.2 \end{bmatrix}, A_{d1} = \begin{bmatrix} 0.1251 & -0.5122 \\ -0.2 & -0.4 \end{bmatrix},$$
$$B_{\omega 1} = \begin{bmatrix} -1.2159 \\ -0.1921 \end{bmatrix}, C_1 = \begin{bmatrix} -0.5215 & 1.4327 \end{bmatrix},$$
$$C_{d1} = \begin{bmatrix} -0.9321 & 0.1252 \end{bmatrix}, D_{\omega 1} = 2.2121,$$
$$L_1 = \begin{bmatrix} -0.9800 & -1.1210 \end{bmatrix}.$$

For mode 2, the system is described as

$$A_2 = \begin{bmatrix} -2.2111 & -1.4321 \\ -1 & -1.5 \end{bmatrix}, A_{d2} = \begin{bmatrix} 1.200 & 1.5421 \\ -0.35 & -0.3 \end{bmatrix},$$
$$B_{\omega 2} = \begin{bmatrix} 1.500 \\ -0.1200 \end{bmatrix}, C_2 = \begin{bmatrix} -0.2197 & 1.5421 \end{bmatrix},$$
$$C_{d2} = \begin{bmatrix} -0.2727 & 0.2100 \end{bmatrix}, D_{\omega 2} = 2.5490,$$
$$L_2 = \begin{bmatrix} 0.9721 & -1.5412 \end{bmatrix}.$$

Suppose the transition rate matrix is given by

$$\begin{bmatrix} \pi_{11} & \pi_{12} \\ \pi_{21} & \pi_{22} \end{bmatrix} = \begin{bmatrix} -0.3 & 0.3 \\ 0.4 & -0.4 \end{bmatrix}.$$

In this example, we assume

$$E = \begin{bmatrix} 1 & 0 \\ 0 & 0 \end{bmatrix},$$

$d_1 = 0.5$ and $\mu = 0.39$. The purpose is the design of a delay-dependent H_∞ filter in the form of (14.2) such that the filtering error system (14.3) achieves mean-square exponentially admissible with H_∞ performance γ. When $d_2 = 3$ and $\gamma = 4$, using Matlab LMI control Toolbox to solve (14.39)-(14.42), we can get the filter parameters to be determined are as follows:

$$\begin{aligned}
A_{f1} &= \begin{bmatrix} -67.2832 & -71.1765 \\ 209.2346 & 233.6312 \end{bmatrix}, \quad B_{f1} = \begin{bmatrix} -3.2529 \\ 9.2785 \end{bmatrix}, \\
C_{f1} &= \begin{bmatrix} -0.9802 & -1.1216 \end{bmatrix}, \\
A_{f2} &= \begin{bmatrix} -27.4747 & -32.5712 \\ 45.6857 & 54.9458 \end{bmatrix}, \quad B_{f2} = \begin{bmatrix} 2.5941 \\ -4.2869 \end{bmatrix}, \\
C_{f2} &= \begin{bmatrix} 1.1343 & -1.3436 \end{bmatrix}, \\
E_f &= \begin{bmatrix} 1 & 0 \\ 0 & 0 \end{bmatrix}.
\end{aligned}$$

On the other hand, we also provide the calculated minimum H_∞ performance γ with different d_2 achieved by the filtering error system (14.3) in Table 14.5. Although the condition in [167] fails to conclude whether or not there exist filters for this system, the result reported in this chapter has less conservatism than the existing one.

Table 14.5. Example 14.8: Minimum H_∞ performance γ for different d_2

d_2	0.9	1.2	1.5	1.8	2.1
Theorem 14.4	2.6569	2.7559	2.8637	2.9268	2.9634

14.5 Conclusion

The problem of delay-dependent H_∞ filtering has been addressed for SMJSs with time-varying delay in a range. A delay-dependent BRL has been proposed in terms of LMI approach. Based on this, a desired H_∞ filter has been designed such that the corresponding filtering error system is delay-dependent mean-square exponentially admissible with H_∞ performance γ. The numerical examples have demonstrated the effectiveness and reduced conservatism of the given methods.

Appendix

Proof of Lemma 1.6

Proof. It can be found that

$$\begin{bmatrix} Z & Y \\ * & Y^{\mathrm{T}}Z^{-1}Y \end{bmatrix} = \begin{bmatrix} Z^{1/2} & 0 \\ Y^{\mathrm{T}}Z^{-1/2} & 0 \end{bmatrix} \begin{bmatrix} Z^{1/2} & Z^{-1/2}Y \\ 0 & 0 \end{bmatrix} \geqslant 0,$$

where $Y = \begin{bmatrix} Y_1^{\mathrm{T}} & Y_2^{\mathrm{T}} & T_1^{\mathrm{T}} \end{bmatrix}$. Hence,

$$\sum_{\alpha=k-d}^{k-1} \begin{bmatrix} \eta(\alpha) \\ \xi(k) \end{bmatrix}^{\mathrm{T}} \begin{bmatrix} Z & Y \\ * & Y^{\mathrm{T}}Z^{-1}Y \end{bmatrix} \begin{bmatrix} \eta(\alpha) \\ \xi(k) \end{bmatrix} \geqslant 0.$$

After some simple manipulation, the above inequality yields (1.10). This completes the proof.

Proof of Lemma 1.7

Proof. According to Jensen inequality, we have

$$\begin{aligned} -d \int_{t-d}^{t} & \begin{bmatrix} x(\alpha) \\ E\dot{x}(\alpha) \end{bmatrix}^{\mathrm{T}} \begin{bmatrix} Z_1 & Z_2 \\ * & Z_3 \end{bmatrix} \begin{bmatrix} x(\alpha) \\ E\dot{x}(\alpha) \end{bmatrix} \mathrm{d}\alpha \\ & \leqslant - \int_{t-d}^{t} \begin{bmatrix} x(\alpha) \\ E\dot{x}(\alpha) \end{bmatrix}^{\mathrm{T}} \mathrm{d}\alpha \begin{bmatrix} Z_1 & Z_2 \\ * & Z_3 \end{bmatrix} \int_{t-d}^{t} \begin{bmatrix} x(\alpha) \\ E\dot{x}(\alpha) \end{bmatrix} \mathrm{d}\alpha. \end{aligned}$$

Rewriting the right-hand side of the above inequality, we can obtain (1.11). This completes the proof.

References

1. Abdollahi, F., Khorasani, K.: A decentralized Markovian jump H_∞ control routing strategy for mobile multi-agent networked systems. IEEE Transactions on Control Systems Technology 19, 269–283 (2011)
2. Bolzern, P., Colaneri, P., Nicolao, G.D.: Almost sure stability of Markov jump linear systems with deterministic switching. IEEE Transactions on Automatic Control 58, 209–213 (2013)
3. Boukas, E.K.: Singular linear systems with delay: H_∞ stabilization. Optimal Control Applications and Methods 28, 259–274 (2007)
4. Boukas, E.K.: Control of Singular Systems with Random Abrupt Changes. Springer, Berlin (2008)
5. Boukas, E.K., Al-Muthairi, N.F.: Delay-dependent stabilization of singular linear systems with delays. International Journal of Innovative Computing, Information and Control 53, 283–291 (2006)
6. Boukas, E.K., Liu, Z.K.: Robust H_∞ control of discrete-time Markovian jump linear systems with mode-dependent time-delays. IEEE Transactions on Automatic Control 46, 1918–1924 (2001)
7. Boukas, E.K., Liu, Z.K.: Delay-dependent stabilization of singularly perturbed jump linear systems. International Journal of Control 77, 310–319 (2004)
8. Boukas, E.K., Xia, Y.: Descriptor discrete-time systems with random abrupt changes: stability and stabilisation. International Journal of Control 81, 1311–1318 (2008)
9. Boukas, E.K., Xu, S., Lam, J.: On stability and stabilizability of singular stochastic systems with delays. Journal of Optimization Theory and Applications 127, 249–262 (2005)
10. Boyd, S., Ghaoui, L.E., Feron, E., Balakrishnan, V.: Linear Matrix Inequalities in System and Control Theory. SIAM, Philadelphia (1994)
11. Brogliato, B., Lozano, R., Maschke, B., Egeland, O.: Dissipative Systems Analysis and Control: Theory and Applications. Springer, London (2007)
12. Calcev, G., Gorez, R., Neyer, M.D.: Passivity approach to fuzzy control systems. Automatica 34, 339–344 (1998)
13. Chen, L., Zhong, M., Zhang, M.: H_∞ fault detection for linear singular systems with time-varying delay. International Journal of Control, Automation, and Systems 9, 9–14 (2011)

14. Chen, S., Chou, J.: Stability robustness of linear discrete singular time-delay systems with structured parameter uncertainties. IEE Proceedings-Control Theory & Applications 150, 296–302 (2003)
15. Chen, S.H., Chou, J.H.: Robust D-stability analysis for linear uncertain discrete singular systems with state delay. Applied Mathematics Letters 19, 197–205 (2007)
16. Chen, W., Guan, Z., Lu, X.: Delay-dependent output feedback stabilisation of Markovian jump system with time-delay. IEE Proceedings-Control Theory & Applications 151, 561–566 (2004)
17. Chizeck, H.J., Willsky, A.S., Castanon, D.: Discrete-time Markovian-jump linear quadratic optimal-control. International Journal of Control 43, 213–231 (1986)
18. Cobb, D.: Controllability, observability, and duality in singular systems. IEEE Transactions on Automatic Control 29, 1076–1082 (1984)
19. Costa, O.L.V., Fragoso, M.D.: Stability results for discrete-time linear-systems with Markovian jumping parameters. Journal of Mathematical Analysis and Applications 179, 154–178 (1993)
20. Costa, O.L.V., Fragoso, M.D., Marques, R.P.: Discrete-Time Markov Jump Linear Systems. Springer, London (1977)
21. Dai, L.: Singular Control Systems. Springer, Berlin (1989)
22. de Souza, C.E., Trofino, A., Barbosa, K.A.: Mode-independent H_∞ filters for Markovian jump linear systems. IEEE Transactions on Automatic Control 11, 1837–1841 (2006)
23. Ding, Y., Zhu, H., Zhong, S.: Exponential stabilization using sliding mode control for singular systems with time-varying delays and nonlinear perturbations. Communications in Nonlinear Science and Numerical Simulation 16, 4099–4107 (2011)
24. Dong, H., Wang, Z., Gao, H.: Robust H_∞ filtering for a class of nonlinear networked systems with multiple stochastic communication delays and packet dropouts. IEEE Transactions on Signal Processing 58, 1957–1966 (2010)
25. Dong, H., Wang, Z., Ho, D.W.C., Gao, H.: Robust H_∞ fuzzy output-feedback control with multiple probabilistic delays and multiple missing measurements. IEEE Transactions on Fuzzy Systems 58, 712–725 (2010)
26. Dong, H., Wang, Z., Ho, D.W.C., Gao, H.: Variance-constrained H_∞ filtering for a class of nonlinear time-varying systems with multiple missing measurements: The finite-horizon case. IEEE Transactions on Signal Processing 58, 2534–2543 (2010)
27. Dong, H., Wang, Z., Ho, D.W.C., Gao, H.: Robust H_∞ filtering for Markovian jump systems with randomly occurring nonlinearities and sensor saturation: The finite-horizon case. IEEE Transactions on Signal Processing 59, 3048–3057 (2011)
28. Dong, X.: Robust strictly dissipative control for discrete singular systems. IET Control Theory and Applications 1, 1060–1067 (2007)
29. Du, Z., Zhang, Q., Liu, L.: Delay-dependent robust H_∞ control for uncertain singular systems with multiple state delays. IET Control Theory and Applications 3, 731–740 (2009)
30. Fang, M.: Delay-dependent stability analysis for discrete singular systems with time-varying delays. Acta Automatica Sinica 36, 751–755 (2010)

31. Fei, Z., Gao, H., Shi, P.: New results on stabilization of Markovian jump systems with time delay. Automatica 45, 2300–2306 (2009)
32. Feng, Z., Lam, J.: Stability and dissipativity analysis of distributed delay cellular neural networks. IEEE Transactions on Neural Networks 22, 976–981 (2011)
33. Feng, Z., Lam, J.: Robust reliable dissipative filtering for discrete delay singular systems. Signal Processing 92, 3010–3025 (2012)
34. Feng, Z., Lam, J., Gao, H.: α-dissipativity analysis of singular time-delay systems. Automatica 47, 2548–2552 (2011)
35. Fridman, E.: Stability of linear descriptor systems with delay: a lyapunov-based approach. Journal of Mathematical Analysis and Applications 273, 24–44 (2002)
36. Fridman, E., Shaked, U.: New bounded real lemma representations for time-delay systems and their applications. IEEE Transactions on Automatic Control 46, 1973–1979 (2001)
37. Fridman, E., Shaked, U.: A new H_∞ filter design for linear time delay systems. IEEE Transactions on Signal Processing 49, 2839–2843 (2001)
38. Fridman, E., Shaked, U.: H_∞ control of linear state-delay descriptor systems: an LMI approach. Linear Algebra and its Applications 352, 271–302 (2002)
39. Fridman, E., Shaked, U., Xie, L.: Robust H_∞ filtering of linear systems with time varying delay. IEEE Transactions on Automatic Control 48, 159–165 (2002)
40. Gao, H., Chen, T., Chai, T.: Passivity and passification for networked control systems. SIAM Journal on Control and Optimization 46, 1299–1322 (2007)
41. Gao, H., Lam, J., Chen, T., Wang, C.: Stability analysis of uncertain discrete-time systems with time-varying state delay: a parameter-dependent lyapunov function approach. Asian Journal of Control 8, 433–440 (2006)
42. Gao, H., Lam, J., Wang, C.: Robust energy-to-peak filter design for stochastic time-delay systems. Systems & Control Letters 55, 101–111 (2006)
43. Gao, H., Meng, X., Chen, T.: Stabilization of networked control systems with a new delay characterization. IEEE Transactions on Automatic Control 53, 2142–2148 (2008)
44. Gao, H., Wang, C.: Robust $L_2 - L_\infty$ filtering for uncertain systems with multiple time-varying state delays. IEEE Transactions on Circuits and Systems-I: Fundamental Theory and Applications 50, 594–599 (2003)
45. Gao, H., Wang, C.: A delay-dependent approach to robust H_∞ filtering for uncertain discrete-time state-delayed systems. IEEE Transactions on Signal Processing 52, 1631–1640 (2004)
46. Gao, H., Zhao, Y., Lam, J., Chen, K.: H_∞ fuzzy filtering of nonlinear systems with intermittent measurements. IEEE Transactions Fuzzy Systems 17, 291–300 (2009)
47. Gopalsamy, K.: Stability and Oscillations in Delay Differential Equations of Population Dynamics. Kluwer Academic Publishers, Dordrecht (1992)
48. Gu, K., Han, Q.-L., Luo, A.C.J., Niculescu, S.I.: Discretized lyapunov functional for systems with distributed delay and piecewise constant coefficients. International Journal of Control 74, 737–744 (2001)
49. Gu, K., Kharitonov, V.K., Chen, J.: Stability of Time-Delay Systems. Birkhauser, Boston (2003)

50. Haidar, A., Boukas, E.K.: Exponential stability of singular systems with multiple time-varying delays. Automatica 45, 539–545 (2009)
51. Haidar, A., Boukas, E.K., Xu, S., Lam, J.: Exponential stability and static output feedback stabilisation of singular time-delay systems with saturating actuators. IET Control Theory and Applications 3, 1293–1305 (2009)
52. Hale, J.K., Lunel, S.M.V.: Introduction to Functional Differential Equations. Springer, New York (1993)
53. Han, C., Zhang, G., Wu, L., Zeng, Q.: Sliding mode control of T-S fuzzy descriptor systems with time-delay. Journal of the Franklin Institute 349, 1430–1444 (2011)
54. Han, Q.: Absolute stability of time-delay systems with sector-bounded nonlinearity. Automatica 41, 2171–2176 (2005)
55. He, X., Wang, Z., Ji, Y., Zhou, D.: Fault detection for discrete-time systems in a networked environment. International Journal of Systems Science 41, 937–945 (2010)
56. He, X., Wang, Z., Zhou, D.: Robust fault detection for networked systems with communication delay and data missing. Automatica 45, 2634–2639 (2009)
57. He, Y., Liu, G., Rees, D.: New delay-dependent stability criteria for neural networks with time-varying delay. IEEE Transactions on Neural Networks 18, 310–314 (2007)
58. He, Y., Liu, G., Rees, D., Wu, M.: Stability analysis for neural networks with time-varying interval delay. IEEE Transactions on Neural Networks 18, 1850–1854 (2007)
59. He, Y., Wang, Q., Wu, M., Lin, C.: Delay-dependent state estimation for delayed neural networks. IEEE Transactions on Neural Networks 17, 1077–1081 (2006)
60. Hemami, H., Wyman, B.F.: Modeling and control of constrained dynamic systems with application to biped locomotion in the frontal plane. IEEE Transactions on Automatic Control 24, 526–535 (1979)
61. Hu, L., Shi, P., Cao, Y.: Delay-dependent filtering design for time-delay systems with Markovian jumping parameters. International Journal of Adaptive Control and Signal Processing 21, 434–448 (2007)
62. Huang, M., Dey, S.: Stability of kalman filtering with Markovian packet losses. Automatica 43, 598–607 (2007)
63. Ji, X., Su, H., Chu, J.: Delay-dependent robust stability of uncertain discrete singular time-delay systems. In: Proceedings of the 2006 American Control Conference, Minneapolis, Minnesota, USA, pp. 3843–3848 (2006)
64. Ji, X., Su, H., Chu, J.: An LMI approach to robust stability of uncertain discrete singular time-delay systems. Asian Journal of Control 8, 56–62 (2006)
65. Ji, X., Su, H., Chu, J.: Robust state feedback H_∞ control for uncertain linear discrete singular systems. IET Control Theory & Applications 1, 195–200 (2007)
66. Ji, Y., Chizeck, H.J., Feng, X., Loparo, K.A.: Stability and control of discrete-time jump linear-systems. Control-Theory and Advanced Technology 7, 247–270 (1991)
67. Jiang, X., Han, Q.-L., Yu, X.: Stability criteria for linear discrete-time systems with interval-like time-varying delay. In: Proceedings of the 2005 American Control Conference, Portland, OR, USA, pp. 2817–2822 (2005)

68. Karan, M., Shi, P., Kaya, C.Y.: Transition probability bounds for the stochastic stability robustness of continuous- and discrete-time Markovian jump linear systems. Automatica 42, 2159–2168 (2006)
69. Kim, J.H.: New design method on memoryless H_∞ control for singular systems with delayed state and control using LMI. Journal of Franklin Institute 342, 321–327 (2005)
70. Kim, J.H.: Delay-dependent robust and non-fragile guaranteed cost control for uncertain singular systems with time-varying state and input delays. International Journal of Control, Automation, and Systems 7, 357–364 (2009)
71. Kolmanovskii, V.B., Myshkis, A.D.: Applied Theory of Functional Differential Equations. Kluwer Academic Publishers, Dordrecht (1992)
72. Kumar, A., Daoutidis, P.: Feedback control of nonlinear differential-algebraic equation systems. AIChE Journal 41, 619–636 (1995)
73. Lam, J., Gao, H., Wang, C.: Stability analysis for continuous systems with two additive time-varying delay components. Systems & Control Letters 56, 16–24 (2007)
74. Lam, J., Shu, Z., Xu, S., Boukas, E.K.: Robust H_∞ control of descriptor discrete-time Markovian jump systems. International Journal of Control 80, 374–385 (2007)
75. Li, H., Gao, H., Shi, P.: New passivity analysis for neural networks with discrete and distributed delays. IEEE Transactions on Neural Networks 21, 1842–1847 (2010)
76. Li, Q., Zhang, Q., Yi, N., Yuan, Y.: Robust passive control for uncertain time-delay singular systems. IEEE Transactions on Circuits and Systems I: Regular Papers 56, 653–663 (2009)
77. Li, X., de Souza, C.E.: Criteria for robust stability and stabilisation of uncertain linear systems with state-delay. Automatica 33, 1657–1662 (2007)
78. Liang, J., Shen, B., Dong, H., Lam, J.: Robust distributed state estimation for sensor networks with multiple stochastic communication delays. International Journal of Systems Science 42, 1459–1471 (2011)
79. Liang, J., Wang, Z., Liu, X.: On passivity and passification of stochastic fuzzy systems with delays: the discrete-time case. IEEE Transactions on Systems, Man, and Cybernetics-Part B 40, 964–969 (2010)
80. Lin, C., Wang, Q., Lee, T.H.: Robust normalization and stabilization of uncertain descriptor systems with norm-bounded perturbations. IEEE Transactions on Automatic Control 50, 515–520 (2005)
81. Lin, C., Wang, Q., Lee, T.H.: Stability and stabilization of a class of fuzzy time-delay descriptor systems. IEEE Transactions on Fuzzy Systems 14, 542–551 (2006)
82. Lin, C., Wang, Q., Lee, T.H., He, Y.: LMI Approach to Analysis and Control of Takagi-Sugeno Fuzzy Systems with Time Delay. Springer, New York (2007)
83. Lin, C., Wang, Z., Yang, F.: Observer-based networked control for continuous-time systems with random sensor delays. Automatica 45, 578–584 (2009)
84. Lin, J., Fei, S., Shen, J.: Delay-dependent H_∞ filtering for discrete-time singular Markovian jump systems with time-varying delay and partially unknown transition probabilities. Signal Processing 91, 277–289 (2011)

85. Lin, J., Fei, S., Shen, J., Long, F.: Exponential estimates and H_∞ control for singular Markovian jump systems with interval time-varying delay. Proceedings of the Institution of Mechanical Engineers, Part I: Journal of Systems and Control Engineering 224, 437–456 (2010)
86. Lin, J., Fei, S., Wu, Q.: Reliable H_∞ filtering for discrete-time switched singular systems with time-varying delay. Circuits, Systems, and Signal Processing 31, 1191–1214 (2012)
87. Liu, H., Ho, D.W.C., Sun, F.: Design of H_∞ filter for Markov jumping linear systems with non-accessible mode information. Automatica 44, 2655–2660 (2008)
88. Liu, M., Shi, P., Zhang, L., Zhao, X.: Fault-tolerant control for nonlinear Markovian jump systems via proportional and derivative sliding mode observer technique. IEEE Transactions on Circuits and Systems Part I: Regular paper 58, 2755–2764 (2011)
89. Liu, M., You, J.: Observer-based controller design for networked control systems with sensor quantisation and random communication delay. International Journal of Systems Science 43, 1901–1912 (2012)
90. Liu, Y., Wang, Z., Liang, J., Liu, X.: Stability and synchronization of discrete-time Markovian jumping neural networks with mixed mode-dependent time delays. IEEE Transactions on Neural Networks 20, 1102–1116 (2009)
91. Liu, Y., Wang, Z., Liu, X.: Exponential synchronization of complex networks with Markovian jump and mixed delays. Physics Letters A 372, 3986–3998 (2008)
92. Liu, Y., Wang, Z., Liu, X.: State estimation for discrete-time Markovian jumping neural networks with mixed mode-dependent delays. Physics Letters A 372, 7147–7155 (2008)
93. Lu, R., Li, H., Zhu, Y.: Quantized H_∞ filtering for singular time-varying delay systems with unreliable communication channel. Circuits, Systems, and Signal Processing 31, 521–538 (2012)
94. Lu, R., Xu, Y., Xue, A.: H_∞ filtering for singular systems with communication delays. IET Control Theory and Applications 90, 1240–1248 (2010)
95. Luan, X., Liu, F., Shi, P.: Neural-network-based finite-time H_∞ control for extended markov jump nonlinear systems. International Journal of Adaptive Control and Signal Processing 24, 554–567 (2010)
96. Luenberger, D.G.: Dynamic equations in descriptor form. IEEE Transactions on Automatic Control 22, 312–321 (1977)
97. Luenberger, D.G., Arbel, A.: Singular dynamic Leontief systems. Econometrica 45, 991–995 (1977)
98. Ma, L., Wang, Z., Bo, Y., Guo, Z.: Finite-horizon H_2/H_∞ control for a class of nonlinear Markovian jump systems with probabilistic sensor failures. International Journal of Control 84, 1847–1857 (2011)
99. Ma, L., Wang, Z., Niu, Y., Bo, Y., Guo, Z.: Sliding mode control for a class of nonlinear discrete-time networked systems with multiple stochastic communication delays. International Journal of Systems Science 42, 661–672 (2011)
100. Ma, S., Boukas, E.K.: Robust H_∞ filtering for uncertain discrete Markov jump singular systems with mode-dependent time delay. IET Control Theory and Applications 3, 351–361 (2009)

101. Ma, S., Boukas, E.K.: Stability and robust stabilisation for uncertain discrete stochastic hybrid singular systems with time delay. IET Control Theory & Applications 3, 1217–1225 (2009)
102. Ma, S., Boukas, E.K., Chinniah, Y.: Stability and stabilization of discrete-time singular Markov jump systems with time-varying delay. International Journal of Robust and Nonlinear Control 20, 531–543 (2010)
103. Ma, S., Cheng, Z., Zhang, C.: Delay-dependent robust stability and stabilisation for uncertain discrete singular systems with time-varying delays. IET Control Theory and Applications 1, 1086–1095 (2007)
104. Ma, S., Liu, X., Zhang, C.: Delay-dependent stability and stabilization of uncertain discrete-time Markovian jump singular systems with time delay. Anziam Journal 49, 111–129 (2007)
105. Ma, S., Zhang, C.: Robust stability and H_∞ control for uncertain discrete Markovian jump singular systems with mode-dependent time-delay. International Journal of Robust and Nonlinear Control 19, 965–985 (2009)
106. Mahmoud, M.S.: Delay-dependent dissipativity of singular time-delay systems. IMA Journal of Mathematical Control and Information 26, 45–58 (2009)
107. Mahmoud, M.S., Xia, Y.: Design of reduced-order l_2-l_∞ filter design for singular discrete-time systems using strict linear matrix inequalities. IET Control Theory and Applications 4, 509–519 (2010)
108. Mao, X., Koroleva, N., Rodkina, A.: Robust stability of uncertain stochastic differential delay equations. Systems & Control Letters 35, 325–336 (1998)
109. Mao, X., Yuan, C.: Stochastic Differential Equations With Markovian Switching. World Scientific Pub. Co. Inc. (2006)
110. Mao, Z., Jiang, B., Shi, P.: H_∞ fault detection filter design for networked control systems modelled by discrete Markovian jump systems. IET Control Theory and Applications 1, 1336–1343 (2007)
111. Masubuchi, I.: Output feedback controller synthesis for descriptor systems satisfying closed-loop dissipativity. Automatica 43, 339–345 (2007)
112. Masubuchi, I.: Dissipativity inequalities for continuous-time descriptor systems with applications to synthesis of control gains. Systems & Control Letters 55, 158–164 (2008)
113. Moon, Y.S., Park, P., Kwon, W.H., Lee, Y.S.: Delay-dependent robust stabilisation of uncertain state-delayed systems. International Journal of Control 74, 1447–1455 (2001)
114. Nguang, S.K., Assawinchaichote, W., Shi, P.: H_∞ filter for uncertain Markovian jump nonlinear systems: An lmi approach. Circuits, Systems and Signal Processing 26, 853–874 (2007)
115. Park, P.: A delay-dependent stability criterion for systems with uncertain time-invariant delays. IEEE Transactions on Automatic Control 44, 876–877 (1999)
116. Park, P., Ko, J.W., Jeong, C.: Reciprocally convex approach to stability of systems with time-varying delays. Automatica 47, 235–238 (2011)
117. Peaucelle, D., Arzelier, D., Henrion, D., Gouaisbaut, F.: Quadratic separation for feedback connection of an uncertain matrix and an implicit linear transformation. Automatica 43, 795–804 (2007)
118. Petersen, I.R.: A stabilization algorithm for a class of uncertain linear systems. Systems & Control Letters 8, 351–357 (1987)

119. Ren, J., Zhang, Q.: Robust normalization and guaranteed cost control for a class of uncertain descriptor systems. Automatica 48, 1693–1697 (2012)
120. Saadni, S.M., Chaabane, M., Mehdi, D.: Robust stability and stabilization of a class of singular systems with multiple time-varying delays. Asian Journal of Control 8, 1–11 (2006)
121. Scott, B.: Power system dynamic response calculations. Proceedings of the IEEE 67, 219–247 (1979)
122. Shao, H.: New delay-dependent stability criteria for systems with interval delay. Automatica 45, 744–749 (2009)
123. Shen, B., Wang, Z., Shu, H., Wei, G.: On nonlinear H_∞ filtering for discrete-time stochastic systems with missing measurements. IEEE Transactions on Automatic Control 53, 2170–2180 (2008)
124. Shen, B., Wang, Z., Shu, H., Wei, G.: H_∞ filtering for nonlinear discrete-time stochastic systems with randomly varying sensor delays. Automatica 45, 1032–1037 (2009)
125. Shu, Z., Lam, J.: Exponential estimates and stabilization of uncertain singular systems with discrete and distributed delays. International Journal of Control 81, 865–882 (2008)
126. Shu, Z., Lam, J., Xiong, J.: Static output-feedback stabilization of discrete-time Markovian jump linear systems: A system augmentation approach. Automatica 46, 687–694 (2010)
127. Siqueira, A.A.G., Terra, M.H., Buosi, C.: Fault-tolerant robot manipulators based on output-feedback H_∞ controllers. Robotics and Autonomous Systems 55, 785–794 (2007)
128. Su, H., Ji, X., Chu, J.: Delay-dependent robust control for uncertain singular time-delay systems. Asian Journal of Control 8, 80–89 (2006)
129. Tian, E., Yue, D., Yang, T., Gu, Z., Lu, G.: T-S fuzzy model-based robust stabilization for networked control systems with probabilistic sensor and actuator failure. IEEE Transactions on Fuzzy Systems 19, 553–561 (2011)
130. Verghese, G.C., Levy, B.C., Kailath, T.: A generalized state-space for singular systems. IEEE Transactions on Automatic Control 26, 811–831 (1981)
131. Vidyasagar, M., Viswandham, N.: Reliable stabilization using a multicontroller configuration. Automatica 21, 599–602 (1985)
132. Wang, H., Yung, C., Chang, F.: H_∞ Control for Nonlinear Descriptor Systems. Springer, London (2006)
133. Wang, Y.Y., Wang, Q.B., Zhou, P.F., Duan, D.P.: Robust guaranteed cost control for singular Markovian jump systems with time-varying delay. ISA Transactions 51, 559–565 (2012)
134. Wang, Z., Ho, D.W.C., Liu, Y., Liu, X.: Robust H_∞ control for a class of nonlinear discrete time-delay stochastic systems with missing measurements. Automatica 45, 684–691 (2009)
135. Wang, Z., Huang, B., Unbehauen, H.: Robust reliable control for a class of uncertain nonlinear state-delayed systems. Automatica 35, 955–963 (1999)
136. Wang, Z., Liu, Y., Liu, X.: Exponential stabilization of a class of stochastic system with Markovian jump parameters and mode-dependent mixed time-delays. IEEE Transactions on Automatic Control 55, 1656–1662 (2010)
137. Wang, Z., Wei, G., Feng, G.: Reliable H_∞ control for discrete-time piecewise linear systems with infinite distributed delays. Automatica 45, 2991–2994 (2009)

138. Wei, G., Wang, Z., Shu, H.: Robust filtering with stochastic nonlinearities and multiple missing measurements. Automatica 45, 836–841 (2009)
139. Wei, G., Wang, Z., Shu, H., Fang, J.: A delay-dependent approach to H_∞ filtering for stochastic delayed jumping systems with sensor non-linearities. International Journal of Control 80, 885–897 (2007)
140. Wo, S., Shi, G., Zou, Y.: Optimal guaranteed cost observer-based control of uncertain singular time-delay systems. Journal of Systems Engineering and Electronics 18, 111–119 (2007)
141. Wu, A.-G., Duan, G.-R., Liu, W.: Proportional multiple-integral observer design for continuous-time descriptor linear systems. Asian Journal of Control 14, 476–488 (2012)
142. Wu, A.-G., Feng, G., Duan, G.-R.: Proportional multiple-integral observer design for discrete-time descriptor linear systems. International Journal of Systems Science 43, 1492–1503 (2012)
143. Wu, H.: Reliable robust H_∞ fuzzy control for uncertain nonlinear systems with Markovian jumping actuator faults. Journal of Dynamic Systems, Measurement, and Control 129, 252–261 (2007)
144. Wu, J., Chen, T., Wang, L.: Delay-dependent robust stability and H_∞ control for jump linear systems with delays. Systems & Control Letters 55, 939–948 (2006)
145. Wu, L., Ho, D.W.C.: Sliding mode control of singular stochastic hybrid systems. Automatica 46, 779–783 (2010)
146. Wu, L., Shi, P., Gao, H.: State estimation and sliding-mode control of Markovian jump singular systems. IEEE Transactions Automatic Control 55, 1213–1219 (2010)
147. Wu, L., Shi, P., Gao, H., Wang, C.: H_∞ filtering for 2-D Markovian jump systems. Automatica 7, 1849–1858 (2008)
148. Wu, L., Su, X., Shi, P.: Sliding mode control with bounded $\mathscr{L}_2$ gain performance of Markovian jump singular time-delay systems. Automatica 48, 1929–1933 (2012)
149. Wu, L., Zheng, W.: Passivity-based sliding mode control of uncertain singular time-delay systems. Automatica 45, 2120–2127 (2009)
150. Wu, M., He, Y., She, J., Liu, G.: Delay-dependent criteria for robust stability of time-varying delay systems. Automatica 40, 1435–1439 (2004)
151. Wu, M., He, Y., She, J., Liu, G.: Delay-dependent robust stability criteria for uncertain neutral systems with mixed delays. Systems & Control Letters 51, 57–65 (2004)
152. Wu, Z., Jiang, Y., Shi, P.: Adaptive tracking for stochastic nonlinear systems with Markovian switching. IEEE Transactions on Automatic Control 55, 2135–2141 (2010)
153. Wu, Z., Lam, J., Su, H., Chu, J.: Stability and dissipativity analysis of static neural networks with time delay. IEEE Transactions on Neural Networks and Learning Systems 23, 199–210 (2012)
154. Wu, Z., Park, J.H., Su, H., Chu, J.: Dissipativity analysis for singular systems with time-varying delays. Applied Mathematics and Computation 218, 4605–4613 (2011)
155. Wu, Z., Shi, P., Su, H., Chu, J.: Passivity analysis for discrete-time stochastic Markovian jump neural networks with mixed time delays. IEEE Transactions on Neural Networks 22, 1566–1575 (2011)

156. Wu, Z., Shi, P., Su, H., Chu, J.: $l_2 - l_\infty$ filter design for discrete-time singular Markovian jump systems with time-varying delays. Information Sciences 181, 5534–5547 (2012)
157. Wu, Z., Shi, P., Su, H., Chu, J.: Dissipativity analysis for discrete-time stochastic neural networks with time-varying delays. IEEE Transactions on Neural Networks and Learning Systems 24, 345–355 (2013)
158. Wu, Z., Su, H., Chu, J.: Robust stabilization for uncertain discrete singular systems with state delay. International Journal of Robust and Nonlinear Control 18, 1532–1550 (2008)
159. Wu, Z., Su, H., Chu, J.: Delay-dependent H_∞ control for singular Markovian jump systems with time delay. Optimal Control Application and Methods 30, 443–461 (2009)
160. Wu, Z., Su, H., Chu, J.: Delay-dependent stabilization of singular Markovian jump systems with state delay. Journal of Control Theory and Applications 7, 231–236 (2009)
161. Wu, Z., Su, H., Chu, J.: Delay-dependent H_∞ filtering for singular Markovian jump time-delay systems. Signal Processing 90, 1815–1824 (2010)
162. Wu, Z., Su, H., Chu, J.: H_∞ filtering for singular Markovian jump systems with time delay. International Journal of Robust and Nonlinear Control 20, 939–957 (2010)
163. Wu, Z., Su, H., Chu, J.: H_∞ filtering for singular systems with time-varying delay. International Journal of Robust and Nonlinear Control 20, 1269–1284 (2010)
164. Wu, Z., Su, H., Chu, J.: Output feedback stabilization for discrete singular systems with random abrupt changes. International Journal of Robust and Nonlinear Control 20, 1945–1957 (2010)
165. Wu, Z., Xie, X., Shi, P., Xia, Y.: Backstepping controller design for a class of stochastic nonlinear systems with Markovian switching. Automatica 45, 997–1004 (2009)
166. Wu, Z., Zhou, W.: Delay-dependent robust H_∞ control for uncertain singular time-delay systems. IET Control Theory & Applications 1, 1234–1241 (2007)
167. Xia, J.: Robust H_∞ filter design for uncertain time-delay singular stochastic systems with Markovian jump. Journal of Control Theory and Applications 5, 331–335 (2007)
168. Xia, J., Xu, S., Song, B.: Delay-dependent $\mathscr{L}_2$-$\mathscr{L}_\infty$ filter design for stochastic time-delay systems. Systems & Control Letters 56, 579–587 (2007)
169. Xia, Y., Boukas, E.K., Shi, P., Zhang, J.: Stability and stabilization of continuous-time singular hybrid systems. Automatica 45, 1504–1509 (2009)
170. Xia, Y., Li, L., Mahmoud, M.S., Yang, H.: H_∞ filtering for nonlinear singular Markovian jumping systems with interval time-varying delays. International Journal of Systems Science 43, 272–284 (2012)
171. Xia, Y., Zhang, J., Boukas, E.K.: Control for discrete singular hybrid systems. Automatica 44, 2635–2641 (2008)
172. Xie, L., Fu, M., Li, H.: Passivity analysis and passification for uncertain signal processing systems. IEEE Transactions on Signal Processing 46, 2394–2403 (1998)
173. Xu, H., Zou, Y.: H_∞ control for 2-D singular delayed systems. International Journal of Systems Science 42, 609–619 (2011)

174. Xu, S., Chen, T., Lam, J.: Robust H_∞ filtering for uncertain Markovian jump systems with mode-dependent time delays. IEEE Transactions on Automatic Control 48, 900–907 (2003)
175. Xu, S., Dooren, P.V., Stefan, R., Lam, J.: Robust stability and stabilization for singular systems with state delay and parameter uncertainty. IEEE Transactions on Automatic Control 47, 1122–1128 (2002)
176. Xu, S., Lam, J.: Robust stability for uncertain discrete singular systems with state delay. Asian Journal of Control 5, 339–405 (2003)
177. Xu, S., Lam, J.: Robust stability and stabilization of discrete singular systems: An equivalent characterization. IEEE Transactions on Automatic Control 49, 568–574 (2004)
178. Xu, S., Lam, J.: Robust Control and Filtering of Singular Systems. Springer, Berlin (2006)
179. Xu, S., Lam, J.: Reduced-order H_∞ filter for singular systems. Systems & Control Letters 56, 48–57 (2007)
180. Xu, S., Lam, J.: A survey of linear matrix inequality techniques in stability analysis of delay systems. International Journal of Systems Science 39, 1095–1113 (2012)
181. Xu, S., Lam, J., Liu, W., Zhang, Q.: H_∞ model reduction for singular systems: continuous-time case. IEE Proceedings-Control Theory and Applications 150, 637–641 (2003)
182. Xu, S., Lam, J., Mao, X.: Delay-dependent H_∞ control and filtering for uncertain Markovian jump systems with time-varying delays. IEEE Transactions on Circuits and Systems Part I: Regular paper 54, 2070–2077 (2007)
183. Xu, S., Lam, J., Mao, X.: Delay-dependet H_∞ control and filtering for uncertain Markovian jump systems with time-varying delays. IEEE Transactions on Circuits and Systems-I: Regular paper 54, 2070–2077 (2007)
184. Xu, S., Lam, J., Yang, C.: Robust H_∞ control for discrete singular systems with state delay and parameter uncertainty. Dynamics of Continuous, Discrete and Impulsive Systems, Series B: Applications and Algorithms 9, 539–554 (2002)
185. Xu, S., Lam, J., Yang, C.: H_∞ filter for singular systems. IEEE Transactions on Automatic Control 48, 2217–2222 (2003)
186. Xu, S., Lam, J., Yang, C.: Robust H_∞ control for uncertain singular systems with state delay. International Journal of Robust and Nonlinear Control 13, 1213–1223 (2003)
187. Xu, S., Lam, J., Zou, Y.: An improved characterization of bounded realness for singular delay systems and its applications. International Journal of Robust and Nonlinear Control 18, 263–277 (2008)
188. Xu, S., Lam, L., Zhang, L.: Robust D-stability analysis for uncertain discrete singular systems with state delay. IEEE Transactions on Circuit and Systems I: Fundamental Theory and Applications 49, 551–555 (2002)
189. Xu, S., Yang, C.: Stabilization of discrete-time singular systems: a matrix inequalities approach. Automatica 35, 1613–1617 (1999)
190. Xu, S., Zheng, W., Zou, Y.: Passivity analysis of neural networks with time-varying delays. IEEE Transactions on Circuits and Systems Part II: Express Briefs 56, 325–329 (2009)
191. Yang, C., Zhang, Q., Zhou, L.: Stability Analysis and Design for Nonlinear Singular Systems, New York, Berlin (2013)

192. Yang, F., Wang, Z., Hung, Y.S., Gani, M.: H_∞ control for networked systems with random communication delays. IEEE Transactions on Automatic Control 51, 511–518 (2006)
193. Yang, F., Zhang, Q.: Delay-dependent H_∞ control for linear descriptor systems with delay in state. Journal of Control Theory and Applications 3, 76–84 (2005)
194. Yang, F., Zhang, Q.: Optimal guaranteed cost observer-based control of uncertain singular time-delay systems. Dynamics of Continuous, Discrete and Impulsive Systems Series B: Applications and Algorithms 12, 397–411 (2005)
195. Yang, G., Wang, J., Soh, Y.: Reliable H_∞ controller design for linear systems. Automatica 37, 717–725 (2001)
196. Yang, G., Ye, D.: Reliable H_∞ control of linear systems with adaptive mechanism. IEEE Transactions on Automatic Control 55, 242–247 (2010)
197. Yao, X., Wu, L., Zheng, W.: Fault detection filter design for Markovian jump singular systems with intermittent measurements. IEEE Transactions on Signal Processing 59, 3099–3109 (2011)
198. Yip, E.L., Sincovec, R.F.: Solvability, controllability and observability of continuous descriptor systems. IEEE Transactions on Automatic Control 26, 702–707 (1981)
199. Yue, D., Han, Q.: Robust H_∞ filter design of uncertain descriptor systems with discrete and distributed delays. IEEE Transactions on Signal Processing 52, 3200–3212 (2004)
200. Yue, D., Han, Q.: Delay-dependent robust H_∞ controller design for uncertain descriptor systems with time-varying discrete and distributed delays. IEE Proceedings-Control Theory & Applications 152, 628–638 (2005)
201. Zhang, B., Zhou, S., Du. Robust, D.: H_∞ filtering of delayed singular systems with linear fractional parametric uncertainties. Circuits, Systems and Signal Processing 25, 627–647 (2006)
202. Zhang, H., Guan, Z., Feng, G.: Reliable dissipative control for stochastic impulsive system. Automatica 44, 1004–1010 (2008)
203. Zhang, L., Boukas, E.K.: H_∞ control for discrete-time Markovian jump linear systems with partly unknown transition probabilities. International Journal of Robust and Nonlinear Control 19, 868–883 (2009)
204. Zhang, L., Boukas, E.K.: Stability and stabilization of Markovian jump linear systems with partly unknown transition probabilities. Automatica 45, 463–468 (2009)
205. Zhang, L., Boukas, E.K., Lam, J.: Analysis and synthesis of Markov jump linear systems with time-varying delays and partially known transition probabilities. IEEE Transactions Automatic Control 53, 2458–2464 (2008)
206. Zhang, L., Huang, B., Lam, J.: H_∞ model reduction of Markovian jump linear systems. Systems & Control Letters 50, 103–118 (2003)
207. Zhang, L., Lam, J.: Necessary and sufficient conditions for analysis and synthesis of Markov jump linear systems with incomplete transition descriptions. IEEE Transactions on Automatic Control 55, 1695–1701 (2010)
208. Zhang, Q., Liu, W., Hill, D.: A Lyapunov approach to analysis of discrete singular systems. Systems & Control Letters 45, 237–247 (2002)
209. Zhang, W., Branicky, M.S., Phillips, S.M.: Stability of networked control systems. IEEE Control Systems Magazine 21, 84–99 (2001)

210. Zhang, X., Han, Q.: Robust H_∞ filtering for a class of uncertain linear systems with time-varying delay. Automatica 44, 157–166 (2008)
211. Zhang, X., Lu, G., Zheng, Y.: Observer design for descriptor Markovian jumping systems with nonlinear perturbations. Circuits, Systems, and Signal Processing 27, 95–112 (2008)
212. Zhao, Y., Gao, H., Lam, J., Che, K.: Fault detection for fuzzy systems with intermittent measurements. IEEE Transactions on Neural Networks 17, 398–410 (2009)
213. Zhong, R., Yang, Z.: Delay-dependent robust control of descriptor systems with time delay. Asian Journal of Control 8, 36–44 (2006)
214. Zhou, S., Zhang, B., Zheng, W.: Gain-scheduled H_∞ filtering of parameter-varying systems. International Journal of Robust and Nonlinear Control 16, 397–411 (2007)
215. Zhou, W., Fang, J.: Delay-dependent robust H_∞ admissibility and stabilization for uncertain singular system with Markovian jumping parameters. Circuits, Systems and Signal Processing 28, 433–450 (2009)
216. Zhou, W., Lu, H., Duan, C., Li, M.: Delay-dependent robust control for singular discrete-time Markovian jump systems with time-varying delay. International Journal of Robust and Nonlinear Control 20, 1112–1128 (2010)
217. Zhou, Y., Li, J.: Reduced-order $\mathcal{L}_2$-$\mathcal{L}_\infty$ filtering for singular systems: a linear matrix inequality approach. IET Control Theory and Applications 2, 773–781 (2008)
218. Zhu, S., Zhang, C., Cheng, Z., Feng, J.: Delay-dependent robust stability criteria for two classes of uncertain singular time-delay systems. IEEE Transactions on Automatic Control 52, 880–885 (2007)
219. Zhu, X., Wang, Y., Gan, Y.: H_∞ filtering for continuous-time singular systems with time-varying delay. International Journal of Adaptive Control and Signal Processing 25, 137–154 (2011)

Index